GUIDE TO FLUORINE NMR FOR ORGANIC CHEMISTS

GUIDE TO FLUORINE NMR FOR ORGANIC CHEMISTS

Second Edition

WILLIAM R. DOLBIER, JR.

Published by John Wiley & Sons, Inc., Hoboken, New Jersey
Published simultaneously in Canada

For general information on our other products and services or for technical support, please contact our Customer Care Department within the United States at (800) 762-2974, outside the United States at (317) 572-3993 or fax (317) 572-4002.

Wiley also publishes its books in a variety of electronic formats. Some content that appears in print may not be available in electronic formats. For more information about Wiley products, visit our web site at www.wiley.com.

Library of Congress Cataloging-in-Publication Data:

Names: Dolbier, William R.
Title: Guide to fluorine NMR for organic chemists / William R. Dolbier, Jr.
Other titles: Guide to fluorine nuclear magnetic resonance for organic
    chemists
Description: 2nd edition. | Hoboken, New Jersey : John Wiley & Sons, Inc.,
    [2016] | Includes bibliographical references and index.
Identifiers: LCCN 2016013711 (print) | LCCN 2016014534 (ebook) | ISBN
    9781118831083 (cloth) | ISBN 9781118831113 (pdf) | ISBN 9781118831090
    (epub)
Subjects: LCSH: Fluorine compounds–Spectra. | Nuclear magnetic resonance
    spectroscopy.
Classification: LCC QD412.F1 D65 2016 (print) | LCC QD412.F1 (ebook) | DDC
    547/.02–dc23
LC record available at https://lccn.loc.gov/2016013711

Typeset in 11/13pt, TimesTenLTStd by SPi Global, Chennai, India.

Printed in the United States of America

10 9 8 7 6 5 4 3 2 1

CONTENTS

PREFACE

Fluorine's unique polar and steric properties as a substituent and the influence that fluorinated substituents can have upon the physical and chemical properties of molecules have induced increasing number of synthetic organic chemists to incorporate fluorine into target compounds of synthetic interest. In preparing compounds that contain fluorine, one first faces the daunting task of learning the intricacies of fluorine's often unique synthetic methodologies.

Then, once the desired fluorine-containing compounds have been synthesized, the real fun begins as the world of fluorine NMR is entered. However, one's first encounter with fluorine NMR can also present a problem because although most synthetic organic chemists are thoroughly familiar with the use of proton and carbon NMR for compound characterization, few have much experience with the use of fluorine NMR for that purpose. Thus, there is a need for a place where a person can turn to obtain a concise but thorough introduction to fluorine NMR itself and, just as importantly, to learn how the presence of fluorine substituents can enhance the efficacy of both proton and carbon NMR as tools for structure characterization.

Simply speaking, that is the continued purpose of the second edition of this book – to provide you, the working organic chemist, with virtually everything you need to know about fluorine NMR, including an understanding of the impact of fluorine substituents upon proton and carbon NMR, and new in the second edition, upon ^{31}P and ^{15}N NMR chemical shifts and coupling constants, to the extent that such data are available.

This book is primarily intended for use by academic and industrial organic chemists, most of whom have interest in fluorinated compounds of potential pharmaceutical and agrochemical interest. Such compounds are for the most part what I will call "lightly" fluorinated – that is containing one or at most a few fluorine-containing substituents, with the emphasis on isolated fluorine substituents, CF_2 groups, and trifluoromethyl substituents. However, virtually all fluorine-containing substituents that might be of interest, including C_2F_5 and SF_5, are discussed. More heavily fluorinated compounds are not totally ignored, but the emphasis is upon the lightly fluorinated species. Data for covalently bound inorganic fluorides are also presented.

Hopefully, this book will work both to provide an introduction to the novice and as a resource for those chemists who are more experienced in working with fluoroorganic compounds. As you will soon notice, the book has not been written by an NMR "specialist," but rather has been written for working organic chemists *by* a working organic chemist.

I thank all of the readers of the first edition, especially those who pointed out errors or omissions and those who have made suggestions for additions and improvements in the second edition. Hopefully, the second edition does a better job of covering the great diversity of compounds that contain fluorine.

This book would not have been possible without the continued encouragement of my wife, Jing, the critical technical assistance of Dr Ion Ghiviriga in obtaining and interpreting NMR spectra, and the able assistance of my current and past research group members who synthesized key model compounds and who, along with Dr Ghiviriga, obtained all spectra that appear in this book. They include Dr Ying Chang, Dr Wei Xu, Dr Oleksandre Kanishchev, Dr Simon Lopez, Dr Xiao-Jun Tang, Lianhao Zhang, Henry Martinez, Seth Thomoson, Xianjin Yang, Masamune Okamoto, and Zuxiao Zhang. I also thank my son, Stephen, for his assistance in creating figures of spectra for the second edition.

William R. Dolbier, Jr.
July 30, 2016
Gainesville, Florida

CHAPTER 1

GENERAL INTRODUCTION

1.1. WHY FLUORINATED COMPOUNDS ARE INTERESTING?

The reason that organic chemists are interested in compounds that contain fluorine is simple. Because of fluorine's steric and polar characteristics, even a *single* fluorine substituent, placed at a propitious position within a molecule, can have a remarkable effect upon the physical and chemical properties of that molecule. Discussions of the impact of fluorine on physical and chemical properties of compounds have appeared in numerous reviews and monographs.[1–13] There are also a number of recent reviews on the subject of fluorine in medicinal, agrochemical, and materials chemistry.[14–23]

1.1.1. Steric Size

In terms of its steric impact, fluorine is the smallest substituent that can replace a hydrogen in a molecule, other than an isotope of hydrogen. Table 1.1 provides insight into the comparative steric impact of various fluorinated substituents on the equilibrium between axial and equatorial substitution in cyclohexane.[24]

Guide to Fluorine NMR for Organic Chemists, Second Edition. William R. Dolbier, Jr.
© 2016 John Wiley & Sons, Inc. Published 2016 by John Wiley & Sons, Inc.

TABLE 1.1. Values of A for Some Common Substituents

$$-\Delta G^0 \ (\text{kcal/mol}) = A$$

X	A value	X	A value
H	[0]	F	0.2
OH	0.5	OCF_3	0.8
OCH_3	0.6	SCF_3	1.2
CH_3	1.7	CH_2F	1.6
C_2H_5	1.8	CHF_2	1.9
i-C_3H_7	2.2	CF_3	2.4
Ph	2.8	C_2F_5	2.7

TABLE 1.2. Substituent Effects: σ_p-Values and F-Values

Substituent	σ_p	F	Substituent	σ_p	F
H	[0]	[0]	CH_2F	0.11	0.15
F	0.06	0.45	CHF_2	0.32	0.29
Cl	0.23	0.42	CF_3	0.54	0.38
OH	−0.37	0.33	C_2F_5	0.52	0.44
NH_2	−0.66	0.08	OCF_3	0.35	0.39
NO_2	0.78	0.65	SCF_3	0.50	0.36
CH_3	−0.17	0.01	SF_5	0.68	0.56
			CH_2CF_3	0.09	0.15

1.1.2. Polar Effects

Fluorine is, of course, the most electronegative atom on the periodic table. σ_p-Values and F-values (the "pure" field inductive effect) provide indications of the electron-withdrawing influence of substituents, and it can be seen that fluorine itself has the largest F value of an atomic substituent. The values for σ_P and F for various other fluorinated (and nonfluorinated) substituents provide insight into the relative electron-withdrawing power of fluorinated substituents (Table 1.2).[25]

1.1.3. Effect of Fluorine Substituents on Acidity and Basicity of Compounds

The strong electronegativity of the fluorinated substituents is reflected in the effect that this group has upon the acidity of alcohols, carboxylic acids, and sulfonic acids, as well as the effect it has on the basicity of amines (Tables 1.3–1.6).[1, 26]

TABLE 1.3. Carboxylic Acid Acidity

XCH_2CO_2H	pK_a
X = H	4.8
X = F	2.59
X = NO$_2$	1.32
X = CF$_3$	2.9
X = CF$_3$CH$_2$	4.2
CF$_3$CO$_2$H	0.2

TABLE 1.4. Sulfonic Acids

	pK_a
CH_3SO_3H	−2.6
CF_3SO_3H	−12

TABLE 1.5. Alcohol Acidity

	pK_a
CH_3CH_2OH	15.9
CF_3CH_2OH	12.4
$(CF_3)_2CH–OH$	9.3
$(CF_3)_3C–OH$	5.4

TABLE 1.6. Amine Basicity

XCH_2NH_2	pK_b			
X = CH$_3$	3.3			
X = CH$_2$CF$_3$	5.1			
X = CF$_3$	8.3	10.0	12.8	27.0 (DMSO)

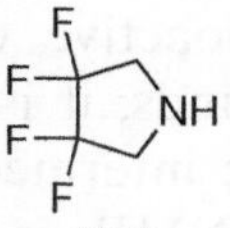

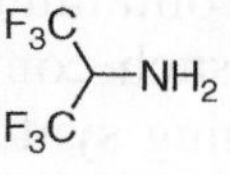

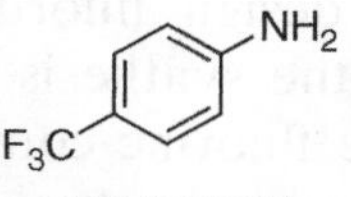

(all in H_2O, unless otherwise indicated)

1.1.4. Effect of Fluorinated Substituents on Lipophilicity of Molecules

Lipophilicity is an important consideration in the design of biologically active compounds because it often controls absorption, transport, or receptor binding; that is, it is a property that can enhance the bioavailability of a compound. The presence of fluorine in a substituent gives rise to enhanced lipophilicity.

For substituents on benzene, lipophilicities are given by values of π_X, as measured by the equation in Scheme 1.1, where P values are the octanol/water partition coefficients.

Scheme 1.1

$$\pi_X = \log P_{C6H5X} - \log P_{C6H6}$$

$$SO_2CH_3 < OH < NO_2 < OCH_3 < H < F < Cl < SO_2CF_3 < CH_3 < SCH_3 < CF_3 < OCF_3$$

$$< SF_5 < SCF_3 < C_2F_5$$

Representative π values: CH_3 (0.56), CF_3 (0.88), OCF_3 (1.04), SF_5 (1.23), SCF_3 (1.44)

As a measure of the impact of fluorine on a molecule's lipophilicity, the π-value of a CF_3 group is 0.88, as compared to 0.56 for a CH_3 group.

1.1.5. Other Effects

There is also evidence that single, carbon-bound fluorine substituents, particularly when on an aromatic ring, can exhibit specific polarity influences, including H-bonding, that can strongly influence binding to enzymes.[16, 27]

These and other insights regarding structure–activity relationships for fluorinated organic compounds allow researchers interested in exploiting the effects of fluorine substitution on bioactivity to more effectively design fluorine-containing bioactive compounds. In the process of the synthesis of such compounds, it is necessary to characterize the fluorine-containing synthetic intermediates and ultimate target compounds. Knowledge of ^{19}F NMR is essential for such characterization.

1.1.6. Analytical Applications in Biomedicinal Chemistry

Over the last decade or so, NMR spectroscopy has emerged as a screening tool to facilitate the drug discovery process, and nowhere has this been more the case than with ^{19}F NMR spectroscopy (more about this in Chapter 2).

1.2. INTRODUCTION TO FLUORINE NMR[28]

Aside from carbon and hydrogen, fluorine-19 is probably the most studied nucleus in NMR. The reasons for this include both the properties of the fluorine nucleus and the importance of molecules containing fluorine. The nucleus ^{19}F has the advantage of 100% natural abundance and a high magnetogyric ratio, about 0.94 times that of ^{1}H. The chemical shift range is very large compared to that of hydrogen, encompassing a range of >350 ppm for organofluorine compounds. Thus, resonances of different fluorine nuclei in a multifluorine-containing compound are usually well separated and the spectra are usually of first order. The nuclear spin quantum number for fluorine is $\frac{1}{2}$ and thus fluorine couples to proximate protons and carbons in a manner similar to hydrogen, and relaxation times are sufficiently long for spin–spin splittings to be resolved. Moreover, long-range spin–spin coupling constants to fluorine can have substantial magnitude, which can be particularly useful in providing extensive connectivity information, especially in ^{13}C NMR spectra.

Although it is of less general importance because of the limited number of phosphorous-containing fluoroorganics, ^{31}P also has a nuclear spin quantum number of $\frac{1}{2}$, its natural abundance is 100%, and it couples strongly to neighboring fluorine. When present, it can therefore have a significant influence on fluorine NMR spectra. ^{15}N also has a nuclear spin quantum number of $\frac{1}{2}$. However, its couplings to fluorine are almost never measured directly because of the very low natural abundance of ^{15}N (0.366%), combined with its small gyromagnetic ratio (-4.314 MHz/T), which is about 1/10th that of ^{1}H. Thus, indirect methods are almost always used to determine both the chemical shifts and any F–N coupling constants.

As hopefully demonstrated by the many examples in this book, a judicious use of fluorine NMR in combination with proton, carbon, phosphorous, and nitrogen NMR can provide unique advantages in the art of structure characterization. This is particularly true when one brings to the task a knowledge of the impact of fluorine substituents on the chemical shifts of and coupling constants to neighboring H, C, P, and N atoms.

1.2.1. Chemical Shifts

Fluorotrichloromethane ($CFCl_3$) has become the accepted, preferred internal reference for the measurement of ^{19}F NMR spectra, and, as such it is assigned a shift of zero. Signals upfield of the $CFCl_3$ peak are assigned negative chemical shift values, whereas those downfield of $CFCl_3$ are assigned positive values for their chemical shifts. When reporting fluorine chemical shifts, it is advised to report them relative to $CFCl_3$.

Other compounds that are commonly encountered as internal standards, particularly in the earlier literature are as follows:

CF_3CO_2H: −76.2 ppm
Hexafluorobenzene: −162.2 ppm
Trifluoromethylbenzene: −63.2 ppm
Ethyl trifluoroacetate: −75.8 ppm.

However, $CFCl_3$ has the advantage that its presence will not have any influence upon a compound's chemical shifts, plus its observed signal lies substantially downfield of most signals deriving from carbon-bound fluorine. Therefore, most fluorine chemical shifts (δ) are negative in value.

Nevertheless, one must be aware that some significant fluorine-containing functional groups, such as acylfluorides ($\delta \sim +20$ ppm), sulfonylfluorides ($\sim +60$ ppm), and pentafluorosulfanyl (SF_5) substituents (up to +85 ppm), have signals in the region *downfield* from $CFCl_3$. Signals deriving from aliphatic CH_2F groups lie at the high field end of the range, with *n*-alkyl fluorides absorbing at about −218 ppm. Methyl fluoride has one the highest field chemical shifts of an organofluorine compound at −268 ppm, tetrakis(fluoromethyl)silane having perhaps the highest (−277 ppm). Chapter 2 provides an overview of fluorine chemical shifts, with subsequent chapters providing details for each type of fluorinated substituent.

All chemical shift data presented in this book come either from the primary literature or from spectra obtained in the author's laboratory. All spectra actually depicted in the book, unless otherwise noted, derive from spectra obtained by the author at the University of Florida. All data from the literature were obtained via searches using Reaxys or SciFinder. Persons interested in accessing such primary literature can do so readily via these databases by simply searching for the specific compound mentioned in the text.

It should be noted that there are variations in reported chemical shifts for particular compounds in the literature, as would be expected. *Usually these variations are less than ±2 ppm,* and they can usually be attributed to concentration and solvent effects (as well as simple experimental error!). When given a choice, data reported using $CDCl_3$

as solvent have been preferred, with chemical shifts being reported to the nearest ppm (except occasionally when comparisons within a series from a common study are reported). When multiple values have been reported in the literature, the author has used his judgment regarding choice of the value to use in the book.

Increasingly effective efforts have been made to calculate fluorine chemical shifts. Some leading references to such theoretical work are provided.[29–32]

1.2.2. Coupling Constants

Fluorine spin–spin coupling constants to other fluorine nuclei, to neighboring hydrogen nuclei and to carbons, phosphorous, or nitrogen in the vicinity of the fluorine substituents are highly variable in magnitude but are also highly characteristic of their environment. The magnitude of such characteristic coupling constants is discussed in each of the subsequent chapters that describe the different structural situations of fluorine substitution.

Spin–spin coupling constants are reported throughout this book as absolute values of $|J|$ in hertz (Hz), and they have all been obtained either from the primary literature or from spectra obtained in the author's laboratory.

REFERENCES

Regarding the multitude of NMR chemical shifts of specific compounds that are provided within the text, references for chemical shifts of individual compounds will for the most part not be cited. It is assumed that if such references are required, the reader can find them by a quick search using either Reaxys or SciFinder. The author found Reaxys much the superior database for locating specific NMR data.

1. Chambers, R. D. *Fluorine in Organic Chemistry*; Blackwell Publishing: Oxford, **2004**.

2. Chambers, R. D. *Fluorine in Organic Chemistry*; John Wiley and Sons: New York, **1973**.

3. Uneyama, K. *Organofluorine Chemistry*; Blackwell Publishing: Oxford, **2006**.

4. Hiyama, T. *Organofluorine Compounds. Chemistry and Applications*; Springer: Berlin, **2000**.

5. Welch, J. T.; Eswarakrishnan, S. *Fluorine in Bioorganic Chemistry*; John Wiley and Sons: New York, **1991**.

6. Begue, J.-P.; Bonnet-Delpon, D. *Bioorganic and Medicinal Chemistry of Fluorine*; Wiley: Hoboken, NJ, **2008**.

7. Smart, B. E. In *Organofluorine Chemistry – Principles and Commercial Applications*; Banks, R. E., Smart, B. E., Tatlow, J. C., Eds.; Plenum Press: New York, **1994**, p 57.

8. O'Hagan, D. *Chem. Soc. Rev.* **2008**, *37*, 308.

9. Kirsch, P. *Modern Fluoroorganic Chemistry*; Wiley-VCH: Weinheim, **2004**.

10. Berger, R.; Resnati, G.; Metrangolo, P.; Weber, E.; Hullinger, J. *Chem. Soc. Rev.* **2011**, *40*, 3496.

11. Bissantz, C.; Kuhn, B.; Stahl, M. *J. Med. Chem.* **2010**, *53*, 5061.

12. Ojima, I. *Fluorine in Medicinal Chemistry and Chemical Biology*; Wiley-Blackwell: Chichester, UK, **2009**.

13. Huchet, Q. A.; Kuhn, B.; Wagner, B.; Fischer, H.; Kansy, M.; Zimmerli, D.; Carreira, E. M.; Mueller, K. *J. Fluorine Chem.* **2013**, *152*, 119.

14. Böhm, H.-J.; Banner, D.; Bendels, S.; Kansy, M.; Kuhn, B.; Müller, K.; Obst-Sander, U.; Stahl, M. *ChemBioChem* **2004**, *5*, 637.

15. Müller, K.; Faeh, C.; Diederich, F. *Science* **2007**, *317*, 1881.

16. Purser, S.; Moore, P. R.; Swallow, S.; Gouverneur, V. *Chem. Soc. Rev.* **2008**, *37*, 320.

17. Kirk, K. L. *Org. Proc. Res. Dev.* **2008**, *12*, 305.

18. Isanbor, C.; O'Hagan, D. *J. Fluorine Chem.* **2006**, *127*, 303.

19. Begue, J.-P.; Bonnet-Delpon, D. *J. Fluorine Chem.* **2006**, *127*, 992.

20. Kirk, K. L. *J. Fluorine Chem.* **2006**, *127*, 1013.

21. Fujiwara, T.; O'Hagan, D. *J. Fluorine Chem.* **2014**, *167*, 16.

22. Ampt, K. A. M.; Aspers, R. L. E. G.; Jaeger, M.; Geutjes, P. E. T. J.; Honing, M.; Wijmenga, S. S. *Magn. Res. Chem.* **2011**, *49*, 221.

23. Wang, J.; Sanchez-Rosello, M.; Acena, J. L.; del Pozo, C.; Sorochinsky, A. E.; Fustero, S.; Soloshonok, V. A.; Liu, H. *Chem. Rev.* **2014**, *114*, 2432.

24. Carcenac, Y.; Tordeux, M.; Wakselman, C.; Diter, P. *New J. Chem.* **2006**, *30*, 447.

25. Hansch, C.; Leo, A.; Taft, R. W. *Chem. Rev.* **1991**, *91*, 165.

26. Roberts, R. D.; Ferran, H. E., Jr.; Gula, M. J.; Spencer, T. A. *J. Am. Chem. Soc.* **1980**, *102*, 7054.

27. Champagne, P. A.; Desroches, J.; Paquin, J.-F. *Synthesis* **2015**, *47*, 306.

28. Hesse, M.; Meier, H.; Zeeh, B. *Spectroscopic Methods in Organic Chemistry*; Georg Thieme Verlag: Stuttgart, **1997**.

29. Wiberg, K. B.; Zilm, K. W. *J. Org. Chem.* **2001**, *66*, 2809.

30. Fukaya, H.; Ono, T. *J. Comput. Chem.* **2004**, *25*, 51.

31. Saielli, G.; Bini, R.; Bagno, A. *Theor. Chem. Acc.* **2012**, *131*, 1140.

32. Raimer, B.; Jones, P. G.; Lindel, T. *J. Fluorine Chem.* **2014**, *166*, 8.

CHAPTER 2

AN OVERVIEW OF FLUORINE NMR

2.1. INTRODUCTION

If one wishes to obtain a fluorine NMR spectrum, one must of course first have access to a spectrometer with a probe that will allow observation of fluorine nuclei. Fortunately, most modern high-field NMR spectrometers that are available in industrial and academic research laboratories today have this capability. Probably the most common NMR spectrometers in use today for taking routine NMR spectra are 300, 400, and 500 MHz instruments, which measure proton spectra at 300, 400, and 500 MHz, carbon spectra at 75.5, 100.6, and125.8 MHz and fluorine spectra at 282, 376, and 470 MHz, respectively. For the most part, and unless otherwise mentioned, the spectra that are depicted in this book are 500 MHz for proton, 125.8 for carbon, and 282 for fluorine, and all have been obtained within the University of Florida Chemistry Department NMR facility.

Before obtaining and attempting to interpret fluorine NMR spectra, it would be advisable to become familiar with some of the fundamental concepts related to fluorine chemical shifts and spin–spin coupling constants that are presented in this book. There is also a very nice introduction to fluorine NMR by Brey and Brey.[1]

Guide to Fluorine NMR for Organic Chemists, Second Edition. William R. Dolbier, Jr.
© 2016 John Wiley & Sons, Inc. Published 2016 by John Wiley & Sons, Inc.

For those new to the field of fluorine NMR, there are a number of convenient aspects about fluorine NMR that make the transition from proton NMR to fluorine NMR relatively easy. With a nuclear spin of $1/2$ and having almost equal sensitivity to hydrogen along with sufficiently long relaxation times to provide reliable integration values, ^{19}F nuclei provide NMR spectra that very much resemble proton spectra, with the additional benefit of having a much broader range of chemical shifts, which means that one usually will not encounter overlapping signals in compounds that contain multiple fluorine-containing substituents, and thus most spectra will be of first order. Also, since it is not usual to employ proton decoupling when obtaining fluorine NMR spectra, one will observe not only coupling between proximate fluorine substituents but also between fluorine nuclei and proton nuclei, with the magnitude of geminal and vicinal F–F and F–H coupling constants generally being larger than the respective H–H spin–spin coupling constants.

As is the case for ^{1}H spectra, but not for ^{13}C spectra, the intensities of individual signals in ^{19}F NMR spectra constitute an accurate measure of the relative number of fluorine atoms responsible for such signals.

Because today the majority of organic chemists who make fluoroorganic compounds work in pharmaceutical and agrochemical industries, and such people are primarily interested in lightly fluorinated molecules, the emphasis in this book will be the NMR analysis of compounds containing one, two, or three fluorine atoms or bearing substituents containing a limited number of fluorines, with the goal of understanding how the chemical shifts and spin–spin couplings of such substituents are affected by the structural environment in which they exist.

2.2. FLUORINE CHEMICAL SHIFTS

The observed resonance frequency of any NMR-active nucleus depends in a characteristic manner upon the magnetic environment of that nucleus. The effective field strength (B_{eff}) felt by the nucleus of an atom that has magnetic moment differs from the imposed field (B_0) in the following manner (Equation 2.1):

$$B_{eff} = B_0 - \sigma B_0, \quad \text{where } \sigma \text{ is the dimensionless shielding constant}$$

$$(2.1)$$

This shielding constant, σ, is made up of three terms (Equation 2.2):

$$\sigma = \sigma_{dia} + \sigma_{para} + \sigma^i \qquad (2.2)$$

The diamagnetic term, σ_{dia}, corresponds to the opposing field resulting from the effect of the imposed field upon the electron cloud surrounding the nucleus. In this case, electrons closer to the nucleus give rise to greater shielding than distant ones.

The paramagnetic term, σ_{para}, derives from the excitation of p-electrons by the external field, and its impact is opposite to that of diamagnetic shielding. The term, σ^i, derives from the effect of neighboring groups, which can increase or decrease the field at the nucleus. σ can also be affected by intermolecular effects, in most cases deriving from interaction of the solvent.

In the case of proton spectra, only s-orbitals are present. Thus, only σ_{dia} is important, whereas, in contrast, the paramagnetic term, σ_{para}, is dominant in determining the relative shielding of fluorine nuclei. Thus, the "normal" intuitions regarding "shielding" that most chemists have acquired while working with ^1H NMR generally do not apply when it comes to predicting relative chemical shifts in ^{19}F NMR. For example, the fluorine nucleus of $ClCH_2CH_2F$ is more highly shielded ($\delta_F = -220$) than that of CH_3CH_2F ($\delta_F = -212$).

There are other notable differences between fluorine and proton NMR spectra. For example, the effects of anisotropic magnetic fields, such as those generated by ring currents, are relatively much less important for fluorine than for proton NMR. Thus, the ranges of vinylic and aromatic fluorine chemical shifts overlap completely. Also notable is the much greater sensitivity of single carbon-bound F-substituents to environment than carbon-bound CF_2 or CF_3 substituents. Single fluorine chemical shifts, which encompass vinylic, aryl, and saturated aliphatic fluorine substituents, range between about -70 and -238 ppm, whereas the similar range for CF_2 groups is -80 to -130 ppm, and that of the CF_3 group is even smaller, between about -52 and -87 ppm.

In general, and all other things being equal, the fluorines of a trifluoromethyl group are more deshielded than those of a CF_2H or $R-CF_2-R'$ group, which are themselves more deshielded than a single fluorine substituent (Scheme 2.1).

2.2.1. Some Aspects of Shielding/Deshielding Effects on Fluorine Chemical Shifts

2.2.1.1. Influence of α-Halogen or Chalcogen. As shown in Tables 2.1 and 2.2, the effects of α-halogen or α-chalcogen substitution on fluorine chemical shifts are variable, depending on whether this substitution is on CF_3, CF_2H, and CH_2F groups. Thus, CF_3 and CF_2H groups are observed to be increasingly deshielded by $F < Cl < Br < I$

Scheme 2.1

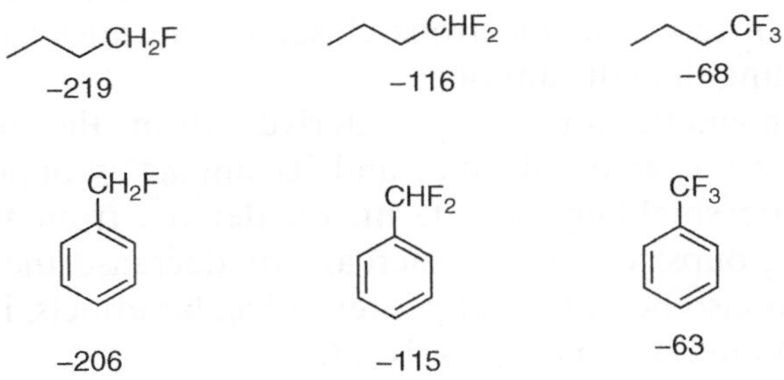

CH$_2$F	CHF$_2$	CF$_3$
−219	−116	−68

CH$_2$F	CHF$_2$	CF$_3$
−206	−115	−63

TABLE 2.1. Impact of α-Halogen Substitution on Fluorine Chemical Shifts

X	CH$_3$	F	Cl	Br	I
δ, CF$_3$X	−65	−62	−33	−21	−5
δ, HCF$_2$X	−110	−78	−73	−70	−68
δ, H$_2$CFX	−212	−143	−169	–	−191

TABLE 2.2. Impact of α-Chalcogen Substitution on Fluorine Chemical Shifts

Y	F	OPh	SPh	SePh
CF$_3$Y, δ	−62	−58	−43	−37
HCF$_2$Y, δ	−79	−87	−96	−94
H$_2$CFY, δ	−143	−149	−180	–

(opposite to the trend observed for proton chemical shifts), with the trend being more pronounced for CF$_3$. In contrast, CH$_2$F groups are increasingly *shielded* going from F to I substitution (Table 2.1).

Similarly, the analogous α-chalcogen substitution only exhibits a consistent deshielding trend for the CF$_3$ group (F <OPh <SPh < SePh), with shielding effects being observed for both CF$_2$H and CH$_2$F groups (Table 2.2).

2.2.1.2. Substituent Effects of Aryl Fluorides versus Benzylic Fluorides.
A remarkable difference is observed for the effect of *para* substituents on the chemical shifts of aryl fluorides versus benzyl fluorides. In the case of aryl fluorides, electron-donor substituents cause *shielding* of the fluorine nucleus, whereas such substituents give rise to *deshielding* of the fluorine nuclei of benzylic fluorides (Scheme 2.2).[2]

Scheme 2.2 Contrasting effects of substituents on fluorine chemicals shifts of aryl and benzyl fluorids

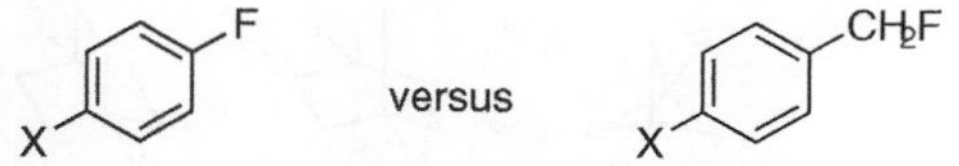

versus

X = NO$_2$	−103.6		−216.2	Most shielded
CN	−104.0		−215.7	
H	−113.8		−207.3	
Cl	−116.7		−208.0	
Me	−119.2		−204.3	
OMe	−125.2	Most shielded	−199.8	

With regard to the aryl fluorides, the degree of shielding is considered to be related to the π charge density of the fluorine atom.[3] On the other hand, hyperconjugative π–σ_{CF}^{*} interactions are thought to be responsible for the observed trend for the series of benzyl fluorides.

2.2.1.3. Homohyperconjugative Effects in 7-Norbornenyl Fluorides.
Related to the observed trend of p-substituted benzyl fluorides is the observation that the *anti*-isomer of norbornen-7-yl fluoride exhibits a large deshielding when compared to either the *endo* isomer or the saturated system (Scheme 2.3).[4] This was attributed to homohyperconjugative π–σ_{CF}^{*} effects, in this case because the π-bond is antiperiplanar to the C—F bond.

Scheme 2.3 7-Norbornenyl fluorides

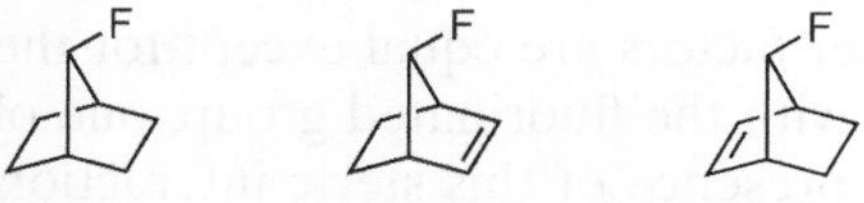

−202.9   −200.8   −178.2

Such shift variations, as in the cases of benzylic and 7-norbornenyl fluorides, where the presence of donor groups leads to *deshielding* appear to be counterintuitive.

2.2.1.4. Transmission of Polar Substituent Effects in Bridgehead Fluorides.
There is a considerable literature, mainly from the group of Adcock, dealing with the transmission of polar substituent effects with

Scheme 2.4 Substituent effects on chemical shifts of bridgehead bicyclic fluorides

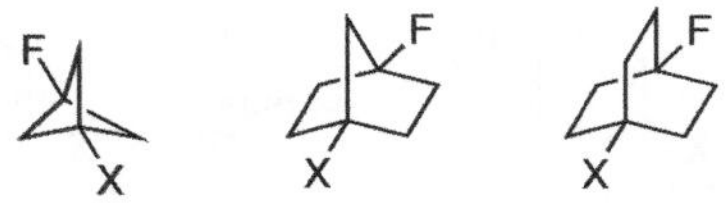

X = NO$_2$	−181.2	−174.8	−158.3
H	**−133.2**	**−182.7**	**−148.4**
NH$_2$	−171.2	−171.4	−155.9
CH$_3$	−145.4	−176.1	−152.3

bridgehead bicyclic fluorides. In such systems, fluorine chemical shifts act as sensitive probes for monitoring σ-electron delocalization effects (Scheme 2.4).

The observed, sometimes disparate, effects shown in Scheme 2.4 are considered to be derived predominantly from electric field and electronegativity effects, with σ-electron delocalization mechanisms such as "through-bond" and "through-space (TS)" effects being invoked. Those readers interested in such effects should consult the primary literature in this area.[4–6]

2.2.1.5. Steric Deshielding of Fluorine Substituents. Another significant and not infrequently encountered impact on fluorine chemical shifts is the deshielding influence of a physically proximate alkyl group upon CF$_3$ groups, CF$_2$ groups, and aromatic C−F (based upon limited data, there does not appear to be significant effect on CH$_2$F groups).[7] Under circumstances such as those depicted in Scheme 2.5, where structurally all other factors are equal except for the steric interaction of the alkyl group with the fluorinated group, one observes significant deshielding in the presence of this steric interaction. This deshielding is understood to occur only when there is direct overlap of the van der Waals radii of the alkyl group and that of the fluorine, and the deshielding is thought to be the result of van der Waals forces of the alkyl group restricting the motion of electrons on the fluorine and thus making the fluorine nucleus respond to the magnetic field as if the electron density were lowered.

The most common situation where this effect is seen is in a comparison of E- and Z- isomers of trifluoromethyl- or difluoromethyl-substituted alkenes, but as the naphthalene examples indicate, the effect is not unique to that situation.

Scheme 2.5

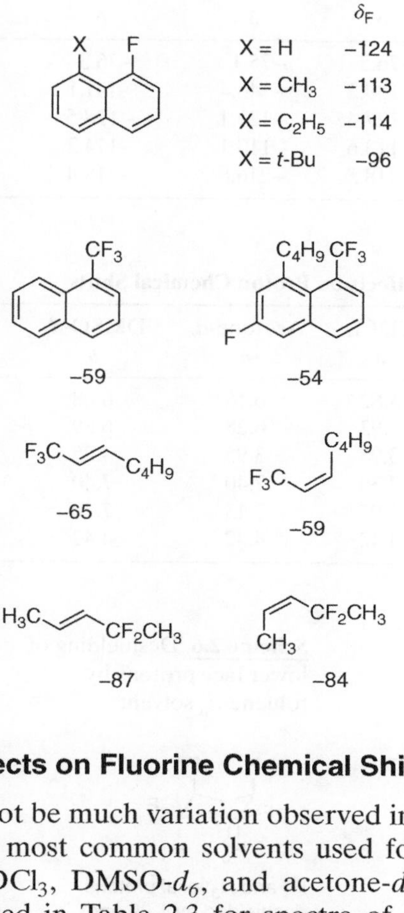

$$\delta_F$$

X = H   −124
X = CH$_3$ −113
X = C$_2$H$_5$ −114
X = t-Bu   −96

CF$_3$

−59

C$_4$H$_9$ CF$_3$

F

−54

F$_3$C⌐⌐C$_4$H$_9$

−65

C$_4$H$_9$

F$_3$C

−59

H$_3$C⌐⌐CF$_2$CH$_3$

−87

CF$_2$CH$_3$

CH$_3$

−84

2.2.2. Solvent Effects on Fluorine Chemical Shifts

There will usually not be much variation observed in fluorine chemical shifts for the three most common solvents used for obtaining NMR spectra, that is, CDCl$_3$, DMSO-d_6, and acetone-d_6, as can be seen in the data presented in Table 2.3 for spectra of a series of typical fluorine-containing compounds in various solvents. The variation in fluorine chemical shifts for these three solvents is no more than ±1 ppm. Thus, in reporting chemical shifts in this book, no mention of specific solvent is made, although the vast majority of spectra will have been measured in CDCl$_3$.

Larger solvent effects can be observed for proton spectra, particularly when using benzene-d_6. As can be seen from the data in Table 2.4, proton chemical shifts in the other solvents, particularly CDCl$_3$ and acetone-d_6, are reasonably consistent.

These types of dramatic shifts are indicative of C—H/π electrostatic interactions, when molecules have hydrogens that are strongly polarized in an electropositive fashion. A notable example has been noted

TABLE 2.3. Solvent Effects on Fluorine Chemical Shifts

Compound	CDCl$_3$ δ	DMSO-d_6 δ	Acetone-d_6 δ	Benzene-d_6 δ	CD$_3$OD δ
CF$_3$CHClBr	−76.5	−75.1	−76.3	−76.6	−77.5
HCF$_2$CF$_2$CH$_2$OH	−139.2	−140.4	−141.1	−139.7	−141.8
	−127.4	−127.1	−128.5	−127.8	−129.4
Fluorobenzene	−113.6	−113.1	−114.2	−113.3	−115.2
1-Fluorooctane	−218.5	−216.8	−218.4	−218.2	−219.7

TABLE 2.4. Solvent Effects on Proton Chemical Shifts

Compound	CDCl$_3$ δ	Acetone-d_6 δ	DMSO-d_6 δ	Benzene-d_6 δ	CD$_3$OD δ
CF$_3$CHClBr	5.82	6.16	6.08	*4.63*	5.77
HCF$_2$CF$_2$CH$_2$OH	5.93	6.28	6.89	*5.34*	6.12
	3.98	3.95	4.25	*3.29*	3.86
Fluorobenzene	7.30	7.40	7.40	*6.88*	7.33
	7.07	7.15	7.20	*6.78*	7.09
1-Fluorooctane	4.42	4.42	4.42	*4.12*	4.44

Scheme 2.6 Deshielding of lower face protons by toluene-d_6 solvent

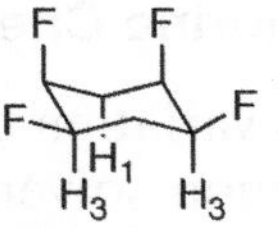

H$_1$ and H$_3$ axial protons deshielded by 1.69 and 1.31 ppm, respectively

by O'Hagan for multifluorine-substituted cyclohexanes, such as the all-*syn* isomer of 1,2,4,5-tetrafluorocyclohexane, that is polarized to have electronegative and electropositive faces (Scheme 2.6).[8] All of the lower-face protons are significantly deshielded in toluene-d$_6$, relative to CD$_2$Cl$_2$, but the axial protons are particularly so.

2.2.3. Overall Summary of Fluorine Chemical Shift Ranges

Figure 2.1 provides a quick overview of the basic chemical shift ranges for carbon-bound F, CF$_2$, and CF$_3$ substituents. Specific details

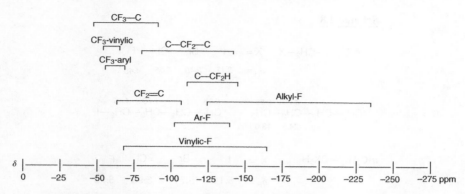

FIGURE 2.1. Overview of fluorine chemical shifts (relative to $CFCl_3$ ($\delta = 0$))

regarding the effect of environment on such chemical shifts are found in Chapters 3–5, respectively.

Although the chemical shifts of most commonly encountered organofluorine compounds are upfield of $CFCl_3$ and thus have negative values, there are a number of structural situations for fluorine that lead to positive chemical shifts (downfield from $CFCl_3$). These include acyl and sulfonyl fluorides as well as the fluorines of SF_5 substituents.

2.3. THE EFFECT OF FLUORINE SUBSTITUENTS ON PROTON CHEMICAL SHIFTS

In general, fluorine substitution has exactly the impact upon proton chemical shifts that one would expect from a substituent of its high electronegativity. That is, its effect is largely inductive in nature. Schemes 2.7 and 2.8 provide a comparison of the relative effect of fluorine substituents as compared with those of other halogens and oxygen, which is closest in electronegativity to fluorine.

Scheme 2.7

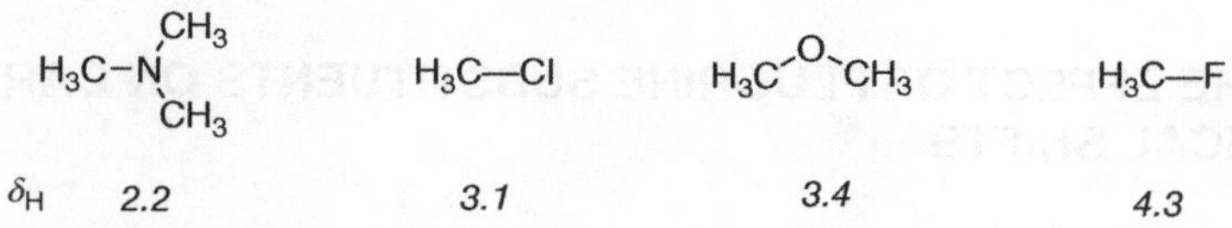

$H_3C-\overset{\underset{\displaystyle CH_3}{\displaystyle |}}{\underset{\displaystyle |}{\overset{\displaystyle CH_3}{N}}}$	H_3C-Cl	$H_3C\overset{O}{\diagup}CH_3$	H_3C-F
δ_H 2.2	3.1	3.4	4.3

Thus, it is seen that the effect of a *single* fluorine substituent upon the chemical shifts of protons attached to the same carbon adheres to

Scheme 2.8

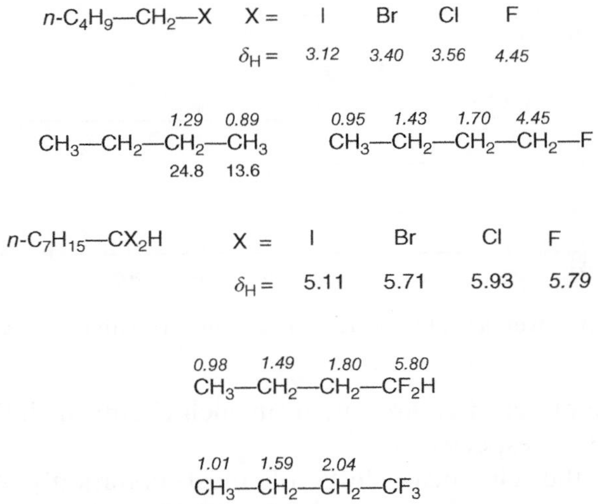

n-C$_4$H$_9$—CH$_2$—X   X =   I   Br   Cl   F

δ_H =   3.12   3.40   3.56   4.45

$\overset{1.29}{}$ $\overset{0.89}{}$
CH$_3$—CH$_2$—CH$_2$—CH$_3$
$\underset{24.8}{}$ $\underset{13.6}{}$

$\overset{0.95}{}$ $\overset{1.43}{}$ $\overset{1.70}{}$ $\overset{4.45}{}$
CH$_3$—CH$_2$—CH$_2$—CH$_2$—F

n-C$_7$H$_{15}$—CX$_2$H   X =   I   Br   Cl   F

δ_H =   5.11   5.71   5.93   5.79

$\overset{0.98}{}$ $\overset{1.49}{}$ $\overset{1.80}{}$ $\overset{5.80}{}$
CH$_3$—CH$_2$—CH$_2$—CF$_2$H

$\overset{1.01}{}$ $\overset{1.59}{}$ $\overset{2.04}{}$
CH$_3$—CH$_2$—CH$_2$—CF$_3$

expectations based upon its expected inductive effect relative to atoms of lesser electronegativity.

As is consistent with other properties deriving from inductive influences of a substituent, the influence of fluorine substituents on proton chemical shifts drops off dramatically as the fluorine becomes farther removed from the hydrogen in question. Thus, protons on the γ-carbon or farther away are essentially unaffected by a single F.

An interesting phenomenon is noted for the effect of the two fluorines of a CF$_2$H group upon the chemical shift of its hydrogen. As seen by the comparative data given in Scheme 2.8, the observed degree of deshielding due to the two fluorines is *less* than that of two chlorines, and barely more than two bromines.

Lastly, as can be seen from the data at the bottom of Scheme 2.8, the inductive effect of a trifluoromethyl group affects chemical shifts of protons as distant as three carbons away.

2.4. THE EFFECT OF FLUORINE SUBSTITUENTS ON CARBON CHEMICAL SHIFTS

Carbon NMR has been considered an essential characterization tool in spite of the fact that its major isotope, the ^{12}C nucleus, is NMR inactive. In spite of its low natural abundance (1.1%) and its low magnetic moment, which lead to a significantly low overall sensitivity relative to hydrogen (1.76×10^{-4}), ^{13}C has a $1/2$ spin quantum number, and carbon NMR has become essential to structure characterization. The acquisition of ^{13}C NMR spectra has remained a priority in the development of commercial NMR instruments from the very beginning.

As is the case with proton chemical shifts, ^{13}C NMR chemical shifts are affected by proximate fluorine substituents in a manner that would be expected by a substituent of its electronegativity (Scheme 2.9). This inductive effect appears to apply mainly to carbon-1 and carbon-2. Carbon-3 appears to actually be shielded by the nearby fluorine, whereas carbon-4 and carbon-5 are unaffected. Although one must be careful to be aware of exceptions to the trend, ^{13}C nuclei and the protons attached to them often show parallel behavior with respect to the effect of substituents on their chemical shifts.

Scheme 2.9

$$CH_3—CH_2—CH_2—CH_2—CH_3 \qquad CH_3—CH_2—CH_2—CH_2—CH_2—F$$
14.1 22.4 34.2 22.4 14.1 $\qquad$ 14.0 22.8 27.9 30.8 82.8

$$CH_3—CH_2—CH_2—CF_2H$$
13.8 15.8 36.2 117.6

$$CH_3—CH_2—CH_2—CF_3$$
13.4 15.7 35.9 127.5

2.5. THE EFFECT OF FLUORINE SUBSTITUENTS ON ^{31}P CHEMICAL SHIFTS

^{31}P is an NMR active nucleus with a nuclear spin of $1/2$. Its sensitivity (relative to H) is only 0.066, but because of its 100% natural abundance, the overall relative sensitivity of ^{31}P is about 375 times that of ^{13}C. Thus, acquisition of phosphorous NMR spectra is relatively easy. The internal standard generally used for measuring ^{31}P chemical shifts is 85% H_3PO_4.

When in the structural vicinity of a fluorine substituent, phosphorous chemical shifts can be significantly affected by the presence of fluorine substituents but not in the way one might think. In contrast to the intuitive expectation based upon the inductive effect of fluorine, which results in the deshielding of protons and carbons in their vicinity, the effect of proximal fluorine upon the chemical shifts of phosphorous compounds is to *shield* the phosphorous nuclei. Scheme 2.10 provides a few examples.

<u>**Scheme 2.10**</u>

Ph_3P versus Ph_3PF_2

$\delta_P = -6$ $\delta_P = -55$

$Ph_3\overset{+}{P}$-CH_3 BF_4^- versus $Ph_3\overset{+}{P}$-CH_2F BF_4^-

$\delta_P = +21.6$ $\delta_P = +18.2$

Ph_2PO-CH_3 versus Ph_2PO-CH_2F

$\delta_P = +30.2$ $\delta_P = +23.4$

$Ph_2PCH_2CH_3$ versus $Ph_2PCH_2CF_3$

$\delta_P = -12$ $\delta_P = -27$

For those examples of phosphorous-containing compounds included in this book, ^{31}P NMR data will be reported, when available.

2.6. THE EFFECT OF FLUORINE SUBSTITUENTS ON ^{15}N CHEMICAL SHIFTS

Nitrogen NMR data for compounds containing fluorine have been relatively scarce in the literature. Although the ^{14}N nucleus is NMR active and is >99% natural abundance, its sensitivity is only 1/1000th that of hydrogen, and it is encumbered by a nuclear spin of 1, which means that it possesses an *electric nuclear quadrupole moment* that produces signal broadening in spectra. Also, the spectra reflect the $2NI + 1$ rule, which means that a single nitrogen will split hydrogens and fluorines into broad triplets. These factors, when combined with its very low sensitivity, means that ^{15}N couplings are not observed under normal conditions of taking proton or fluorine NMR spectra.

On the other hand, nitrogen's minor (0.366%) natural-occurring isotope, ^{15}N, has a nuclear spin of $1/2$ but still only 1/1000th the sensitivity of a proton. This gives it a 3.9×10^{-6} overall sensitivity relative to hydrogen. Nevertheless in recent years, ^{15}N NMR data, deriving from indirect methodology, have begun to appear in the literature.

For compounds that have couplings between N, F, and H, the measurement of ^{15}N chemical shifts and ^{19}F–^{15}N couplings through the use of a ^1H–^{15}N gHSQC experiment can be convenient, since such experiments use common instrument capabilities: signal routing for indirect detection, pulse-field gradients for the suppression of the signals of the

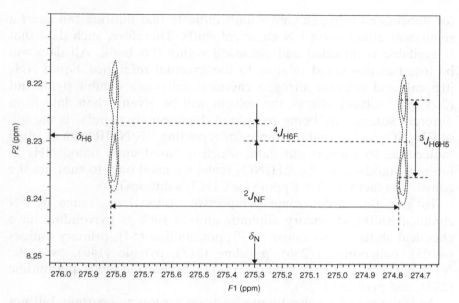

FIGURE 2.2. H6-N cross peak in a ^{1}H–^{15}N gHSQC experiment with 2-fluoropyridine

protons bound to ^{15}N, and a probe tunable to ^{15}N. ^{19}F–^{15}N couplings are measured in F1, where the H–N couplings have been focused. If ^{15}N decoupling is used during acquisition, the two cross peaks are separated by J_{FN} in F1 and by J_{FH} in F2. This method is limited to N atoms that couple with protons. In the example provided in Figure 2.2,[1] both H6 of the 2-fluoropyridine and N couple with the fluorine substituent. Thus, there are two peaks for the alpha and beta states of ^{19}F. Each of the peaks preserves the multiplicity of the signal in the proton spectrum: H6 is a doublet of 4 Hz through coupling with H5. The chemical shifts of H6 (8.23 ppm) and N (275.3 ppm) can be read in the middle of the two peaks on the respective F2 and F1 axes. The coupling constant $^2J_{FN}$ (52.8 Hz) is the distance in hertz between the peaks on the ^{15}N axis (F1). The coupling constant $^4J_{H6F}$ (0.8 Hz) is the distance in hertz between the peaks on the ^{1}H axis (F2).

With many compounds of interest to both pharmaceutical and agrochemical chemists containing nitrogen, and with the increasing capabilities of modern NMR instrumentation, the potential for use of ^{15}N NMR data in structure characterization is beginning to be recognized and utilized. Substituent effects on ^{15}N chemical shifts can be quite dramatic, and the relatively few data that are available

[1] Thanks to Dr. Ion Ghiviriga for providing this Figure.

for fluorinated nitrogen compounds indicate that fluorine can exert a significant effect upon ^{15}N chemical shifts. Therefore, such data that is available is included and discussed within this book. All data will be listed as downfield relative to the external reference, liquid NH_3 [0 ppm], and because nitrogen chemical shifts can exhibit significant (2–5 ppm) solvent effects, the solvent will be given, when data from different sources are being compared. Increasingly popular is the use of CH_3NO_2 as internal reference for reporting ^{15}N NMR data of small molecules. To convert our data, which is based upon liquid NH_3 as internal standard, to the CH_3NO_2 scale, we need only to subtract the conversion factor of 379.8 ppm (for $CDCl_3$ solutions).

To give the reader some perspective of the large range of ^{15}N chemical shifts, secondary aliphatic amines such as pyrrolidine have chemical shifts in the range of 38 ppm, aniline (54), primary amides (~105), benzonitrile (256), pyridine (317), pyrrole (146), pyrazole: $N - 1$ (207), $N - 2$ (399), imidazole (211), pyridazine (400), pyrimidine (295), and pyrazine (333).[9]

As is the case for substituent effects on proton and carbon, but not phosphorous NMR chemical shifts, nitrogen chemical shifts exhibit deshielding when in the vicinity of substituents that are more electronegative than carbon (Scheme 2.11). Presumably, although there are little data to support this assertion, the effect diminishes rapidly with distance of the substituent from the nitrogen, as is the case for proton and carbon chemical shifts. However, in aromatic heterocycles, such as pyridine, the effect of the substituent is transmitted through the π system, as indicated by the significant influence of para substituents of pyridines (Scheme 2.12).[10]

Scheme 2.11 The effect of neighboring fluorine on ^{15}N chemical shifts

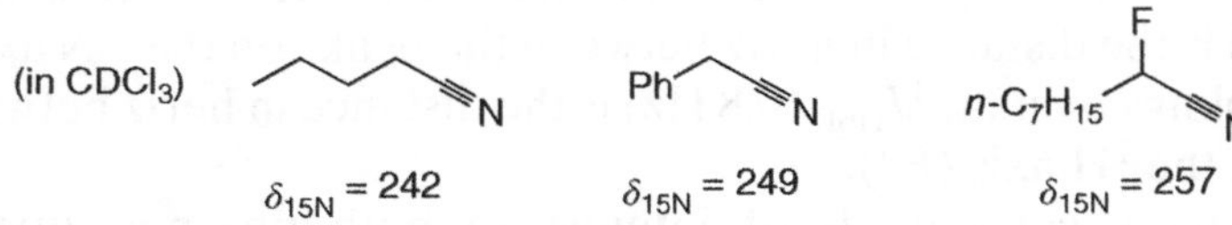

(in $CDCl_3$)

$\delta_{15N} = 242$ $\delta_{15N} = 249$ $\delta_{15N} = 257$

Scheme 2.12 ^{15}N chemical shifts of 4-substituted pyridines

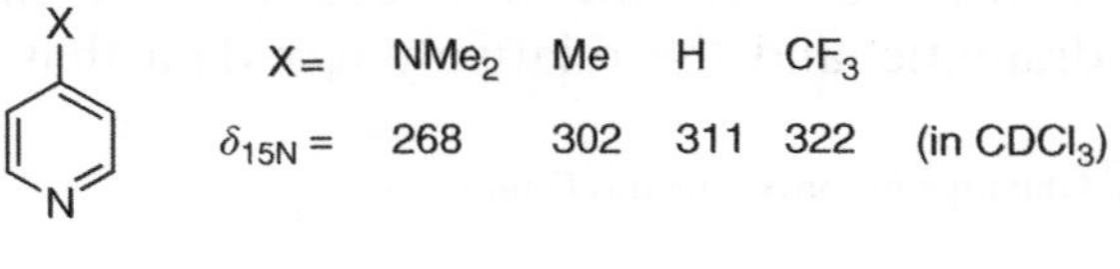

X =	NMe_2	Me	H	CF_3	
$\delta_{15N} =$	268	302	311	322	(in $CDCl_3$)

2.7. SPIN–SPIN COUPLING CONSTANTS TO FLUORINE

Couplings between NMR active nuclei, such as hydrogen, carbon, phosphorous, nitrogen, and fluorine, are generally transmitted through the electrons of the covalent bonds, and therefore decrease rapidly with the increasing number of bonds on the pathway between the two coupling nuclei. The fact that the values of the spin–spin coupling constants become, in most cases, very small after three bonds have been for a long time the basis of structure elucidation by NMR.

Most fluorine NMR spectra are what are considered to be *first order* in nature, which means that, because fluorine, hydrogen, and phosphorous nuclei all are $I = \frac{1}{2}$ nuclei, multiplicities resulting from spin–spin coupling will reflect the $2nI + 1$ rule, that is, the $n + 1$ rule. The relative intensities of the peaks within the multiplets will also correspond to the binomial expansion given by Pascal's triangle for spin-$\frac{1}{2}$ nuclei. Thus, fluorine NMR signals will exhibit multiplets that derive from both fluorine–fluorine and hydrogen–fluorine, and when phosphorous is present, phosphorous–fluorine coupling. Most of the following discussion will relate to fluorine–fluorine and hydrogen–fluorine coupling.

As a simple example, the fluorine and proton NMR spectra of $CF_2Cl\text{-}CHCl_2$ (Figure 2.3) are given in Figures 2.4 and 2.5, respectively. In these spectra, the only coupling is between the two magnetically equivalent fluorines of the CF_2Cl group and the single hydrogen of the $CHCl_2$ group, the former being split into a *doublet* by the single hydrogen, while the latter is split into a *triplet* by the two fluorines, each with requisite identical 10 Hz coupling constants.

In another example, the signal for the trifluoromethyl group of 1,1,1-trifluoropropane ($\delta = -69.1$) will be split into a triplet ($^3J_{HF} = 10.5$) by the neighboring two protons in its fluorine NMR (Figure 2.6), with these same protons ($\delta = 2.10$) being split into a quartet of quartets in the proton NMR spectrum (Figure 2.7) by the CF_3 group ($^3J_{FH} = 10.5$) and by the CH_3 group ($^3J_{HH} = 7.5$ Hz).

There are some general concepts and trends that should be mentioned here regarding the observed magnitude of vicinal F–F and

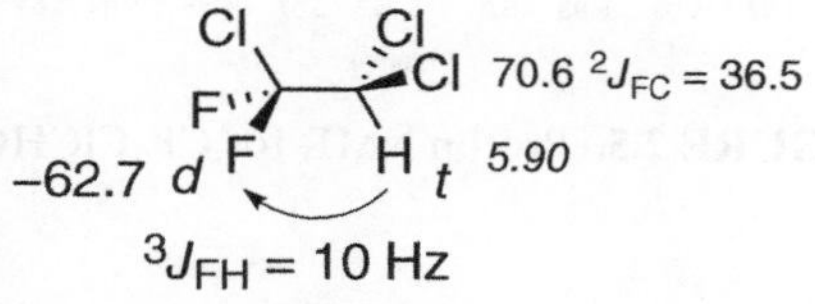

FIGURE 2.3. NMR data for $CF_2ClCHCl_2$

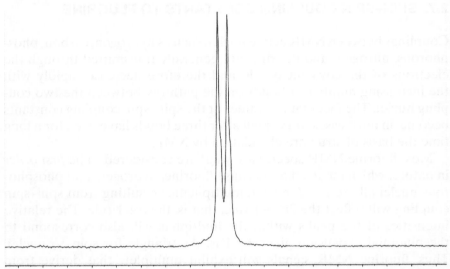

FIGURE 2.4. Fluorine NMR of $CF_2ClCHCl_2$

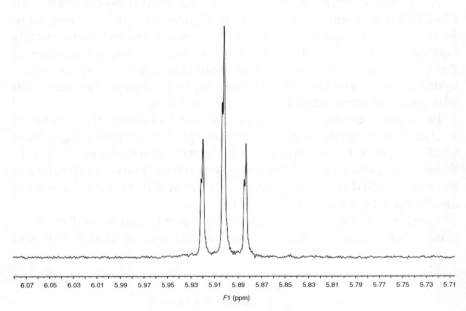

FIGURE 2.5. Proton NMR for $CF_2ClCHCl_2$

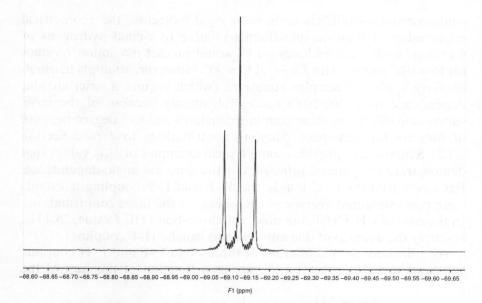

FIGURE 2.6. ^{19}F NMR spectrum of 1,1,1-trifluoropropane

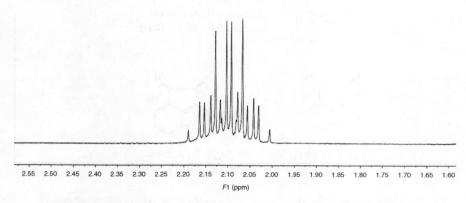

FIGURE 2.7. Expansion of CH$_2$ region of ^{1}H spectrum of CH$_3$CH$_2$CF$_3$

F–H coupling constants. The major influences on vicinal F–F and F–H coupling constants in nonstrained compounds are the torsional angles Ø between the coupled nuclei and the nature (particularly the electronegativities) and position of neighboring substituents. A Karplus-type dependence of the magnitudes of both F–F and F–H three-bond coupling constants upon the dihedral angle between the coupling nuclei was confirmed empirically by Williamson et al.,[11] and thus observed values of J have been able to be used to evaluate conformational equilibria or in more rigid molecules the geometrical relationship of fluorine substituents relative to vicinal hydrogens or fluorines. Such a dependency on Ø would predict maximum J values for Ø = 180° and 0°, with J = ~0 at Ø = 90°. However, attempts to *quantitatively* apply the Karplus equations (which require a strict angular dependence have not been successful, mainly because of the large substituent effects on these coupling constants and to a degree because of fluorine through-space coupling contributions to J (see Section 2.7.2). Scheme 2.13 provides some typical examples of $^3J_{HF}$ values that demonstrate the general principles of the dihedral angle dependence. For freely rotating C–C bonds, the H–F and F–F coupling constants comprise a weighted average of the values for the three conformations. In the case of CH_3CH_2F, the observed three-bond HF J value, 26.4 Hz, is simply the average of one anti and two gauche H–F couplings.

As indicated, the magnitudes of three-bond F–F and F–H coupling constants are also observed to vary as a function of the sum of the

Scheme 2.13

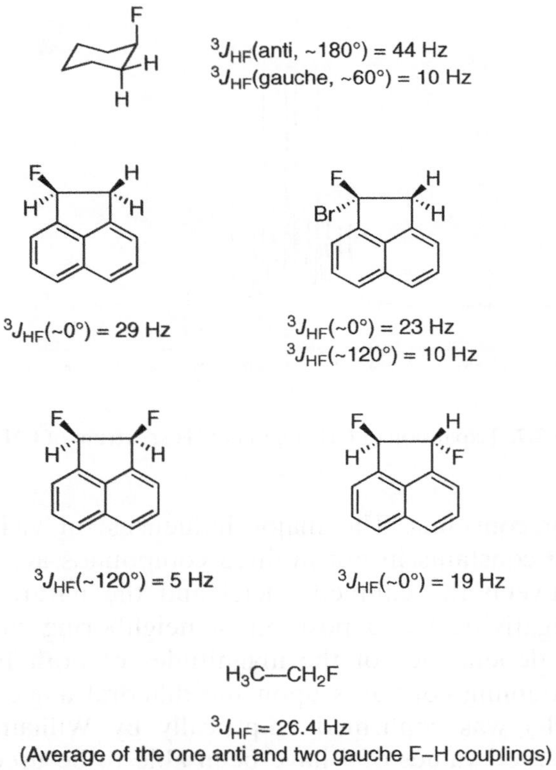

$^3J_{HF}$(anti, ~180°) = 44 Hz
$^3J_{HF}$(gauche, ~60°) = 10 Hz

$^3J_{HF}$(~0°) = 29 Hz

$^3J_{HF}$(~0°) = 23 Hz
$^3J_{HF}$(~120°) = 10 Hz

$^3J_{HF}$(~120°) = 5 Hz

$^3J_{HF}$(~0°) = 19 Hz

H_3C—CH_2F

$^3J_{HF}$ = 26.4 Hz
(Average of the one anti and two gauche F–H couplings)

TABLE 2.5. Vicinal Coupling Constants as a Function of Multiple Electronegative Substituents

$^3J_{FH}$ (Hz)		$^3J_{FH}$ (Hz)		$^3J_{FH}$ (Hz)	
CH_3-CH_2F	27				
CH_3-CHF_2	21	CF_3-CH_3	13	CF_3-CH_2F	16
CH_3-CF_3	13	CF_3-CH_2-Cl	8.5	CF_3-CHF_2	3
CH_2F-CH_2F	17	CF_3-CHCl_2	4.7	$CF_3-CH_2-CR_3$	~0
CHF_2-CHF_2	3			CF_3-CF_2-O-R	~0
CF_3-CHF_2	3			CF_3-CF_2-S-R	3

electronegativities of the other substituents that are on the two carbons in question, with the absolute values of these coupling constants *decreasing* with increasing substituent electronegativity. For example, in the extreme case of adjacent CF_2 groups, the F–F coupling constant can approach zero in magnitude. Some examples are given in Table 2.5 that will demonstrate this principle.

In describing coupling relationships within molecules, nuclei such as the fluorines in the schemes above, that have a first-order coupling relationship are represented by letters that are far away in the alphabet (i.e., AX), according to the Pople notation. Virtually, all of the lightly fluorinated compounds that are discussed in this book exhibit coupling between hydrogen and fluorine and many of them also exhibit fluorine–fluorine coupling. Most are first-order AX or AMX systems.

It should also be mentioned that most modern NMR facilities are capable of doing fluorine–hydrogen decoupling experiments, that is, $^{19}F\{^1H\}$ or $^1H\{^{19}F\}$ decoupled spectra, particularly when the fluorine signals occur over a relatively small range of chemical shifts. Such decoupling can drastically simplify proton NMR spectra. There are specific instrumental requirements for running such experiments, but a laboratory that does significant work with fluorochemicals will at times find this capability to be indispensable. An example of a situation where such decoupling was possible and provided unique insight is provided later in this chapter (Section 2.8, Figures 2.15 and 2.16[2]).

2.7.1. Effect of Molecule Chirality on Coupling

An additional complication regarding splitting patterns and coupling constants that is often encountered when there is chirality associated

[2]The assistance of Prof. David O'Hagan and Dr. Tomas Lehl of the University of St. Andrews, UK in providing this spectrum of *meso*-1,2-difluoro-1,2-diphenylethane is very much appreciated.

with the fluorine-bearing carbon or one proximal to it. A simple, fully analyzable example of this phenomenon is provided by examining the fluorine and proton NMR spectra of 1,2-dibromofluoroethane (Figure 2.8a, b).

Without considering chirality, one might expect that the signal for the fluorine of 1,2-dibromofluoroethane would be comprised of a doublet of triplets (six peaks), but as you can see from the actual fluorine spectrum given in Figure 2.9, the signal is split into eight peaks (doublet of doublets of doublets).

This outcome derives from the fact that the carbon that bears the fluorine is chiral, which makes the two vicinal hydrogens diastereotopic and thus magnetically nonequivalent. In such a case, the two diastereotopic protons will not only appear as separate signals (an AB system), but they usually will also couple to vicinal fluorines (and hydrogens) with different coupling constants. Examining Figure 2.8b, which represents

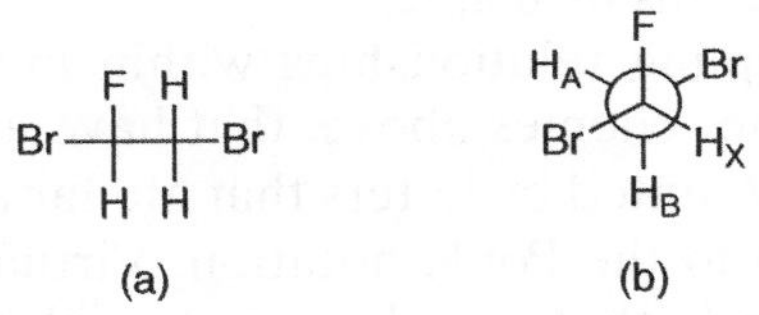

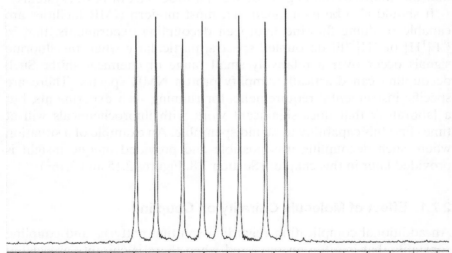

FIGURE 2.8. Diastereotopic protons of 1,2-dibromofluoroethane

FIGURE 2.9. ^{19}F NMR spectrum of 1,2-dibromofluoroethane

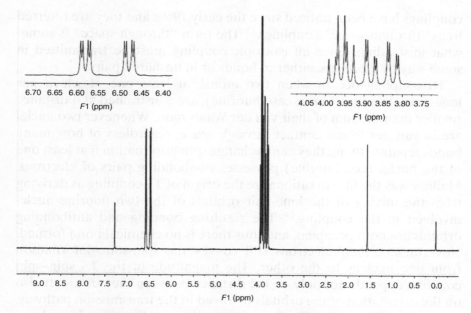

FIGURE 2.10. ^{1}H NMR spectrum of 1,2-dibromofluoroethane

the probable most stable conformation for 1,2-dibromofluoroethane, allows one to rationalize the differences in magnitude of the various three-bond coupling constants in the molecule. Thus, the fluorine signal at δ −136.3 derives from a 50 Hz two-bond coupling with H_X, and three-bond couplings of 11.5 and 28.5 Hz with H_A and H_B, respectively.

The ^{1}H NMR spectrum of this compound, shown in Figure 2.10, exhibits the same type of complications, with H_X appearing at δ 6.53 (ddd, $^2J_{FX} = 50$, $^3J_{BX} = 8.5$, $^3J_{AX} = 11.5$ Hz), H_A appearing at δ 3.96 (dt, $^3J_{FA} = {}^2J_{AB} = 11.5$, $^3J_{AX} = 8.5$ Hz), and H_B appearing at δ 3.85 (ddd, $^3J_{FB} = 28.5$, $^2J_{AB} = 11.5$, $^3J_{BX} = 2.5$ Hz). H_A's *triplet* derives from two coincidentally identical 11.5 Hz coupling constants.

A normal proton-decoupled carbon NMR spectrum of a chiral compound such as 1,2-dibromofluoroethane does not exhibit any complications, and the ^{13}C NMR spectrum of 1,2-dibromofluoroethane is shown and discussed later in this chapter.

2.7.2. Through-Space Coupling

In some particular instances, a large coupling between fluorine and a hydrogen, a carbon, a nitrogen, or another fluorine that may be separated by many bonds (four, five, six, or more) is observed. Such couplings have been noticed since the early 1960s and they are referred to as "through-space" couplings.[13] The term "through space" is somewhat misleading, since all isotropic coupling must be transmitted in some way by electrons, either in bonds or in unshared pairs.

TS coupling occurs when two atoms, at least one of which has lone-pair electrons (in our case fluorine), are constrained at a distance smaller than the sum of their van der Waals radii. Whenever two nuclei are in van der Waals contact through space, regardless of how many bonds separate them, they can exchange spin information if at least one of the nuclei (i.e., fluorine) possesses nonbonding pairs of electrons. Mallory was the first to rationalize the origin of TS coupling as deriving from the mixing of the lone-pair orbitals of the two fluorine nuclei involved in the coupling.[14] The resulting bonding and antibonding orbitals are both occupied, and thus there is no chemical bond formed. Nevertheless, their electrons still convey the spin-state information from one nucleus to the other. The magnitude of the TS spin–spin coupling depends not only on the distance between the nuclei but also on the orientation of the orbitals involved in the transmission pathway. Since this early rationalization, many other explanations have been proposed regarding the transmission mechanism, and this remains an active area of investigation.[15]

An early example of TS fluorine–fluorine coupling is shown in Scheme 2.14, where a "formally" five-bond F–H coupling constant of 167–170 Hz is observed.

<u>**Scheme 2.14**</u>

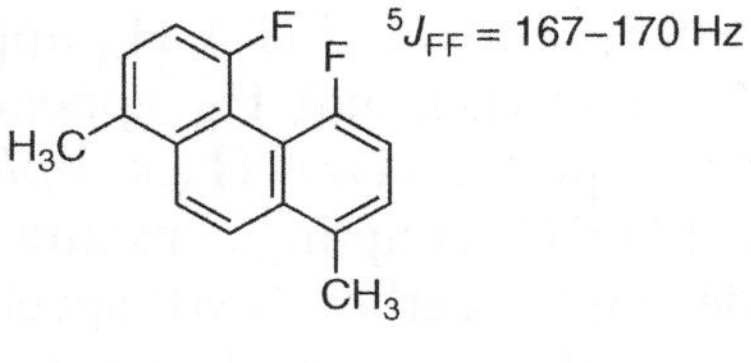

$^5J_{FF} = 167–170$ Hz

Transmission pathways other than two overlapping lone-pair orbitals have been identified, notably the overlap of a lone-pair orbital of fluorine with an occupied *bonding* orbital, which explains the TS coupling between ^{19}F and ^1H or ^{13}C.[15] A classic example of TS coupling between fluorine and hydrogen can be seen in the comparison of the six-bond coupling constants for the two similarly substituted compounds in Scheme 2.15,[16, 17] and other examples of long-range coupling of fluorine

Scheme 2.15

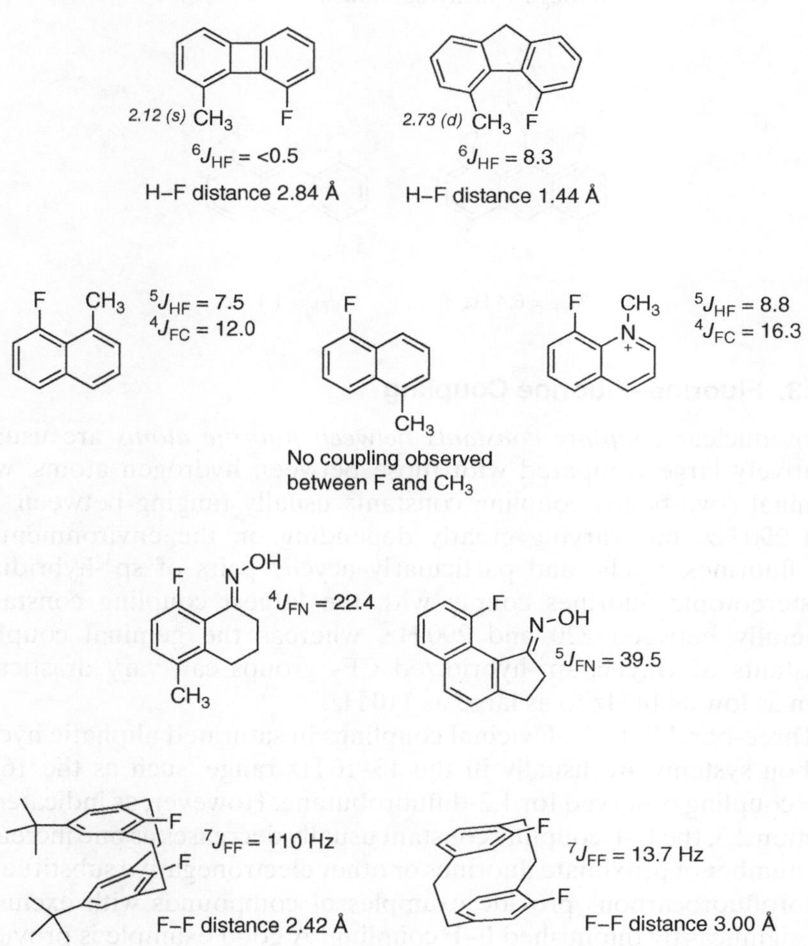

$2.12\ (s)$ CH_3 F
$^6J_{HF} = <0.5$
H–F distance 2.84 Å

$2.73\ (d)$ CH_3 F
$^6J_{HF} = 8.3$
H–F distance 1.44 Å

$^5J_{HF} = 7.5$
$^4J_{FC} = 12.0$

No coupling observed
between F and CH_3

$^5J_{HF} = 8.8$
$^4J_{FC} = 16.3$

$^4J_{FN} = 22.4$

$^5J_{FN} = 39.5$

$^7J_{FF} = 110\ Hz$
F–F distance 2.42 Å

$^7J_{FF} = 13.7\ Hz$
F–F distance 3.00 Å

to hydrogen, carbon, nitrogen, and fluorine are given in the same scheme.[14, 18, 19]

There are also examples, although rare, where F–F coupling appears to be "transmitted" through a phenyl substituent, example of which is given in Scheme 2.16.[20]

Although initially mainly of theoretical interest, through-space couplings are now considered an essential element of structure, in particular stereochemical elucidation, and there was an early study in which the fluorine–fluorine TS couplings in ^{19}F-labeled amino acids were used to elucidate the folding of a protein.[21]

Scheme 2.16 F–F coupling transmitted through a phenyl substituent

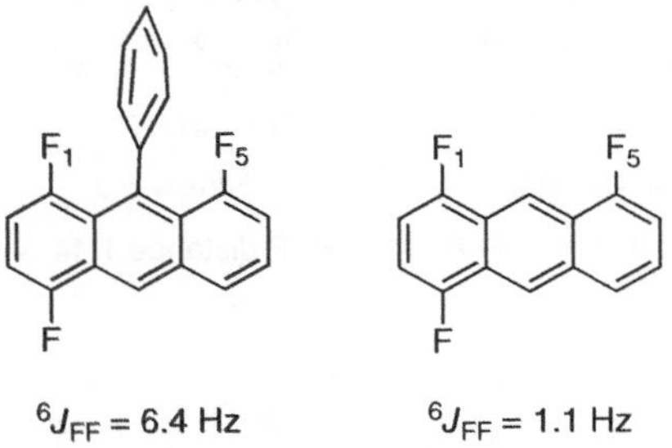

$^6J_{FF} = 6.4$ Hz $^6J_{FF} = 1.1$ Hz

2.7.3. Fluorine–Fluorine Coupling

Homonuclear *coupling constants between fluorine atoms* are usually relatively large compared with those between hydrogen atoms, with geminal (two-bond) coupling constants usually ranging between 100 and 290 Hz, but varying greatly depending on the environment of the fluorines. Cyclic and particularly acyclic pairs of sp^3-hybridized diastereotopic fluorines couple with the largest coupling constants, generally between 220 and 290 Hz, whereas the geminal coupling constants of vinylic, sp^2-hybridized CF_2 groups can vary drastically, from as low as 14 Hz to as large as 110 Hz.

Three-bond F–C–C–F vicinal couplings in saturated aliphatic hydrocarbon systems are usually in the 13–16 Hz range, such as the 16 Hz F–F coupling observed for 1,2-difluorobutane. However, as indicated in Section 2.3, the F–F coupling constant usually decreases as one increases the number of proximate fluorines or other electronegative substituents. Chlorofluorocarbons provide examples of compounds with exclusive but significantly diminished F–F coupling. A good example is provided by the fluorine NMR spectrum of 1,1,2-trichloro-1,2,2-trifluoroethane (F113), depicted in Figure 2.11, for which the two signals at −67.6 and −71.8 exhibit a diminished $^3J_{FF}$ coupling constant of 9.6 Hz. The relative lack of resolution in this spectrum most likely derives from the multiple isotopes of chlorine, which will give rise to broadening of the signals due to isotope effects.

The largest three-bond F–F coupling constants are observed between *trans*-vinylic fluorines, where coupling constants can be as large as 145 Hz, which can be compared to the much smaller *cis*-coupling constants (<35 Hz, but sometimes much less). For example, the vicinal fluorines in 1-chloro-1,2-difluoroethenylbenzene (Scheme 2.17) appear as doublets due to the three-bond F–F coupling between the two

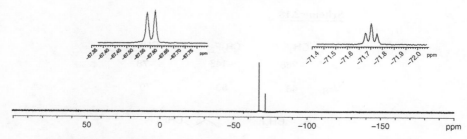

FIGURE 2.11. ^{19}F NMR spectrum of 1,1,2-trichloro-1,2,2-trifluoroethane

Scheme 2.17

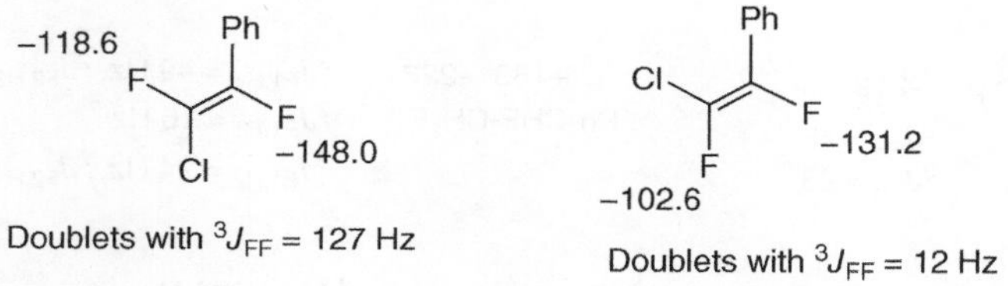

Doublets with $^3J_{FF}$ = 127 Hz

Doublets with $^3J_{FF}$ = 12 Hz

fluorine nuclei. The *trans*-coupling constant is much larger (127 Hz) than the respective *cis*-coupling constant (12 Hz). (Note also that the *cis*-vicinal fluorines *deshield* each other significantly, relative to the *trans*-vicinal pair.)

2.7.4. Coupling Between Fluorine and Hydrogen

Coupling constants between fluorine and hydrogen in saturated compounds are also large and characteristic, depending on whether one is dealing with a single fluorine, a CF_2 group, or a CF_3 group. *Two-bond coupling constants* for a single fluorine range from a low of 46 Hz for CH_3F to an extreme high value of 79 Hz for CF_3H (Scheme 2.18). However, most H–F couplings for R–CH_2–F groups are in the range of 47–55 Hz, and R–CF_2–H couplings are consistently between 57 and 59 Hz.

Three-bond couplings exhibit even greater variation, with the largest coupling constants between vicinal F and H being observed for the single fluorine substituent, 21–27 Hz (Scheme 2.19). In contrast, the range for similar coupling to a CF_2 group is between 14 and 22 Hz, and vicinal coupling of H to a CF_3 group is normally only 7–11 Hz. Thus, as is the case for three-bond F–F coupling constants, the value of vicinal

Scheme 2.18

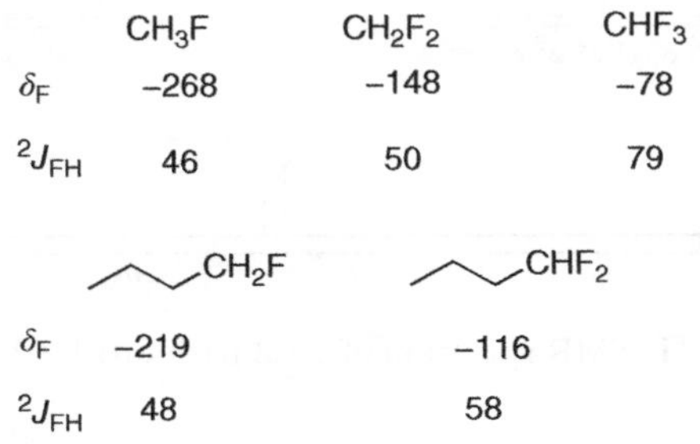

	CH_3F	CH_2F_2	CHF_3
δ_F	−268	−148	−78
$^2J_{FH}$	46	50	79

CH_2F CHF_2

δ_F	−219	−116
$^2J_{FH}$	48	58

Scheme 2.19

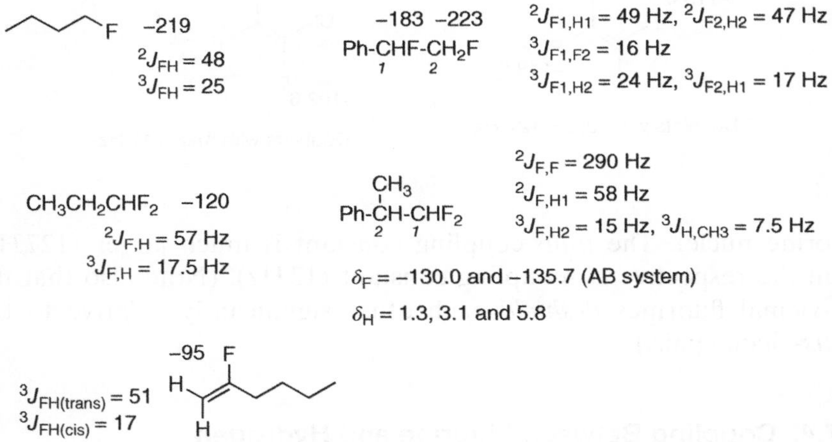

F   −219
$^2J_{FH} = 48$
$^3J_{FH} = 25$

−183 −223
Ph-CHF-CH$_2$F
 1   2

$^2J_{F1,H1} = 49$ Hz, $^2J_{F2,H2} = 47$ Hz
$^3J_{F1,F2} = 16$ Hz
$^3J_{F1,H2} = 24$ Hz, $^3J_{F2,H1} = 17$ Hz

$CH_3CH_2CHF_2$ −120
$^2J_{F,H} = 57$ Hz
$^3J_{F,H} = 17.5$ Hz

CH$_3$
|
Ph-CH-CHF$_2$
 2   1

$^2J_{F,F} = 290$ Hz
$^2J_{F,H1} = 58$ Hz
$^3J_{F,H2} = 15$ Hz, $^3J_{H,CH3} = 7.5$ Hz

$\delta_F = -130.0$ and −135.7 (AB system)
$\delta_H = 1.3, 3.1$ and 5.8

$^3J_{FH(trans)} = 51$
$^3J_{FH(cis)} = 17$

−95 F

F–H coupling constants also decreases as one accumulates additional electronegative substituents on the carbons bearing the coupling nuclei.

Vinylic fluorine can have very large (35–52 Hz) three-bond coupling constants to hydrogen when the fluorine and the hydrogen are *trans* to each other. Analogous *cis*-coupling constants are smaller and generally range from 14 to 20 Hz.

The examples of F–F and H–F coupling constants given in Scheme 2.19 are typical for acyclic compounds of the type described above. Specific coupling constant data are provided for each class of fluorinated molecules as they are discussed in Chapters 3–6.

Observation of H–F coupling has been used in postulating evidence for intramolecular C–F ⋯ H–O hydrogen bonding in specific cases where geometrical conditions allow it (Scheme 2.20).[22–24]

Scheme 2.20 Intramolecular F–HO hydrogen bonding

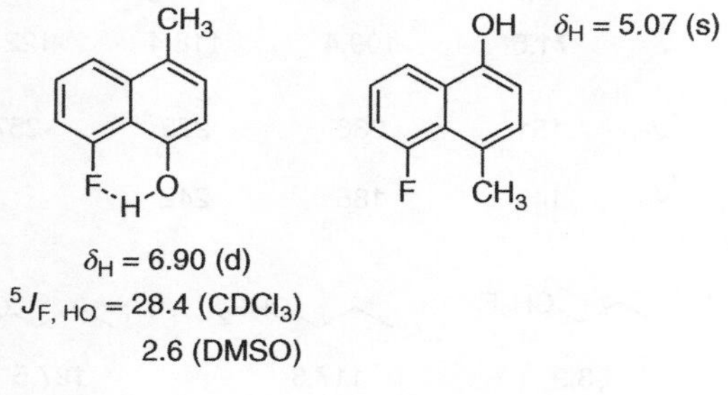

$\delta_H = 6.90$ (d)
$^5J_{F, HO} = 28.4$ (CDCl$_3$)
2.6 (DMSO)

$\delta_H = 5.07$ (s)

Scheme 2.21

versus

More basic

$^6J_{FH} = 43.7$ Hz

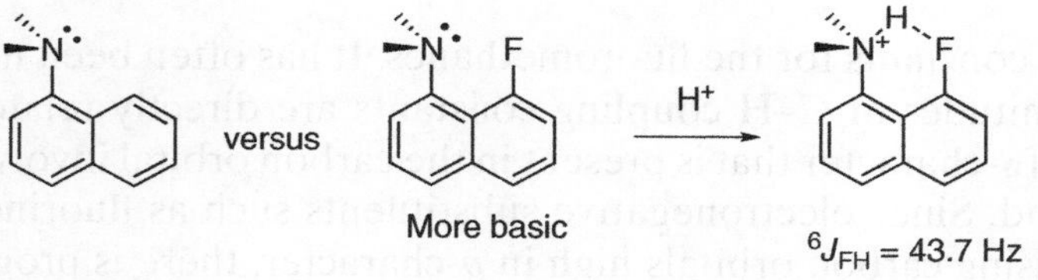

Sometimes a strongly hydrogen-bonded system, such as the fluorinated "proton sponge" compound shown in Scheme 2.21, can have an impact on reactivity.[25]

2.7.5. Coupling Between Fluorine and Carbon

The use of ^{13}C NMR has benefitted from the fact that, because of the relatively small range of proton chemical shifts, techniques for broadband decoupling of protons were developed early on, and proton decoupled spectra, which generally results in singlet signals for each magnetically equivalent carbon in a compound, are what are now generally reported in the literature. When fluorine is introduced into a compound, coupling between fluorine and carbon is thus readily discerned and much information can be gleaned by an understanding of the straightforward relationship between the F–C coupling constant and the number of bonds separating the two nuclei.

The magnitude of one-bond *fluorine coupling to carbon* in a simple fluorohydrocarbon can vary from 151 to 280 Hz, depending again on whether one, two, or three fluorines are bound to the carbon. Scheme 2.22 demonstrates these trends in the fluoromethanes and fluoroalkanes. The scheme also includes mention of the one-bond H–C

Scheme 2.22

	CH_3F	CH_2F_2	CHF_3	CF_4
δ_C	71.6	109.4	118.4	122.4
$^1J_{FC}$	151	236	275	257
$^1J_{HC}$	149	185	242	

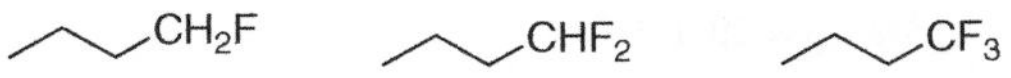

	$\diagdown CH_2F$	$\diagdown CHF_2$	$\diagdown CF_3$
δ_C	83.9	117.6	127.5
$^1J_{FC}$	165	239	276

coupling constants for the fluoromethanes. It has often been noted that the magnitudes of C–H coupling constants are directly related to the degree of s-character that is present in the carbon orbital involved in the C–H bond. Since electronegative substituents such as fluorine bind to carbon using carbon orbitals high in p-character, there is progressively more s-character in the carbon orbitals used for the C–H bonds going from CH_3F to CHF_3, which is reflected in the trend observed in the C–H coupling constants of the fluoromethanes.

In addition, replacement of one of the fluorine atoms with a chlorine on a multifluoro-substituted carbon always increases the one-bond F–C coupling constant, with Br and I giving rise to even greater increases (Table 2.6). This is a consistent trend, regardless of whether the halogens are bound to CF_3, CF_2H, or CH_2F groups.

The effect of replacing a fluorine on a multifluoro-substituted carbon with OR, SR, or SeR groups on one-bond F–C coupling constants can be highly variable depending on the number of fluorines remaining on

TABLE 2.6. Impact of α-Halogen Substitution on Carbon Chemical Shifts and One-Bond F–C Coupling Constants

X	F	Cl	Br	I
CF_3X, δ	122.4	118.0	112.9	78.2
$^1J_{FC}$ (Hz)	257	299	324	344
HCF_2X, δ	118.4	118.0	–	83.4
$^1J_{FC}$ (Hz)	272	288	–	308
H_2CFX, δ	109.4	–	–	–
$^1J_{FC}$ (Hz)	235	–	–	–

TABLE 2.7. Impact of α-Chalcogen Substitution on Carbon Chemical Shifts and One-Bond F–C Coupling Constants

Y	F	OPh	SPh	SePh
CF_3Y, δ	122.4	121.0	130.0	123.0
$^1J_{FC}$ (Hz)	259	251	308	333
HCF_2Y, δ	118.4	116.0	–	–
$^1J_{FC}$ (Hz)	274	260	–	–
H_2CFY, δ	109.4	100.5	88.2	–
$^1J_{FC}$ (Hz)	235	217	219	–

the carbon, but the trend of increasing coupling constants for the series O–S–Se is still observed (Table 2.7).

Although no explanation for this trend has yet been offered, it is consistent with another empirical "rule" that seems to be generally followed relating fluorine chemical shifts to one-bond F–C coupling constants: "The coupling constants tend to decrease with increasing (i.e., more positive) fluorine chemical shifts"[26] (Scheme 2.23).

Scheme 2.23

	CH_3F	n-C_4H_9F	CH_2F_2	n-$C_3H_7CHF_2$	CHF_3	n-$C_3H_7CF_3$	CF_4
δ_F	−268	−219	−148	−116	−78	−68	−62
$^1J_{FC}$	151	165	236	239	275	276	257

Two-bond F–C coupling remains quite large, usually about 20 Hz, and significant coupling can be observed for three- and sometimes even four-bond interactions. Very important are the consistent one-, two-, three-, and four-bond F–C coupling constants that are observed in fluorinated aliphatic and aromatic systems (Scheme 2.24). The strength of coupling persists for greater distances in aromatic systems. The consistency of such coupling constants provides extremely useful insights regarding the connectivity of carbons with respect to the fluorine substituent(s).

Scheme 2.24

$|J|_{FC}$ values (Hz)

$$\underset{5}{CH_3}-CH_2-CH_2-CH_2-\underset{167}{CH_2}-F \qquad 20$$

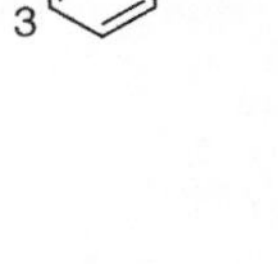

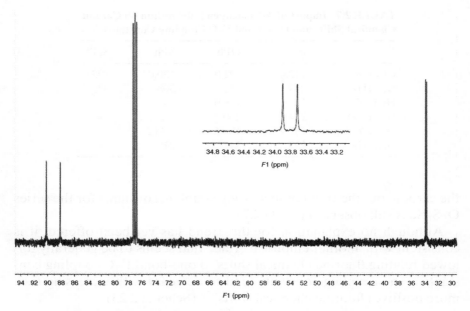

FIGURE 2.12. ^{13}C NMR spectrum of 1,2-dibromofluoroethane

Figure 2.12 shows the carbon NMR spectrum for 1,2-dibromofluoro-ethane. As usual, the protons are decoupled, so that the only couplings that one can see are those between fluorine and carbon. Two signals are observed at 33.8 and 89.1 ppm, with one-bond and two-bond coupling constants of 257 and 23.5 Hz, respectively. (The multiplet at ~77 ppm derives from the solvent, CDCl$_3$.)

2.7.6. Coupling Between Fluorine and Phosphorous

^{31}P is an NMR active nucleus with a nuclear spin of $^1/_2$. When in the structural vicinity of a fluorine substituent, it exhibits relatively strong coupling. A few examples are given in Scheme 2.25. When they are available, P–F coupling constants to fluorine are reported for the phosphorous-containing compounds discussed in this book.

<u>**Scheme 2.25**</u>

Ph$_3$PF$_2$ $^1J_{FP}$ = 668   (One-bond coupling)

Ph$_2$PO—CH$_2$F $^2J_{FP}$ = 58   (Two-bond coupling)

Ph$_2$P—CH$_2$CH$_2$F $^2J_{FP}$ = 15   (Three-bond coupling)

2.7.7. Coupling Between Fluorine and Nitrogen

Because of the low natural abundance and low sensitivity of the ^{15}N nucleus, one never sees such couplings, even as sidebands, under normal conditions for running fluorine NMR spectra. Essentially, all data regarding NF coupling constants are obtained using indirect methodology, most commonly through the use of ^{1}H–^{15}N gHSQC experiments (see Section 2.6 for additional discussion).

For fluorinated nitrogen-containing compounds, one-bond N–F couplings can be quite large, especially in inorganic compounds: $F_2N{=}O$ (356 Hz) and NF_3 (196 Hz). Few one-bond N–F couplings of organic compounds have been measured, Selectfluor (84 Hz) being one of these (Scheme 2.26).

Scheme 2.26 Typical F–N spin–spin couplings

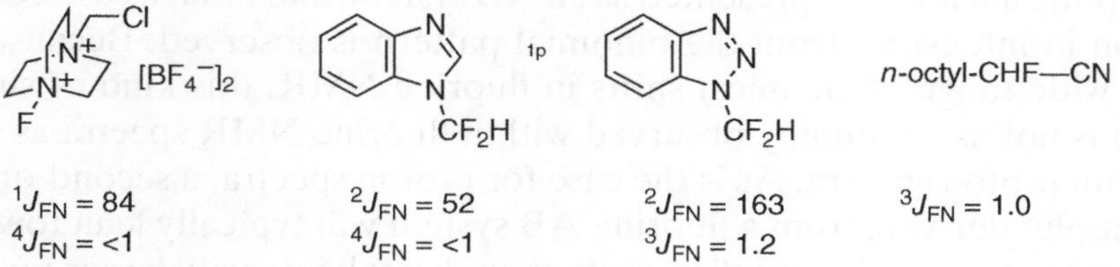
Cl
N+
N+
[BF4−]2
F
N
N
CF2H
1p
N
N
N
CF2H
n-octyl-CHF—CN
1JFN = 84
4JFN = <1
2JFN = 52
4JFN = <1
2JFN = 163
3JFN = 1.2
3JFN = 1.0

Two-bond couplings can vary considerably (Scheme 2.27). In fluoroheterocyclics, such as in 2-fluoropyridine, the three-bond coupling constants are much larger than those for compounds, where only sigma bonds link the F and the N. In the absence of through-space interactions, three-bond or greater distance couplings between F and N are always small but are a little larger when transmitted through a π system (Scheme 2.28).

Scheme 2.27 Typical two-bond F–N coupling constants

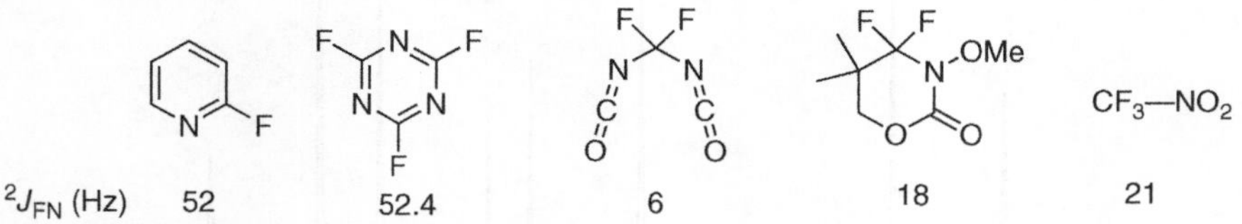
N
F
F
N
F
N
N
F
F
F
N
C
O
N
C
O
F
F
N
OMe
O
O
CF3—NO2
2JFN (Hz)
52
52.4
6
18
21

Larger long-distance F–N coupling constants are observed when through-space interactions are involved (see Section 2.7.2, Scheme 2.15).

Scheme 2.28 Typical three-bond F–N coupling constants

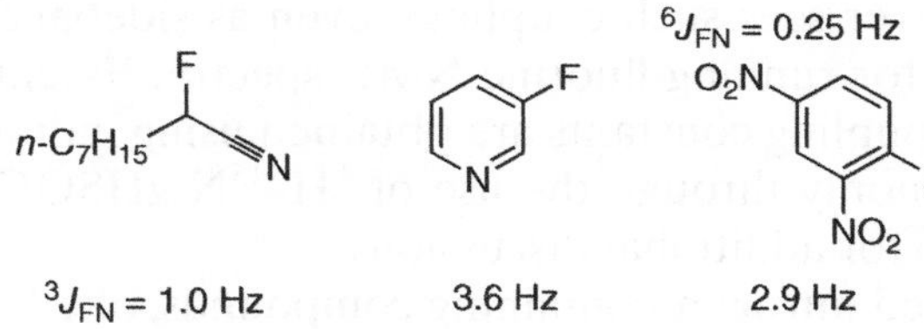

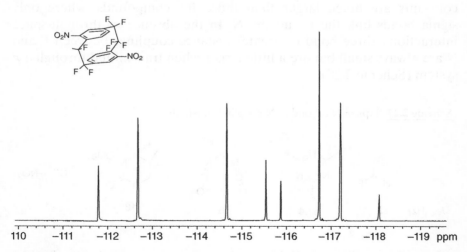

$^6J_{FN} = 0.25$ Hz

n-C$_7$H$_{15}$

$^3J_{FN} = 1.0$ Hz   3.6 Hz   2.9 Hz

2.8. SECOND-ORDER SPECTRA[27]

Second-order effects begin to appear in a spectrum when the chemical shift difference (in hertz) between the coupling nuclei is less than about 10 times the value of the coupling constant (i.e., $\Delta v/J \leq 10$). Such coupling nuclei are represented as an AB system, and in such cases, deviation in intensities from the binomial pattern is observed. Because of the wide range of chemical shifts in fluorine NMR, this kind of situation is not as commonly observed within fluorine NMR spectra as it is within proton spectra. As is the case for proton spectra, a second-order multiplet deriving from a fluorine AB system will typically lean toward the resonances of its coupling partner, with peak intensity larger for the inner peaks and smaller for the outer peaks. Figure 2.13 provides an example of the two AB systems in a fluorine NMR spectrum, that of pseudo *para*-dinitrooctafluoro[2.2]paracyclophane.

FIGURE 2.13. ^{19}F NMR spectrum of pseudo-*p*-dinitro-1,1,2,2,9,9,10,10-octafluoro-[2.2]paracyclophane

There is a second, more complicated and for fluorine NMR spectra more common situation that will lead to second-order spectra, that in which *chemically equivalent* fluorines (same chemical shift) are *magnetically nonequivalent*. This occurs when the chemically equivalent fluorines do not have the same coupling constants to specific other nuclei in the molecule.

Both homotopic fluorines such as those in difluoromethane and 2,2-difluoropropane and 1,1-difluoroethene, and enantiotopic fluorines such as those in chlorodifluoromethane and 2,2-difluorobutane (Scheme 2.29) would be chemically equivalent.

Scheme 2.29

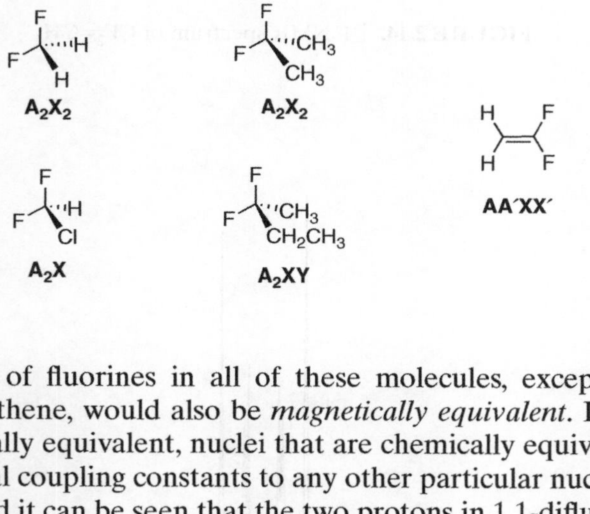

The pairs of fluorines in all of these molecules, except those in 1,1-difluoroethene, would also be *magnetically equivalent*. In order to be magnetically equivalent, nuclei that are chemically equivalent must have identical coupling constants to any other particular nucleus in the molecule, and it can be seen that the two protons in 1,1-difluoroethene do not have the same spatial relationship with respect to a given fluorine substituent. For example, the F_a substituent has a *cis* relationship to H_a, but a *trans* relationship to H_b (Scheme 2.30). A spin system such as this one is represented as an AA′XX′ system, which contrasts with the A_2X_2, A_2X, and A_2XY systems in Scheme 2.29 wherein both fluorines in each of these systems have identical $^2J_{HF}$ coupling constants.

Scheme 2.30

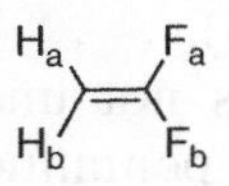

CF$_2$=CH$_2$

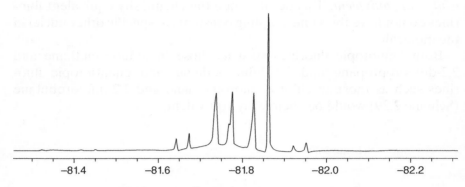

FIGURE 2.14. ^{19}F NMR spectrum of CF$_2$=CH$_2$

CF$_2$=CH$_2$

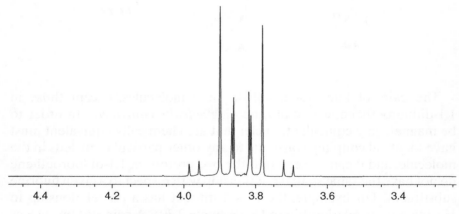

FIGURE 2.15. ^{1}H NMR spectrum of CF$_2$=CH$_2$

Any spin system that contains fluorine substituents that are chemi-cally equivalent but not magnetically equivalent is, by definition, second order. Such spectra can appear deceptively simple, or more commonly they can be amazingly complex. The fluorine and proton spectra of the simple, symmetrical compound, 1,1-difluoroethene exemplify the latter situation (Figures 2.14 and 2.15).

Magnetic nonequivalence is not uncommon, often deriving from the constraints of a ring, as in pentafluorophenyl derivatives or other

Scheme 2.31

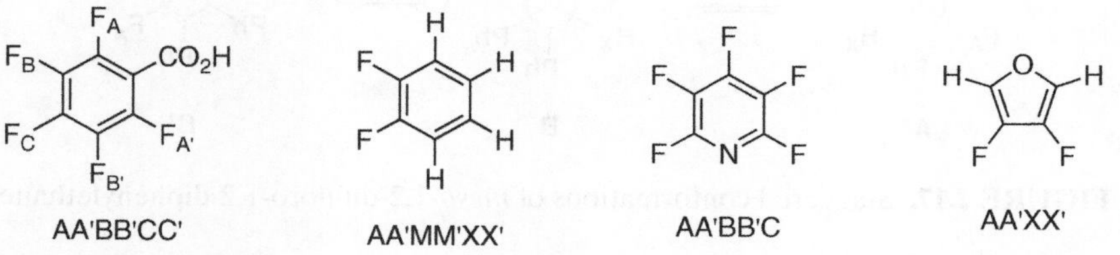

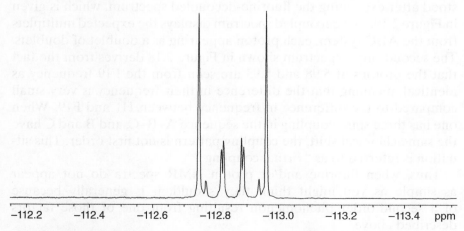

AA'BB'CC'   AA'MM'XX'   AA'BB'C   AA'XX'

symmetrically fluorine substituted ring systems such as those shown in Scheme 2.31. The fluorine and proton NMR spectra of all three difluorobenzenes are representative of the appearance of second-order spectra of polyfluoroaromatics. They can be found in Section 3.9.3, Figures 3.17–3.22.

Another common situation that can lead to second-order spectra is an open-chain system such as *meso*-1,2-difluoro-1,2-phenylethane whose magnetically nonequivalent spin system and resultant second-order fluorine NMR spectrum (Figure 2.16) can only be understood by examination of the contributing conformations about its fluorine bearing carbons.[12]

The symmetry of this molecule makes the fluorines chemically equivalent but not magnetically equivalent. Examination of the three staggered conformations of AA'XX' spin system (Figure 2.17) helps one understand this situation.

meso-PhCHFCHFPh

-112.2   -112.4   -112.6   -112.8   -113.0   -113.2   -113.4   ppm

FIGURE 2.16. ^{19}F NMR spectrum of meso-1,2-difluoro-1,2-diphenylethane.[12]

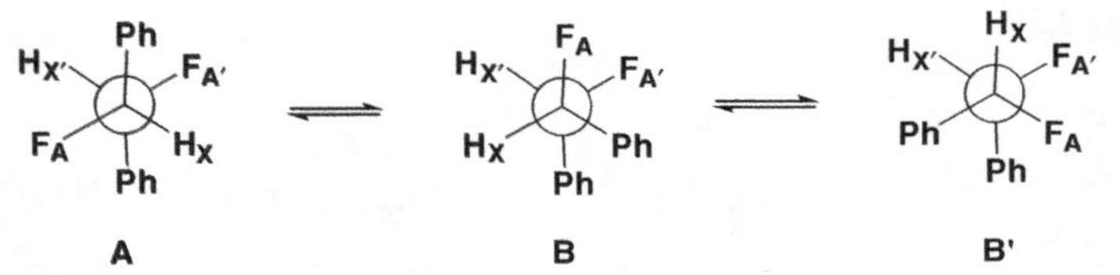

FIGURE 2.17. Staggered conformations of *meso*-1,2-difluoro-1,2-diphenylethane

Determination of individual coupling constants from a second-order spectrum such as those in the above figures cannot be accomplished by simple inspection of the spectrum. Such an analysis requires simulation of spectra via an intuitive fitting of coupling constant values to specific coupling relationships.[28] On the basis of such analyses, it is sometimes possible to determine the relative contribution of individual conformations based upon the estimated values for a full *anti* $^3J_{HF}$ coupling constant of 32 Hz and a full *gauche* $^3J_{HF}$ coupling constant of approximately 8 Hz (in vicinal difluoro systems).[29]

Based on these values, if the three conformations in Figure 2.17 contributed equally, the vicinal F–H coupling constant should be 16 Hz. Since the actual value was estimated to be 14 Hz, this would indicate that conformer A (with only *gauche* three-bond F–H interactions) must be slightly favored.

There is still another situation that leads to second-order spectra and this one usually cannot be anticipated. For example, take a look at the proton spectrum of 3,3,3-trifluoropropene in Figure 2.18. This spectrum is not the simple one that one would expect for a monosubstituted ethylene. However, the second-order nature of this spectrum can be understood after examining the fluorine-decoupled spectrum, which is given in Figure 2.19. The decoupled spectrum displays the expected multiplets from the ABC system, each proton appearing as a doublet of doublets. The second-order spectrum shown in Figure 2.18 derives from the fact that the protons at 5.98 and 5.93 are seen from the F19 frequency as identical, meaning that the difference in their frequency is very small compared to the difference in frequency between H1 and F19. When one has three spins coupling in the sequence A–B–C, and B and C have the same chemical shift, the coupling pattern is not first order. This situation is referred to as "virtual coupling."

Thus, when fluorine and/or proton NMR spectra do not appear as simple as you might think they should, it is generally because of a second-order phenomenon resulting from one of those factors described above.

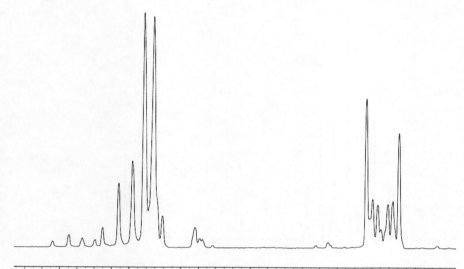

FIGURE 2.18. ^{1}H NMR spectrum of 3,3,3-trifluoropropene

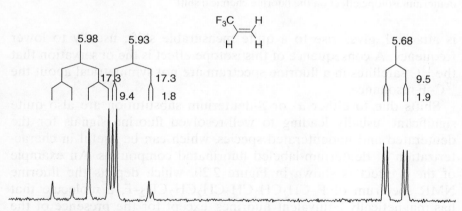

FIGURE 2.19. Fluorine decoupled ^{1}H NMR spectrum of 3,3,3-trifluoropropene

2.9. ISOTOPE EFFECTS ON CHEMICAL SHIFTS

Because fluorine is relatively sensitive to its environment and has such a large range of chemical shifts, considerable changes in chemical shift can be observed when a nearby atom is replaced by an isotope. For example, replacement of ^{12}C by ^{13}C for the atom to which the fluorine

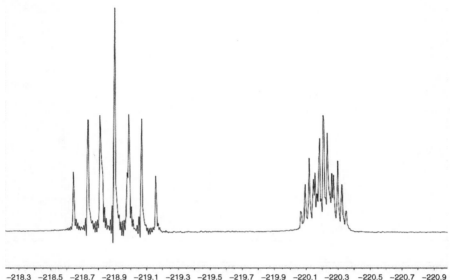

-218.3 -218.5 -218.7 -218.9 -219.1 -219.3 -219.5 -219.7 -219.9 -220.1 -220.3 -220.5 -220.7 -220.9

F1 (ppm)

FIGURE 2.20. ^{19}F NMR spectrum of 1,6-difluorohexane-1,1-d_2, demonstrating the deuterium isotope effect on the fluorine chemical shift

is attached, gives rise to a quite measurable shift, usually to lower frequency. A consequence of this isotope effect is the observation that the ^{13}C satellites in a fluorine spectrum are not symmetrical about the ^{12}C–F resonance.

Shifts due to either α- or β-deuterium substitution are also quite significant, usually leading to well-resolved fluorine signals for the deuterated and undeuterated species, which can be useful in characterization of deuterium-labeled fluorinated compounds. An example of the α-effect is shown in Figure 2.20, which depicts the fluorine NMR spectrum of F–CH$_2$CH$_2$CH$_2$CH$_2$CH$_2$CD$_2$–F, a molecule that has magnetically equivalent fluorines, except for the presence of the deuteriums. The observed isotope effect on the chemical shift is a 1.31 ppm *upfield* shift (or 0.65 ppm upfield per deuterium). This effect is an α-deuterium isotope effect, in this case a two-bond effect on fluorine.

Scheme 2.32 provides further insight regarding both α- and β-deuterium isotope effects on fluorine chemical shifts with a series of deuterated chloro, fluoroethylenes.[30]

For the *cis*-1-chloro-2-fluoroethylene, an α-deuterium isotope effect (one D) of 0.6 ppm is observed, along with a *trans*-β-deuterium isotope effect of 0.4 ppm. Looking at the *trans*-1-chloro-2-fluoroethylene system, the α-deuterium isotope effect is 0.5 ppm and the *cis*-β-isotope

Scheme 2.32

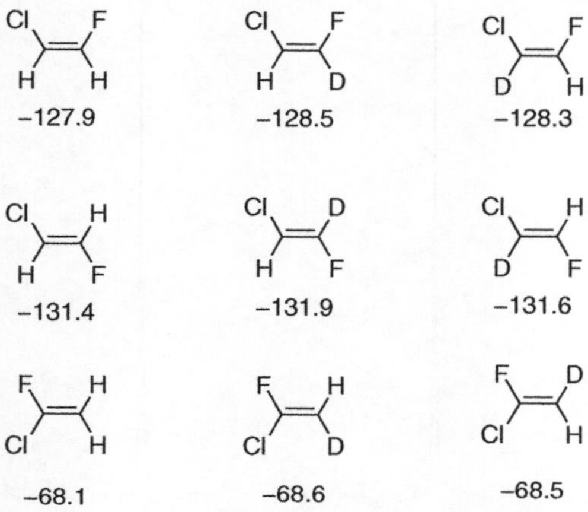

Cl, F / H H   −127.9

Cl, F / H D   −128.5

Cl, F / D H   −128.3

Cl, H / H F   −131.4

Cl, D / H F   −131.9

Cl, H / D F   −131.6

F, H / Cl H   −68.1

F, H / Cl D   −68.6

F, D / Cl H   −68.5

effect is 0.2 ppm. For the 1-chloro-1-fluoroethylene system, *trans-β*-isotope effect is 0.5 ppm, with the *cis-β*-isotope effect being 0.4 ppm.

Other examples of stereoelectronic effects on three-bond transmission of deuterium isotope effects on fluorine chemical shift are provided in the 2-deuteriofluorocyclohexane system, where an *anti*-deuterium gives rise to an incremental shift of 0.35 ppm, whereas *gauche* deuterium has only about half the effect (Scheme 2.33).[31]

Scheme 2.33

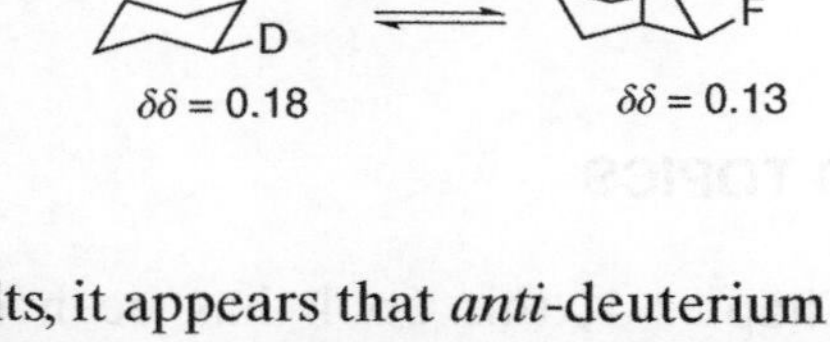

$\delta\delta = 0.35$ $\delta\delta = 0.15$

$\delta\delta = 0.18$ $\delta\delta = 0.13$

From these results, it appears that *anti*-deuterium substitution transmits its isotope effect better than *gauche*-deuterium substitution, the same trend as is observed in transmission of coupling constants.

Figure 2.21 depicts the CH_2F/CD_2F region of the ^{13}C NMR spectrum of 1,6-difluorohexane-d_2, and it provides a measure of the deuterium isotope effect upon the carbon chemical shift of the CH_2F carbon

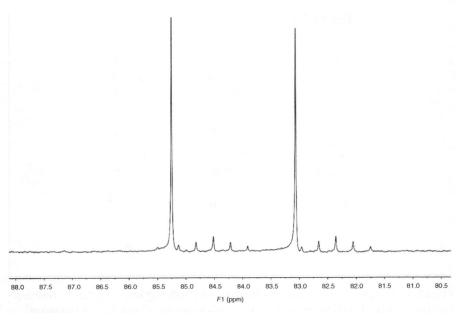

F1 (ppm)

FIGURE 2.21. CH_2F/CD_2F region of the ^{13}C NMR spectrum of 1,6-difluorohexane-1,1-d_2 demonstrating deuterium isotope effect on ^{13}C chemical shifts

(compared to the CD_2F carbon). This isotopic shift is much less easily discernable than that observed in the fluorine NMR, in part because of the D–C coupling that further splits the fluorine-split doublet carbon peak of the CD_2F group. The observed isotope effect on the ^{13}C chemical shift is a 0.72 ppm *upfield* shift. The F–C coupling constants for the two groups appear to be slightly different (165 vs 163 Hz for the CH_2F and CD_2F groups, respectively). This effect is an α-deuterium isotope effect on carbon, in this case a one-bond effect. This is not particularly a fluorine-related effect, since it would be observed for any deuterium-substituted carbon, whether or not a fluorine substituent was present.

2.10. ADVANCED TOPICS

As indicated in Chapter 1, this book has been designed to be an introduction to fluorine NMR and to serve as a practical handbook for use by organic chemists involved in the synthesis of fluorinated organic compounds, with an emphasis on characterization of lightly fluorinated organic compounds. It is meant to serve as a primary source of fluorine, proton, carbon, and to a more limited extent phosphorous and nitrogen chemical shift and coupling constant data of such compounds. It is not designed to be the end regarding all possible applications of fluorine NMR. There are times when the data in this book are not sufficient to address detailed structural issues, in particular those related to stereochemistry, and in such cases the reader may want to consider more advanced NMR methodology, in particular, multidimensional ^{19}F NMR. Although this book does not deal with such techniques in any detail, it is recognized that they will at times become indispensable, and thus an introduction is provided to ^{19}F 2D NMR techniques, with appropriate references to sources of more information. Usually, the use of these techniques will require specific instrumentation and the assistance of an NMR specialist.

Also, one would be remiss if one did not at least mention the various analytical and diagnostic NMR techniques that have come into prominent use over the last decade, particularly with respect to their applications to drug discovery. Because of the sensitivity of its chemical shift to local environment, ^{19}F NMR has increasingly been used as a probe in the study of structure and dynamics of small molecules, proteins, and other biological systems.[32–36]

A couple of examples are provided: (i) The use of an amine-sensing palladium complex (A) (Scheme 2.34) now provides chemists with an easy way to determine the enantiomeric excess of chiral amines[37]; (ii) the diastereotopic nature of the two fluorines in the CF_2H group of difluoromethionine (B) in Scheme 2.34 has been exploited in the investigation of protein structure and function because their chemical shift difference varies depending upon the rotational mobility of the CF_2H group[38]; (iii) labeling small proteins with 3-fluorotyrosine or trifluoromethyl-phenylalanine has allowed chemists to study the structural and dynamic properties of proteins inside live bacteria.[39]

^{19}F NMR-based screening has become widely used as a powerful and reliable tool in the identification of potential drug candidates.[40–42]

Scheme 2.34

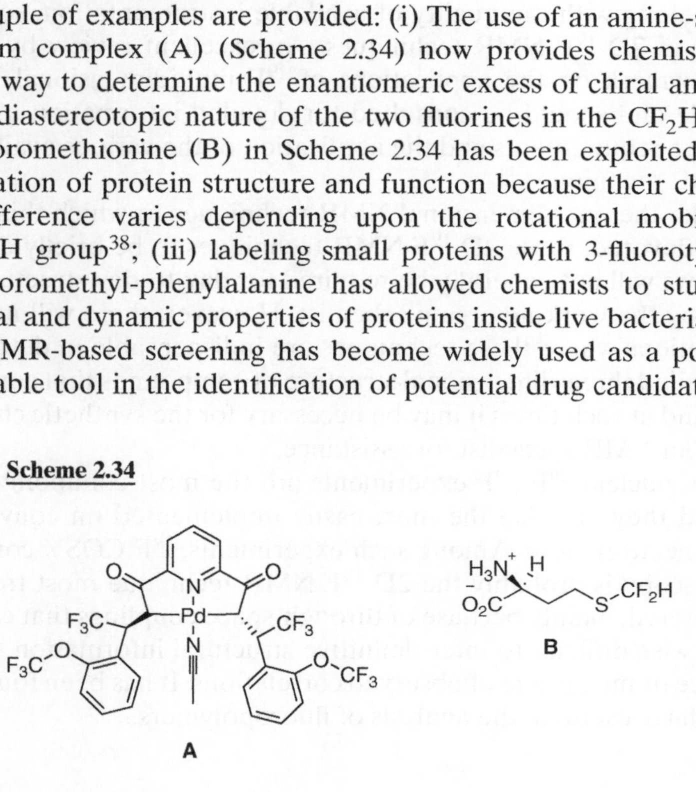

A

B

The use of fluorine labeled ligands or substrates allows one, via ^{19}F NMR, to detect weak intermolecular interactions. One such technique (FAXS – Fluorine chemical shift Anisotropy and eXchange for Screening) utilizes a fluorine-containing "spy" molecule to monitor changes in the transverse relaxation rate of the ^{19}F resonance in the presence of a series of test compounds.[43]

2.10.1. Multidimensional ^{19}F NMR

The use of two-dimensional (2D) NMR techniques has become almost routine for a detailed analysis of complex organic molecules containing carbon and hydrogen. In contrast, 2D ^{19}F NMR methods are not nearly so commonly used in the analysis of fluorine-containing molecules. The reasons for this are generally a combination of instrumental requirements combined with intrinsic differences between fluorine and proton NMR, in particular the wide range of ^{19}F chemical shifts, which to an extent negates the need for 2D, but also can create problems, for example, with respect to uniform excitation of the entire ^{19}F bandwidth.

However, as the interest in compounds that contain fluorine has grown, along with the quality of available instrumentation, so has the interest in 2D ^{19}F NMR techniques. An excellent review by Battiste and Newmark on the applications of ^{19}F multidimensional NMR is available and should be consulted for detailed information about the hardware requirements and the application of the various available ^{19}F 2D NMR techniques.[44]

Unlike the one-dimensional NMR techniques to which this book is largely devoted, those 2D ^{19}F NMR techniques to be briefly discussed as follows will not generally be required for day-to-day structure elucidation by the working organic chemist. However, there will inevitably be situations where these techniques are indispensable in determining the detailed three-dimensional structure of compounds that contain fluorine, and at such times it may be necessary for the synthetic chemist to turn to an NMR specialist for assistance.

Homonuclear ^{19}F–^{19}F experiments are the most commonly carried out, and they are also the most easily implemented on conventional NMR spectrometers. Among such experiments, ^{19}F COSY correlation spectroscopy is probably the 2D ^{19}F NMR technique most frequently encountered, mainly because of through-space couplings that can make it otherwise difficult to infer definitive structural information from the presence or magnitude of observed correlations. It has been found to be particularly useful in the analysis of fluoropolymers.

Homonuclear ¹⁹F-TOCSY and ¹⁹F-NOESY experiments are utilized much less frequently than ¹⁹F-COSY, but as discussed in the Battiste review,[44] they are very much underutilized experiments, which when implemented can provide unique structural and particularly stereochemical information.

For compounds that contain a limited number of fluorine atoms, heteronuclear ¹⁹F–¹H 2D NMR experiments such as ¹⁹F–¹H HETCOR and ¹H–¹⁹F heteronuclear Overhauser spectroscopy (HOESY) can provide considerable assistance distinguishing structural isomers and diastereomers as well as for conformational analysis. HOESY experiments have been frequently used for conformational analysis of biomolecules containing fluorine labels.[33]

Diffusion-ordered spectroscopy (DOSY) "provides a means for virtual separation of compounds by providing a 2D map in which one axis is the chemical shift, while the other is that of the diffusion coefficient."[45] The direct combination of ¹⁹F NMR and DOSY has been shown to be very useful for studying drug formulations with fluorine-containing compounds that are part of a complex mixture.[46]

As an example of the kind of structural information that can be obtained by multidimensional NMR of fluorinated compounds, Ernst has, through a combination of heteronuclear 2D techniques, been able to assign all 16 carbons and all aromatic protons but was not able to establish unambiguous syn/anti assignments for the bridge CH₂ groups of 4-fluoro-[2.2]paracyclophane (**F-PCP**), and he was also able to distinguish the pseudo-gem, pseudo-ortho, pseudo-meta, and pseudo-para isomers (**F2-PCP**) and similarly able to assign all carbons and protons of these isomers, except for the syn/anti protons of the CH₂ bridge carbons (Scheme 2.35).[47] Similarly, Ghiviriga has been able to assign all fluorine and proton chemicals shifts and coupling constants for mono- and di-fluorine ring-substituted octafluoro[2.2]paracyclophanes (F-AF4 and F2-AF4).[48]

Scheme 2.35

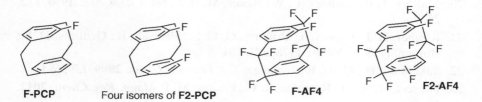

F-PCP   Four isomers of **F2-PCP**   **F-AF4**   **F2-AF4**

REFERENCES

1. Brey, W. S.; Brey, M. L. In *Encyclopedia of Nuclear Magnetic Resonance*; Grant, D. M., Harris, R. K., Eds.; *John Wiley & Sons*: Chichester, **1996**; Vol. *3*, p 2063.

2. Bromilow, J.; Brownlee, R. T. C.; Page, A. V. *Tetrahedron Lett.* **1976**, *17*, 3055.

3. Taft, R. W.; Prosser, F.; Goodman, L.; Davis, G. T. *J. Chem. Phys.* **1963**, *38*, 380.

4. Adcock, W.; Angus, D. L.; Lowe, D. A. *Magn. Reson. Chem.* **1996**, *34*, 675.

5. Adcock, W.; Abeywickrema, A. N.; Kok, G. B. *J. Org. Chem.* **1983**, *49*, 1387.

6. Adcock, W.; Krstic, A. R. *Magn. Reson. Chem.* **2000**, *38*, 115.

7. Gribble, G. W.; Keavy, D. J.; Olson, E. R.; Rae, I. D.; Staffa, A.; Herr, T. E.; Ferrara, M. B.; Contreras, R. H. *Magn. Reson. Chem.* **1991**, *29*, 422.

8. Durie, A. J.; Slawin, A. M. Z.; Lebl, T.; Kirsch, P.; O'Hagan, D. *Chem. Commun.* **2012**, *48*, 9643.

9. Dokalik, A.; Kalchhauser, H.; Mikenda, W.; Schweng, G. *Magn. Reson. Chem.* **1999**, *37*, 895.

10. Kleinmaier, R.; Arenz, S.; Karim, A.; Carlsson, A.-C. C.; Erdelyi, M. *Magn. Reson. Chem.* **2012**, *51*, 46.

11. Williamson, K. L.; Hsu, Y.-F. L.; Hall, F. H.; Swager, S.; Coulter, M. S. *J. Am. Chem. Soc.* **1968**, *90*, 6717.

12. O'Hagan, D.; Rzepa, H. S.; Schuler, M.; Slawin, A. M. Z. *Beilstein J. Org. Chem.* **2006**, *2*, 19.

13. Petrakis, L.; Sederholm, C. H. *J. Chem. Phys.* **1961**, *35*, 1243.

14. Mallory, F. B. *J. Am. Chem. Soc.* **1973**, *95*, 7747.

15. Hierso, J.-C. *Chem. Rev.* **2014**, *114*, 4838.

16. Gribble, G. W.; Douglas, J. R., Jr. *J. Am. Chem. Soc.* **1970**, *92*, 5764.

17. Mallory, F. B.; Mallory, C. W.; Butler, K. E.; Lewis, M. B.; Xia, A. Q.; Luzik, E. D., Jr.; Fedenburgh, L. E.; Ramanjulu, M. M.; Van, Q. N.; Francl, M. M.; Freed, D. A.; Wray, C. C.; Hann, C.; Nerz-Stormes, M.; Carroll, P. J.; Chirlian, L. E. *J. Am. Chem. Soc.* **2000**, *122*, 4108.

18. Gribble, G. W.; Kelly, W. J. *Tetrahedron Lett.* **1981**, *22*, 2475.

19. Ernst, L.; Ibrom, K. *Angew. Chem. Int. Ed. Engl.* **1995**, *34*, 1881.

20. Mallory, F. B.; Mallory, C. W.; Baker, M. B. *J. Am. Chem. Soc.* **1990**, *112*, 2577.

21. Kimber, B. J.; Feeney, J.; Roberts, G. C. K.; Birdsall, B.; Griffiths, D. V.; Burgen, A. S. V. *Nature* **1978**, *271*, 184.

22. Takemura, H.; Ueda, R.; Iwanaga, T. *J. Fluorine Chem.* **2009**, *130*, 684.

23. Fonseca, T. A. O.; Ramalho, T. C.; Freitas, M. P. *Magn. Res. Chem.* **2012**, *50*, 551.

24. Champagne, P. A.; Desroches, J.; Paquin, J.-F. *Synthesis* **2015**, *47*, 306.

25. Scerba, M. T.; Leavitt, C. M.; Diener, M. E.; DeBlase, A. F.; Guasco, T. L.; Siegler, M. A.; Bair, N.; Johnson, M. A.; Lectka, T. *J. Org. Chem.* **2011**, *76*, 7975.

26. Muller, N.; Carr, D. T. *J. Phys. Chem.* **1963**, *67*, 112.

27. Lambert, J. B.; Mazzola, E. P. *Nuclear Magnetic Resonance Spectroscopy*; Pearson Education, Inc.: Upper Saddle River, **2004**.

28. Abraham, R. T.; Loftus, P. *Tetrahedron* **1977**, *33*, 1227.

29. Irig, A. M.; Smith, S. L. *J. Am. Chem. Soc.* **1972**, *94*, 34.

30. Osten, H. J.; Jameson, C. J.; Craig, N. C. *J. Chem. Phys.* **1985**, *83*, 5434.

31. Lambert, J. B.; Greifenstein, L. G. *J. Am. Chem. Soc.* **1973**, *95*, 6150.

32. Gerig, J. T. *Prog. Nucl. Magn. Reson. Spectrosc.* **1994**, *26*, 293.

33. Gakh, Y. G.; Gakh, A. A.; Gronenborn, A. M. *Magn. Reson. Chem.* **2000**, *38*, 551.

34. Cobb, S. L.; Murphy, C. D. *J. Fluorine Chem.* **2009**, *130*, 132.

35. Chen, H.; Viel, S.; Ziarelli, F.; Peng, L. *Chem. Soc. Rev.* **2013**, *42*, 7971.

36. Keita, M.; Kaffy, J.; Troufflard, C.; Morvan, E.; Crousse, B.; Ongeri, S. *Org. Biomol. Chem.* **2014**, *12*, 4576.

37. Zhao, Y.; Swager, T. M. *J. Am. Chem. Soc.* **2015**, *137*, 3221.

38. Vaughan, M. D.; Cleve, P.; Robinson, V.; Durwel, H. S.; Honek, J. F. *J. Am. Chem. Soc.* **1999**, *121*, 8475.

39. Li, C.; Wang, G.-F.; Wang, Y.; Creager-Allen, R.; Lutz, E. A.; Scronce, H.; Slade, K. M.; Ruf, R. A. S.; Mehl, R. A.; Pielak, G. J. *J. Am. Chem. Soc.* **2010**, *132*, 321.

40. Berkowitz, D. B.; Karukurichi, K. R.; de la Salud-Bea, R.; Nelson, D. L.; McCune, C. D. *J. Fluorine Chem.* **2008**, *129*, 731.

41. Shuker, S. B.; Hajduk, P. J.; Meadows, R. P.; Fesik, S. W. *Science* **1996**, *274*, 1531.

42. Dalvit, C.; Vulpetti, A. *Magn. Res. Chem.* **2012**, *50*, 592.

43. Dalvit, C. *Prog. Nucl. Magn. Reson. Spectrosc.* **2007**, *51*, 243.

44. Battiste, J.; Newmark, R. A. *Prog. Nucl. Magn. Reson. Spectrosc.* **2006**, *48*, 1.

45. Cohen, Y.; Avram, L.; Frish, L. *Angew. Chem. Int. Ed.* **2005**, *44*, 520.

46. Poggetto, G. D.; Favaro, D. C.; Nilsson, M.; Morris, G. A.; Tormena, C. F. *Magn. Reson. Chem.* **2013**, *52*, 172.

47. Ernst, L.; Ibrom, K. *Magn. Reson. Chem.* **1997**, *35*, 868.

48. Ghiviriga, I.; Dulong, F.; Dolbier, W. R., Jr. *Magn. Reson. Chem.* **2009**, *47*, 313.

CHAPTER 3

THE SINGLE FLUORINE SUBSTITUENT

3.1. INTRODUCTION

The biological activity of a compound can often be affected dramatically by the presence of even a single fluorine substituent that is placed in a particular position within the molecule. There are diverse reasons for this, which have been discussed briefly in the preface and introduction of this book. A few illustrative examples of bioactive compounds containing *a single fluorine substituent* are given as follows (Figure 3.1). These include what is probably the first example of enhanced bioactivity due to fluorine substitution, that of the corticosteroid **3-1** below wherein Fried discovered, in 1954, that the enhanced acidity of the fluorohydrin enhanced the activity of the compound by increasing its binding affinity to glucocorticoid (GC) receptors and by retarding oxidation of the proximal 11-OH group.[1] Also pictured are the antibacterial β-fluoro amino acid, FA (**3-2**), which acts as a suicide substrate enzyme inactivator, and the well-known antianthrax drug, CIPRO® (**3-3**).

The information and examples presented in this chapter should enable the reader to predict chemical shift and coupling constant values for *a single fluorine substituent* in virtually any possible environment in which it might be encountered.

Guide to Fluorine NMR for Organic Chemists, Second Edition. William R. Dolbier, Jr.
© 2016 John Wiley & Sons, Inc. Published 2016 by John Wiley & Sons, Inc.

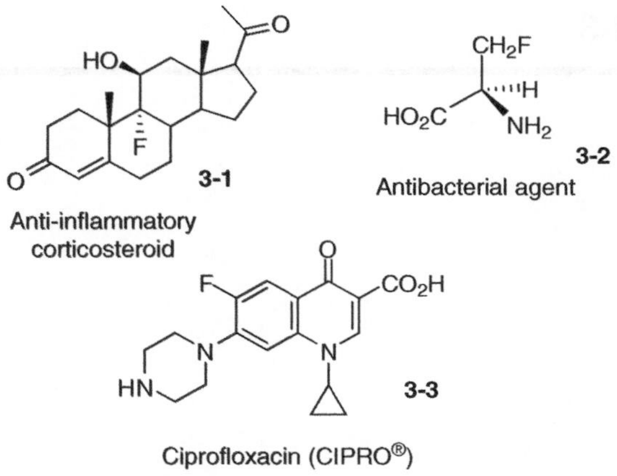

3-1
Anti-inflammatory
corticosteroid

3-2
Antibacterial agent

3-3
Ciprofloxacin (CIPRO®)

FIGURE 3.1. Examples of bioactive compounds containing a single fluorine substituent

3.1.1. Chemical Shifts – General Considerations

As indicated in Chapter 2, the single fluorine substituent has an extremely broad range of observed chemical shifts, which include sulfonyl fluorides and acyl fluorides absorbing downfield in the region of +40 and +25 ppm, respectively, all the way up to fluoromethyltrimethylsilane, with its signal far upfield at −277 ppm.

Even within the different classes of compounds bearing the fluorine substituent, the ranges of chemical shifts are still quite large, but there are predictable trends, both within each class and connecting the different classes of compounds. For example, the range of chemical shifts for a single fluorine within a saturated hydrocarbon goes from the −131 ppm chemical shift observed for *t*-butyl fluoride to the value of −272 ppm observed for methyl fluoride. Primary fluorides absorb at the higher field (more negative) end and tertiary fluorides absorb at the lower field end of the range. Single vinylic and aromatic fluorines absorb at even lower field, within the range of −95 to −130 ppm.

3.1.2. Spin–Spin Coupling Constants – General Considerations

It will also be seen that spin–spin coupling constants between fluorine and hydrogen as well as between fluorine and fluorine, and between fluorine and carbon are quite predictable and thus useful in detailed structure characterization. Compounds containing a single fluorine substituent exhibit the largest three-bond vicinal F–H coupling among saturated hydrofluorocarbon systems, such coupling constants ranging from

21 to 27 Hz. Their two-bond F–H coupling constants (47–51 Hz) are also large, but those for a CH_2F group (48–51 Hz) are smaller than the 56–58 Hz coupling constants exhibited by CF_2H groups.

The one-bond F–C coupling constants observed for $-CH_2F$ and $-CHF-$ groups, which are generally in the 162–170 Hz range, are also much smaller than the 234–250 Hz coupling exhibited by $-CF_2H$ or $-CF_2-$ groups or the 275–285 Hz coupling observed for CF_3 groups. However, both the monofluoro- and the difluoromethyl F–C coupling constants will increase significantly when the carbon is also bound to other electronegative substituents. For example, note the large difference between the one-bond F–C coupling constant of methyl fluoromethyl ether compared to that of 1-fluorobutane (219 vs 165 Hz). These trends probably reflect the relative amount of s-character in the carbon orbital that is used in bonding to fluorine in these compounds, the more s-character, the larger the C–F coupling constant.

3.2. SATURATED HYDROCARBONS[2, 3]

Among the large group of compounds represented as monofluoroalkanes, primary fluorine substituents are the most shielded, with the rule governing relative chemical shifts being quite simple:

Shielding of alkyl fluorides: $CH_3 > 1° > 2° > 3°$
Range of chemical shifts: $-272 \rightarrow -131$ ppm

This trend is consistent with those observed for both proton and carbon chemical shifts, with the proton on the most highly substituted carbon, and the carbon with the most alkyl substituents being the most highly deshielded. Computational work by Wiberg and Zilm has allowed identification of the factors that lead to the observed shielding trends that are observed for alkyl fluorides.[1]

A corollary rule is that branching at either the β- or the γ-position gives rise to increased *shielding* of 1°, 2°, or 3° fluorine nuclei.

3.2.1. Primary Alkyl Fluorides

The typical chemical shift for primary, n-alkyl fluorides is -219, but the values for primary alkyl fluorides vary between -212 for ethyl fluoride and -226 for 2-ethyl-1-fluorobutane (Scheme 3.1). As mentioned earlier, alkyl branching leads to shielding of fluorine nuclei.

<u>**Scheme 3.1**</u>

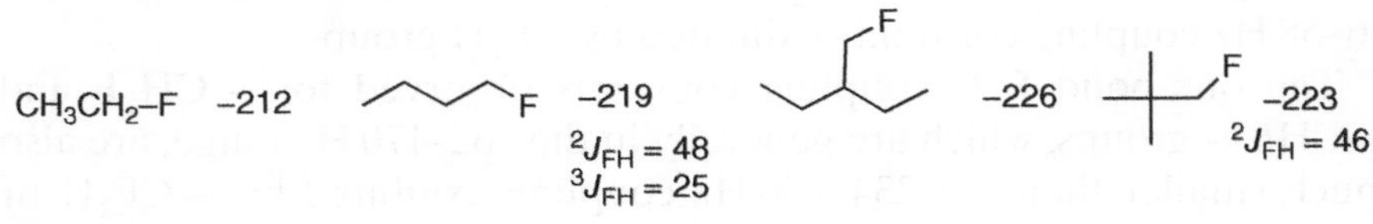

$CH_3CH_2–F$ –212   /\/\F   –219   –226   –223

$^2J_{FH} = 48$ $^2J_{FH} = 46$
$^3J_{FH} = 25$

The coupling constants given here are typical two- and three-bond H–F values for such systems, with the range of two-bond F–H coupling of about 47–49 Hz and that for the three-bond F–H coupling of 21–27 Hz. Since the value of the three-bond H–F coupling constant is approximately half that of the two-bond H–F coupling constant, the net result is that the fluorine signal for an *n*-alkyl fluoride generally has the appearance of a septet, as is exemplified for *n*-pentyl fluoride in Figure 3.2.

When CH_2F is a substituent on most alicyclic rings, such as a cyclohexane ring, the ^{19}F chemical shift of this group is not significantly altered from that of an acyclic system (Scheme 3.2). On the other hand, when it is attached to a *cyclopropane* ring, a unique deshielding influence is

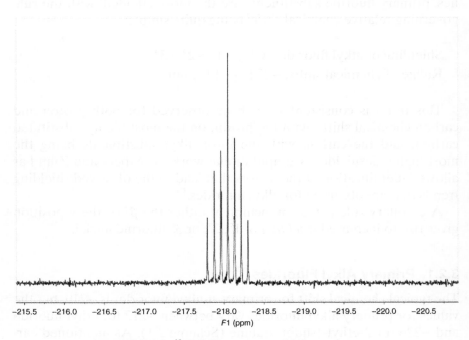

FIGURE 3.2. ^{19}F NMR spectrum of 1-fluoropentane

Scheme 3.2

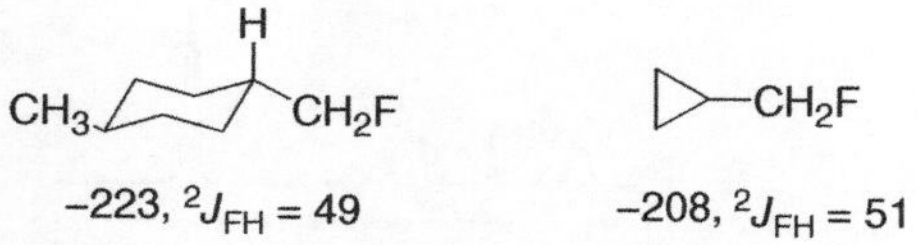

$-223, {}^{2}J_{FH} = 49$ $-208, {}^{2}J_{FH} = 51$

observed. This deshielding is likely a result of a cyclopropane ring's well-recognized ability to stabilize an electron-deficient center via electron donation. It is therefore likely a homohyperconjugative $\pi-\sigma_{CF}{}^{*}$ interaction similar to that discussed as arising in anti-7-norbornenyl fluoride (Section 2.2.1.3).

3.2.1.1. ^{1}H and ^{13}C NMR Data.
The following examples (Scheme 3.3) provide insight into expected proton and carbon chemical shift and coupling constant data for primary alkyl fluorides. It can be seen that the influence on both proton and carbon chemical shifts diminishes rapidly as one moves away from the site of fluorine substitution.

Scheme 3.3

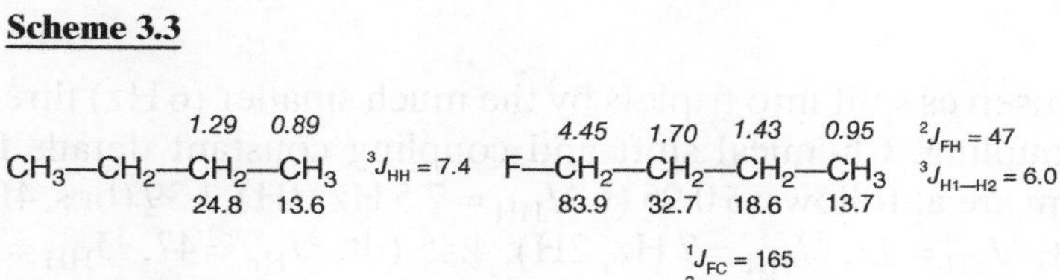

Here and continuing throughout the text, proton chemical shifts are italicized in order to distinguish them from carbon and fluorine chemical shifts.

Figure 3.3 provides a typical example of a proton spectrum of an n-alkyl fluoride. In this spectrum, one can clearly see the doublet signals resulting from the large two-bond F–H coupling (47 Hz), which

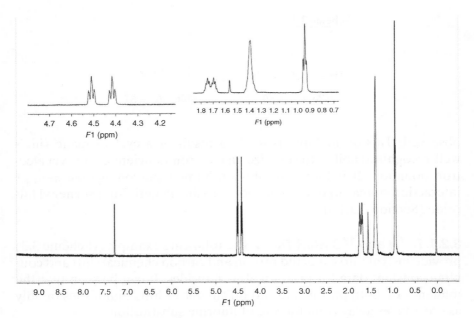

FIGURE 3.3. ^{1}H NMR spectrum of 1-fluoropentane

are themselves split into triplets by the much smaller (6 Hz) three-bond H–H coupling. Chemical shift and coupling constant details for this spectrum are as follows: δ 0.95 (t, $^3J_{HH} = 7.5$ Hz, 3H), 1.39 (br s, 4H), 1.72 (d, pent, $^3J_{FH} = 25$, $^3J_{HH} = 7$ Hz, 2H), 4.45 (dt, $^2J_{HF} = 47$, $^3J_{HH} = 6.0$ Hz, 2H).

Figure 3.4 provides the ^{13}C NMR spectrum of 1-fluoropentane, again a typical spectrum of an *n*-alkyl fluoride. Examining this spectrum allows one to readily distinguish each carbon with respect to its location relative to the fluorine substituent. This can be accomplished not only by comparison of their chemical shifts but also even more definitively by comparison of their fluorine–carbon coupling constants. In the spectrum, one sees the large (164 Hz) one-bond coupling of the signal at 84.2 ppm, the much smaller (20.1 Hz) two-bond F–C coupling of the signal at 30.1 ppm, and the yet smaller (5.0 Hz) three-bond coupling of the signal at 27.3 ppm. The small coupling to the remaining two carbons is not evident in the signals at 22.3 and 14.0 ppm. (The multiplet at ~77.2 ppm in most ^{13}C spectra printed in this book derives from the solvent, CDCl$_3$.)

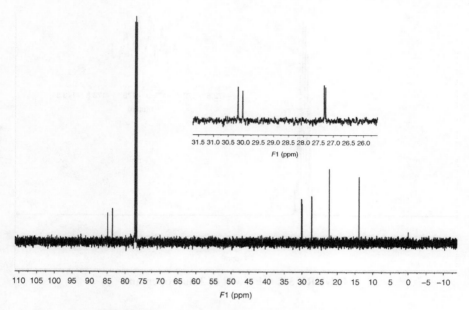

FIGURE 3.4. ^{13}C NMR spectrum of 1-fluoropentane

3.2.2. Secondary Alkyl Fluorides

Secondary alkyl fluorides exhibit a downfield (deshielding) shift of about +35 ppm from their primary analogs, their fluorines typically absorbing at about −183 ppm (Scheme 3.4), and such fluorines will also experience the usual considerable shielding as a result of branching.

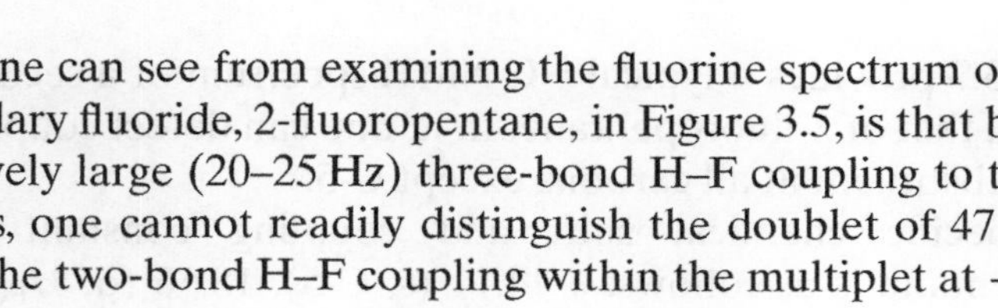

Scheme 3.4

F   F   F

−165   −183   −189

$^{2}J_{FH} = 47$

What one can see from examining the fluorine spectrum of the typical secondary fluoride, 2-fluoropentane, in Figure 3.5, is that because of the relatively large (20–25 Hz) three-bond H–F coupling to the vicinal hydrogens, one cannot readily distinguish the doublet of 47 Hz deriving from the two-bond H–F coupling within the multiplet at −173 ppm.

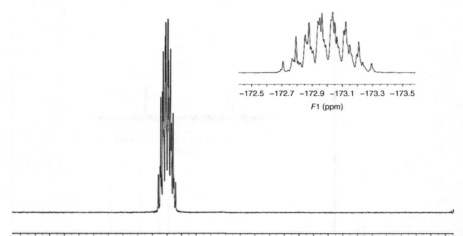

FIGURE 3.5. ^{19}F NMR of 2-fluoropentane

The net result is the multiplet seen in Figure 3.5. The doublet due to the 47 Hz coupling is much more clearly seen in the proton spectrum (Figure 3.6) because the potentially complicating three-bond H–H coupling constants are much smaller (6–7 Hz).

3.2.2.1. Characteristic ^1H and ^{13}C NMR Data. The following examples (Scheme 3.5) provide insight into expected proton and carbon chemical shift and coupling constant data for secondary alkyl fluorides.

As mentioned earlier, the doublet due to the large (48 Hz) two-bond F–H coupling constant can easily be seen in the proton spectrum of 2-fluoropentane (Figure 3.6). Note also the nice doublet of doublets centered at 1.31 ppm ($^3J_{FH} = 24$ Hz, $^3J_{HH} = 6$ Hz) deriving from the C-1 methyl group, which exemplifies the significant difference in magnitude between typical three-bond F–H and three-bond H–H coupling constants.

Figure 3.7 provides a typical ^{13}C NMR spectrum of a secondary fluoroalkane, that of 2-fluoropentane. In it, one can detect coupling to the fluorine substituent at all carbons except C-5.

The specific chemical shift and coupling constant data for 2-fluoropentane is as follows: δ 14.11 (s, C-5), 18.56 (d, $^3J_{FC} = 5.1$ Hz, C-4), 21.20 (d, $^2J_{FC} = 22.3$ Hz, C-1), 39.24 (d, $^2J_{FC} = 20.4$ Hz, C-3), 91.0 (d, $^1J_{FC} = 164$ Hz, C-2).

1.3 1.2 1.1 1.0 0.9
F1 (ppm)
4.80 4.75 4.70 4.65 4.60 4.55 4.50 4.45
F1 (ppm)
7.5 7.0 6.5 6.0 5.5 5.0 4.5 4.0 3.5 3.0 2.5 2.0 1.5 1.0 0.5 0.0
F1 (ppm)

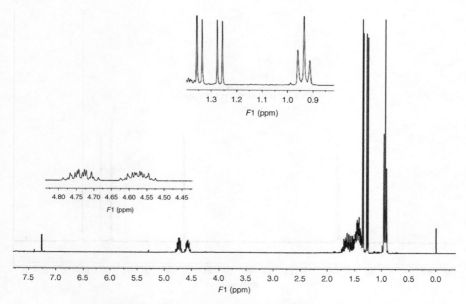

FIGURE 3.6. ^{1}H NMR spectrum of 2-fluoropentane

Scheme 3.5

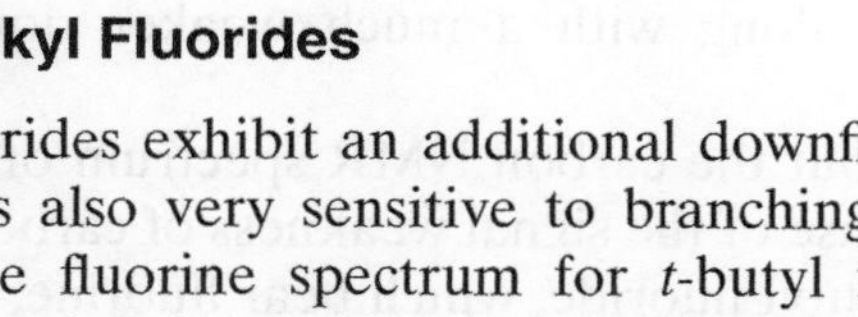

1.33 4.84
CH3-CH-CH3
22.6 87.3
F
3JHH= 7.3
0.96 1.56 4.62
CH3-CH2-CH-CH2-CH3
9.1 27.6 96.5
F
3JHH= 7.3
2JFH= 4.9

3.2.3. Tertiary Alkyl Fluorides

Tertiary alkyl fluorides exhibit an additional downfield shift of about +25 ppm, which is also very sensitive to branching as shown below (Scheme 3.6). The fluorine spectrum for *t*-butyl fluoride is shown in Figure 3.8. The signal at −131 ppm is split into 10 peaks with a three-bond, H–F coupling constant of 21 Hz.

3.2.3.1. *Characteristic ^{1}H and ^{13}C NMR Data.* The following examples provide relevant proton and carbon chemical shift data (Scheme 3.7).

The proton and carbon spectra for *t*-butyl fluoride are provided in Figures 3.9 and 3.10.

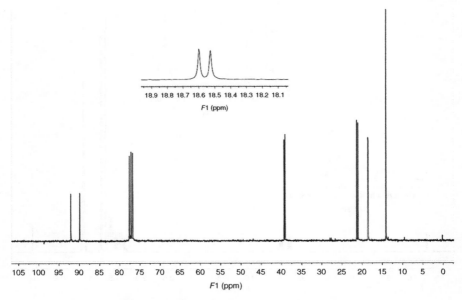

FIGURE 3.7. ^{13}C NMR spectrum of 2-fluoropentane

Scheme 3.6

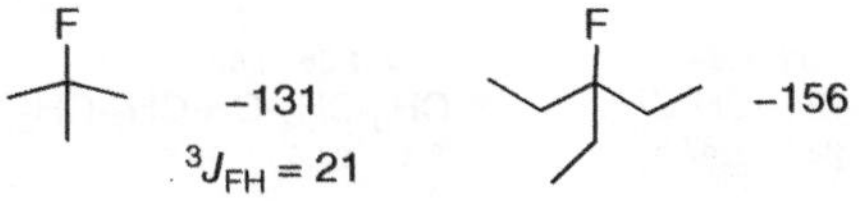

−131 $^3J_{FH} = 21$ −156

The proton spectrum consists of a doublet at δ 1.38 ($^3J_{FH} = 21$), whereas the carbon spectrum exhibits a doublet at 28.7 ppm with $^2J_{FC} = 25$ Hz along with a much weaker doublet at 94.1 ppm ($^1J_{FC} = 162$ Hz).

A comment about the carbon NMR spectrum of *t*-butyl fluoride is appropriate. Because of the signal weakness of carbons such as the tertiary carbon of *t*-butyl fluoride, which bear fluorine, but no hydrogens, many published tabulations of ^{13}C spectra of compounds that contain such structural features fail to report these crucial signals. They can easily be missed, especially if you do not know what you are looking for. Even with relatively concentrated samples, it is usually necessary to run such spectra appropriately in order to accumulate sufficient FT data to see these weak signals.

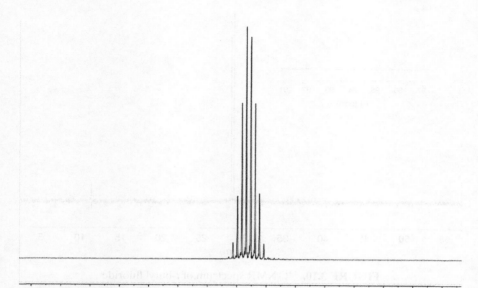

−127.5 −128.0 −128.5 −129.0 −129.5 −130.0 −130.5 −131.0 −131.5 −132.0 −132.5 −133.0 −133.5 −134.0 −134.5
F1 (ppm)

FIGURE 3.8. ^{19}F NMR spectrum of *t*-butyl fluoride

Scheme 3.7

CH3 1.38
94.1
F
$^3J_{FH}$ = 21
H3C
CH3
$^1J_{FC}$ = 162
$^2J_{FC}$ = 25
28.7
F
99.5
1.57
28.5
$^3J_{FH}$ = 26
7.3
0.87

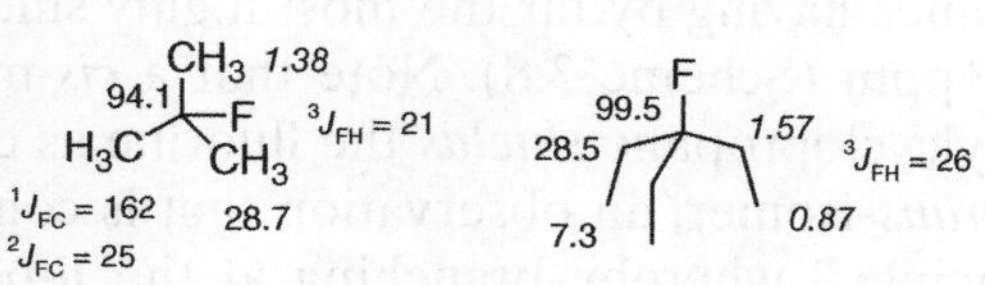

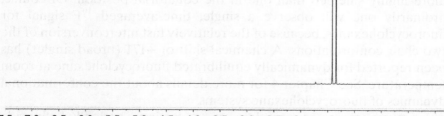

7.5 7.0 6.5 6.0 5.5 5.0 4.5 4.0 3.5 3.0 2.5 2.0 1.5 1.0 0.5 0.0 −0.5
F1 (ppm)

FIGURE 3.9. ^{1}H NMR spectrum of *t*-butyl fluoride

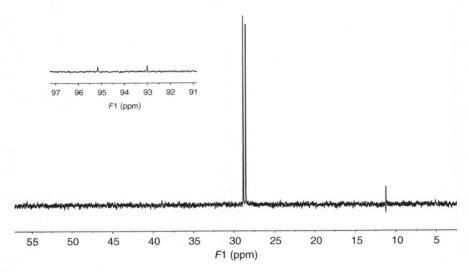

FIGURE 3.10. ^{13}C NMR spectrum of *t*-butyl fluoride

3.2.4. Cyclic and Bicyclic Alkyl Fluorides

Fluorocycloalkanes exhibit a small downfield (+) trend in chemical shifts in going from six- to five- to four-membered rings, but cyclopropanes are unique in exhibiting a strong *upfield* shift, with fluorocyclopropanes having by far the most highly shielded secondary fluorine at −213 ppm (Scheme 3.8). Note that a *cis*-methyl group on 1-fluoro-2-methylcyclopropane *shields* the fluorine as compared to the fluorine of the *trans*-isomer, an observation that is consistent with the "branching principle," whereby branching at the β-position leads to shielding of fluorine substituents.

Looking more closely at fluorocyclohexane systems, it has been observed that a fluorine substituent in the axial position is much more highly shielded than one in the equatorial position. Of course, ordinarily one will observe a single, time-averaged ^{19}F signal for fluorocyclohexanes, because of the relatively fast interconversion of the two chair conformations. A chemical shift of −171 (broad singlet) has been reported for dynamically equilibrated fluorocyclohexane at room temperature. See Chapter 4 for more details about the conformational dynamics of fluorocyclohexane systems.

In going from a secondary to a tertiary cycloalkyl fluoride (Scheme 3.9), one observes the usual deshielding effect as is exemplified by the isomeric 1-fluoro-1-methyl-4-*t*-butylcyclohexanes. Of

Scheme 3.8 Secondary cyclic and bicyclic fluorides

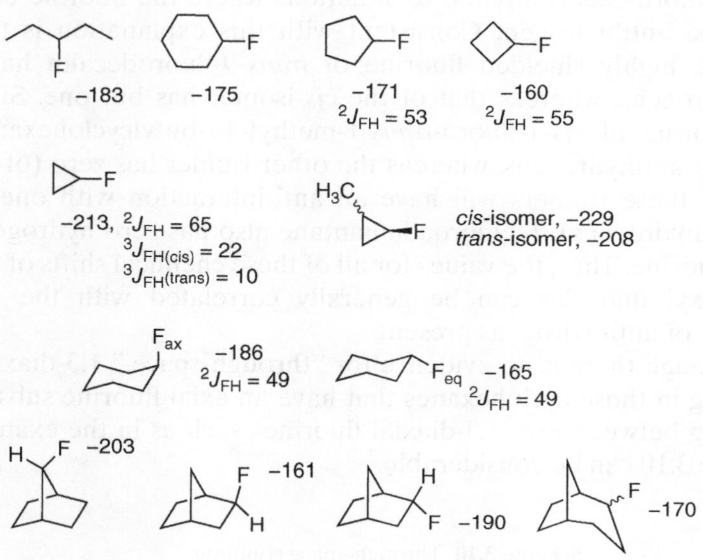

−183   −175   −171   −160
$^2J_{FH} = 53$ $^2J_{FH} = 55$

−213, $^2J_{FH} = 65$
$^3J_{FH(cis)} = 22$
$^3J_{FH(trans)} = 10$

H_3C cis-isomer, −229
trans-isomer, −208

F_{ax} −186 F_{eq} −165
$^2J_{FH} = 49$ $^2J_{FH} = 49$

H, F −203

F −161

H   F −190

F −170

Scheme 3.9 Secondary cyclic and bicyclic tertiary fluorides

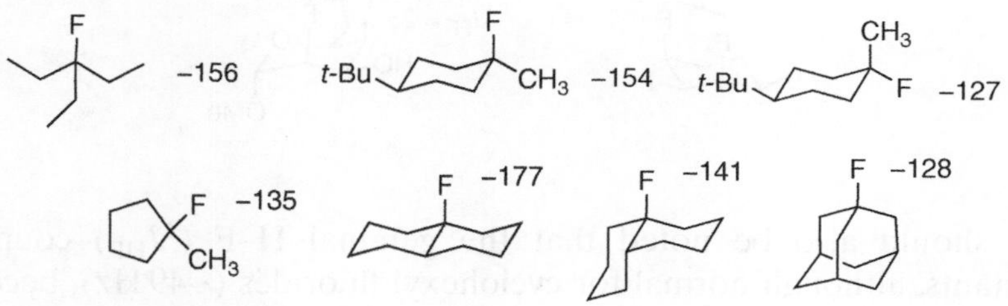

F −156

t-Bu   F CH_3 −154

CH_3
t-Bu   F −127

F −135
CH_3

F −177

F −141

F −128

course, these two isomers exist essentially in the single conformation given, because of the presence of the 4-*t*-butyl substituent.

The chemical shift differences observed for these 1-fluoro-1-methyl-*t*-butylcyclohexanes, the *cis*- and *trans*-9-fluorodecalins, and for 1-fluoroadamantane provide insight regarding the significant influence of conformation upon fluorine chemical shifts in fluorocyclohexanes. The relative chemical shifts of these various cyclohexyl fluorides can be rationalized simply on the basis of what is commonly known as an anomeric effect. That is, a vicinal hydrogen that is rigidly anti to a fluorine substituent will exhibit an "anomeric" double-bond/no-bond resonance (or $\sigma \rightarrow \sigma^*$) interaction, which will lead to relative *shielding* of the fluorine as compared to situations where the fluorine does not have the antihydrogen. Consistent with this explanation is the fact that the highly shielded fluorine of *trans*-9-fluorodecalin has three antihydrogens, whereas that of the *cis*-isomer has but one. Similarly, the fluorine of *cis*-1-fluoro-*trans*-1-methyl-4-*t*-butylcyclohexanes has two ring antihydrogens, whereas the other isomer has zero (of course, both of these isomers will have an anti-interaction with one of the methyl hydrogens). 1-Fluoroadamantane also has zero hydrogens anti to its fluorine. Thus, the values for all of these chemical shifts of tertiary cyclohexyl fluorides can be generally correlated with the relative number of antihydrogens present.

Although there is no evidence for "through-space," 1,3-diaxial F–H coupling in those cyclohexanes that have an axial fluorine substituent, coupling between *two*, 1,3-diaxial fluorines such as in the examples in Scheme 3.10 can be considerable.[2, 3]

Scheme 3.10 Through-space coupling between two 1,3-diaxial fluorine substituents

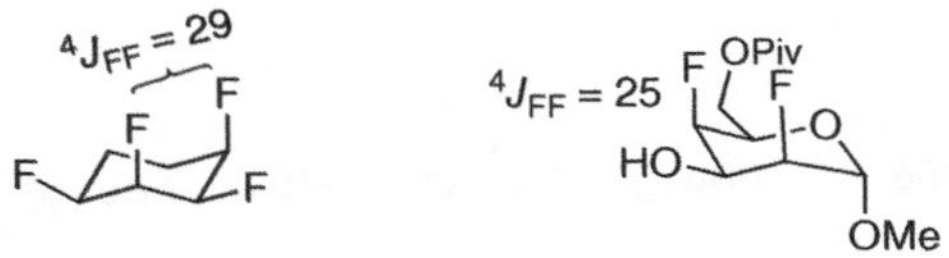

It should also be noted that the geminal H–F ($^2J_{HF}$) coupling constants, although normal for cyclohexyl fluorides ($\sim$49 Hz), become progressively larger with smaller ring size (cyclopentyl fluoride, 53 Hz; cyclobutyl fluoride, 55 Hz), culminating in the characteristically large value ($\sim$65 Hz) for cyclopropyl fluoride. Such an increase is consistent with the presumed increase in s-character of the carbon orbital used to bond to fluorine in the series.

A series of strained, bridgehead fluorinated bicyclic compounds shown in Scheme 3.11 displays a large range of chemical shifts that has no obvious trend. What is surprising is the lack of correlation with s content of the orbital used by carbon to bond to fluorine, which should be decreasing along the series.[4]

3.2.4.1. *^1H and ^{13}C NMR Data.* Pertinent proton and carbon NMR data, in addition to those given above, are provided in Schemes 3.12

Scheme 3.11

−132.5   −157.4   −125.1   −182.0   −147.6   −127.8

Scheme 3.12 Proton and carbon data for cyclic fluorides

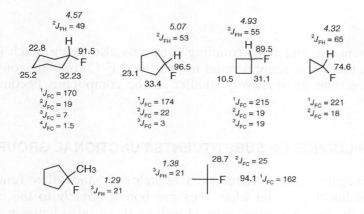

H-F Coupling constants in cyclobutyl fluoride

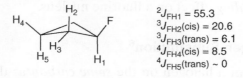

$^2J_{FH1} = 55.3$
$^3J_{FH2}(cis) = 20.6$
$^3J_{FH3}(trans) = 6.1$
$^4J_{FH4}(cis) = 8.5$
$^4J_{FH5}(trans) \sim 0$

and 3.13. Again, the increase in one-bond F–C coupling constant observed in the series, with that for cyclopropyl fluoride being the largest, is noteworthy and is again consistent with the degree of s-character in the F–C bond. A review of ^{13}C NMR spectra of fluorinated cyclopropanes has been published.[5]

When it comes to chemical shifts for bridgehead hydrogens, as was the case with the fluorine shifts, there does not seem to be any understandable trend (Scheme 3.13). On the other hand, the carbon data for 1-fluoro-bicyclo[1.1.1]pentane are unique, with both bridgehead carbons being much more highly shielded than their counterparts

Scheme 3.13 Proton and carbon data for bridgehead fluorides

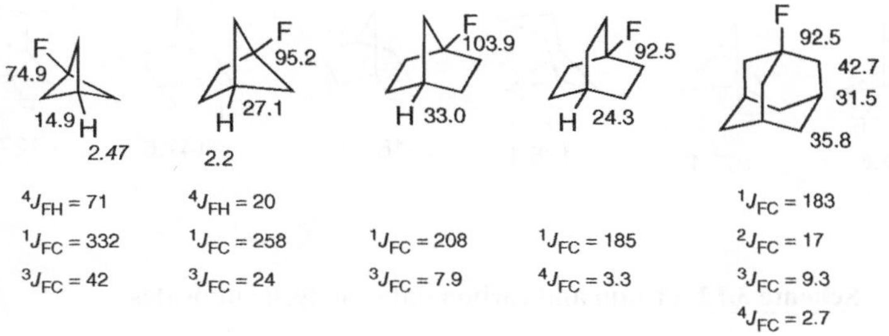
F
74.9
14.9
H
2.47
F
95.2
27.1
H
2.2
F
103.9
H 33.0
F
92.5
H 24.3
F
92.5
42.7
31.5
35.8
$^4J_{FH}$ = 71
$^1J_{FC}$ = 332
$^3J_{FC}$ = 42
$^4J_{FH}$ = 20
$^1J_{FC}$ = 258
$^3J_{FC}$ = 24
$^1J_{FC}$ = 208
$^3J_{FC}$ = 7.9
$^1J_{FC}$ = 185
$^4J_{FC}$ = 3.3
$^1J_{FC}$ = 183
$^2J_{FC}$ = 17
$^3J_{FC}$ = 9.3
$^4J_{FC}$ = 2.7

in the scheme, and their coupling constants also being much larger. Indeed, there is a general trend in one-bond F–C coupling constants, which become progressively smaller as the compounds become less strained.

3.3. INFLUENCE OF SUBSTITUENTS/FUNCTIONAL GROUPS

Electronegative substituents, such as halogens and ether functions, deshield fluorine nuclei when they are bound *directly* to the carbon bearing the fluorine substituent. However, these same halogen, ether, and alcohol substituents, as well as similarly electronegative carbonyl functions, when positioned β to the fluorine substituent, generally give rise to a *shielding* effect on a fluorine nucleus.

3.3.1. Halogen Substitution[6]

Substitution of a halogen on the *same carbon* as that bearing the fluorine substituent gives rise to dramatic incremental deshielding effects (Schemes 3.14 and 3.15).

The deshielding effect of I<Br<Cl, with a greater deshielding effect within the methane series than the primary alkyl series.

A halogen substituent at the β-position to a fluorine generally gives rise to *shielding* of the fluorine nucleus (Scheme 3.16), with fluorine providing the greatest shielding influence and iodine having virtually no effect. Interestingly, addition of second and third β-chlorine substituent leads to progressive *de*shielding.

Scheme 3.14

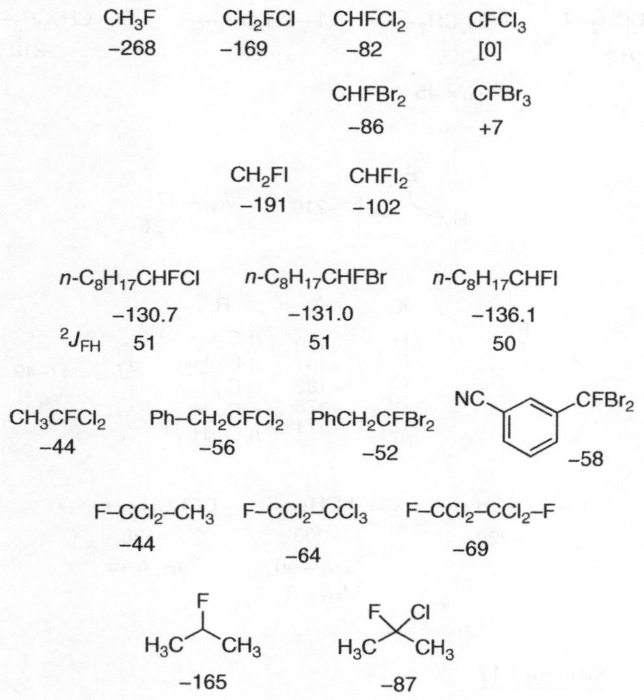

CH₃F	CH₂FCl	CHFCl₂	CFCl₃

$$CH_3F \quad CH_2FCl \quad CHFCl_2 \quad CFCl_3$$
$$-268 \quad\quad -169 \quad\quad -82 \quad\quad [0]$$

$$CHFBr_2 \quad CFBr_3$$
$$-86 \quad\quad +7$$

$$CH_2FI \quad CHFI_2$$
$$-191 \quad\quad -102$$

$$n\text{-}C_8H_{17}CHFCl \quad n\text{-}C_8H_{17}CHFBr \quad n\text{-}C_8H_{17}CHFI$$
$$-130.7 \quad\quad\quad -131.0 \quad\quad\quad -136.1$$
$$^2J_{FH} \quad 51 \quad\quad\quad\quad 51 \quad\quad\quad\quad\quad 50$$

$$CH_3CFCl_2 \quad Ph\text{-}CH_2CFCl_2 \quad PhCH_2CFBr_2$$
$$-44 \quad\quad\quad -56 \quad\quad\quad\quad -52$$

NC⟍ ⟍CFBr₂ (benzene ring)
−58

$$F\text{-}CCl_2\text{-}CH_3 \quad F\text{-}CCl_2\text{-}CCl_3 \quad F\text{-}CCl_2\text{-}CCl_2\text{-}F$$
$$-44 \quad\quad\quad\quad -64 \quad\quad\quad\quad -69$$

H₃C—C(F)—CH₃
−165

H₃C—C(F)(Cl)—CH₃
−87

3.3.1.1. *¹H and ¹³C NMR Data for Halofluoroalkanes.* The relative effect of the four halogens on proton and carbon chemical shifts in haloethanes is shown in Scheme 3.17. The trends in proton and carbon chemical shifts are consistent with relative electronegativity effects.

Scheme 3.18 provides some pertinent proton and carbon chemical shift and coupling constant data for trihalomethanes, including what data are available for fluorodihalomethanes. There are potentially significant solvent effects on proton chemical shifts of all trihalomethanes,

Scheme 3.15

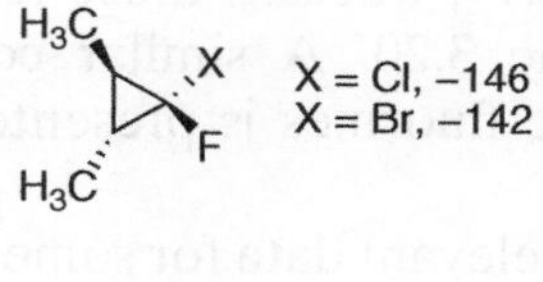

H₃C⟍ (cyclopropane ring with X and F)
H₃C⟋

X = Cl, −146
X = Br, −142

Scheme 3.16

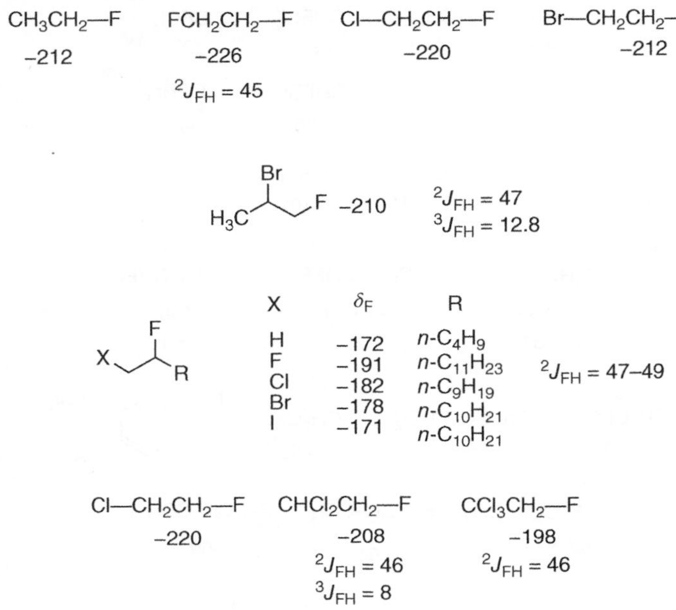

CH_3CH_2-F FCH_2CH_2-F $Cl-CH_2CH_2-F$ $Br-CH_2CH_2-F$

   −212   −226   −220   −212

                    $^2J_{FH} = 45$

H$_3$C—CH(Br)—F   −210 $^2J_{FH} = 47$
                          $^3J_{FH} = 12.8$

X	δ_F	R	
H	−172	n-C$_4$H$_9$	
F	−191	n-C$_{11}$H$_{23}$	$^2J_{FH} = 47\text{--}49$
Cl	−182	n-C$_9$H$_{19}$	
Br	−178	n-C$_{10}$H$_{21}$	
I	−171	n-C$_{10}$H$_{21}$	

X—CH(F)—CH$_2$—R

$Cl-CH_2CH_2-F$ $CHCl_2CH_2-F$ CCl_3CH_2-F

    −220   −208   −198

                      $^2J_{FH} = 46$ $^2J_{FH} = 46$
                      $^3J_{FH} = 8$

Scheme 3.17

	CH$_3$CH$_2$F	CH$_3$CH$_2$Cl	CH$_3$CH$_2$Br	CH$_3$CH$_2$I
δ_H	4.36	3.47	3.37	3.18
δ_C	78.0	38.7	27.9	−1.0

so an attempt was made to list data obtained in CDCl$_3$ as solvent, when available. Note the trends observed in proton and carbon chemical shifts again appear to be entirely consistent with relative electronegativity effects. Thus, the carbon chemical shift of CHI$_3$ is −161 ppm, that of CHBr$_3$ +11.85 ppm, and that of CHFCl$_2$ +105.9 ppm.

Data for a similar series of dihalomethanes and fluoro, halomethanes is presented in Scheme 3.19, whereas those for fluorotrihalomethanes can be found in Scheme 3.20. A similar comparison of methanes containing two and three fluorines is presented in Chapters 4 and 5, respectively.

Scheme 3.21 contains relevant data for some typical 1-fluorohalo and dihaloalkanes.

Scheme 3.22 contains data for 1-fluoro-2-halo- and 3-haloalkanes.

Scheme 3.18

Nonfluorine-containing trihalomethanes

	CHI$_3$	CHI$_2$Br	CHI$_2$Cl	CHBr$_2$I	CHIBrCl	CHBr$_3$	CHICl$_2$	CHBrCl$_2$	CHCl$_3$
δ_H (CDCl$_3$)	4.90	5.74	6.19	6.4	6.71	6.83	7.0	7.2	7.25
δ_H (Acetone-d_6)									7.89
δ_C	−161.4	−99.6	−67.8	−42.9	−15.4	11.85	10.5	54.5	77.1

Fluorodihalomethanes

	CHFI$_2$	CHFIBr	CHFICl	CHFBr$_2$	CHFClBr	CHFCl$_2$
δ_H (CDCl$_3$)	7.32	7.58	7.64		7.83*	
δ_H (Acetone-d_6)				8.32		7.97
$^2J_{FH}$	48	49	50	50		54
δ_C	ua	ua	55.0	75.9	91.4	105.9
$^1J_{FC}$			306	312	304	292

* Solvent unknown; *ua* = unavailable

Scheme 3.19

	CH$_2$I$_2$	CH$_2$Br$_2$	CH$_2$BrCl	CH$_2$Cl$_2$	CH$_2$FCl	CH$_2$FI
δ_H	3.85	4.69	5.14	5.27	5.93	6.35
δ_C	−61.1	18.4	36.7	53.0	*	51.8
$^2J_{FH}$					49	49
$^1J_{FC}$						251

* Data unavailable

Scheme 3.20

	CFCl$_3$	CFCl$_2$Br	CFCl$_2$I	CFClBr$_2$	CFBr$_3$	CFBr$_2$I	CFBrI$_2$	CFI$_3$
δ_F	[0]	+3.2	+5.9	+5.6	+7	−3.8	−9.0	*
δ_C					48.1			
$^1J_{FC}$	337				365			

Scheme 3.21

$^2J_{HF} = 51$
$^3J_{HF} = 5.3$

2.17 6.15
$C_7H_{15}CH_2CHFCl$
39.1 103.0

$^1J_{FC} = 242$
$^2J_{FC} = 20$

$^2J_{HF} = 51$
$^3J_{HF} = 5.3$

2.17 6.45
$C_7H_{15}CH_2CHFBr$
40.7 95.7

$^1J_{FC} = 251$
$^2J_{FC} = 18$

$^2J_{HF} = 50$
$^3J_{HF} = 5.5$

2.17 6.82
$C_7H_{15}CH_2CHFI$
43.2 75.6

$^1J_{FC} = 253$
$^2J_{FC} = 18$

CH_3CFCl_2
118.8

$^1J_{FC} = 294$

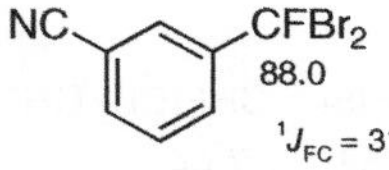

NC—⟨benzene⟩—$CFBr_2$
88.0

$^1J_{FC} = 314$

Scheme 3.22

$\overset{2}{X}-CH_2-\overset{1}{CH_2}-F$

X	δ_{H1}	δ_{H2}	$^2J_{HF}$	$^3J_{HF}$	$^3J_{HH}$
Cl	4.58	3.67	47	23	5.7
Br	4.61	3.49	46	18	4.9
I	3.88	2.54	47	19	6.7

$I-CH_2-CH_2-F$
1.2 82.7

$^1J_{FC} = 174$
$^2J_{FC} = 22$

$^3J_{HH} = 6.9$

Br–CH$_2$–CH$_2$–CH$_2$–F
3.55   4.56

$^2J_{HF} = 47$
$^3J_{HH} = 5.7$

$^3J_{HF} = 26$
$^3J_{HH} = 6.9$

$^2J_{HF} = 47$
$^3J_{HH} = 5.3$

3.25 2.15 4.48
I–CH$_2$–CH$_2$–CH$_2$–F
1.0   33.9 83.0

$^1J_{FC} = 167$
$^2J_{FC} = 20$
$^3J_{FC} = 5.7$

$^3J_{HH} = 7.5$

$^2J_{HF} = 48$
$^3J_{HH} = 5.3$

$^3J_{HF} = 20$
$^3J_{HH} = 5.3$

0.97 1.75 4.54 3.45
CH$_3$–CH$_2$–CH–CH$_2$–Br

$^3J_{FC} = 5.1$ 8.9 26.4 F 33.5 $^2J_{FC} = 25$

$^2J_{FC} = 20$ 93.1

$^1J_{FC} = 176$

3.3.1.2. Vicinal Fluorine Substituents. The examples provided in this section exemplify the specific effect of vicinal fluorine substituents on both chemical shift and upon spin–spin coupling constants (Scheme 3.23). As indicated in Section 3.3.1 (Scheme 3.16), each of the fluorines is somewhat shielded by the presence of the other.

Scheme 3.23

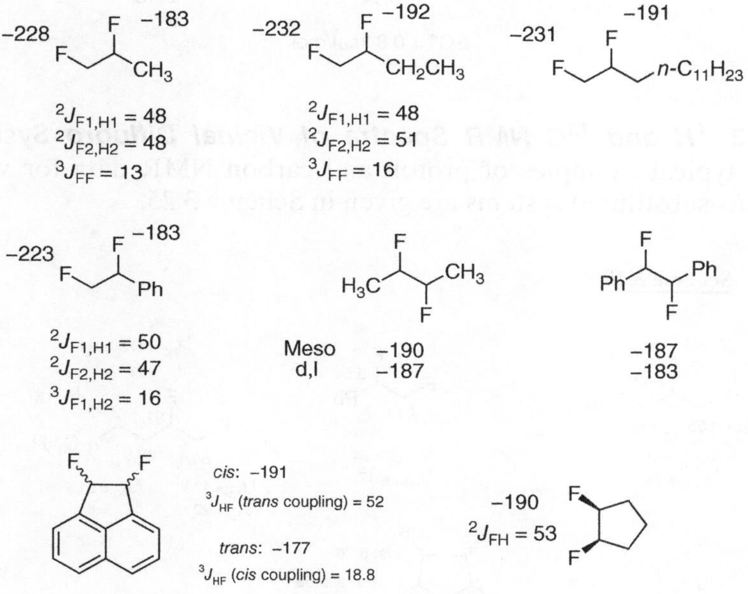

For the above examples, vicinal F–F coupling constants are in the vicinity of 15 Hz, whereas, as seen in Scheme 3.25, the vicinal F–H coupling constants are larger, generally in the range of 20–25 Hz.

In the case of a number of vicinal difluoro systems, such as 2,3-difluoro-2,3-diphenylethane or 2,3-difluorosuccinic acid derivatives, the coupling systems are AA'XX', which means that they will produce second-order spectra (see Section 2.3.7). A case in point are the fluorine and proton spectra of 1,2-difluoroethane, which have been analyzed carefully both experimentally and computationally in order to determine details of the conformational distribution of this molecule (Scheme 3.24).[7] As is commonly known, the gauche conformation is preferred thermodynamically over the anticonformation.

Scheme 3.24

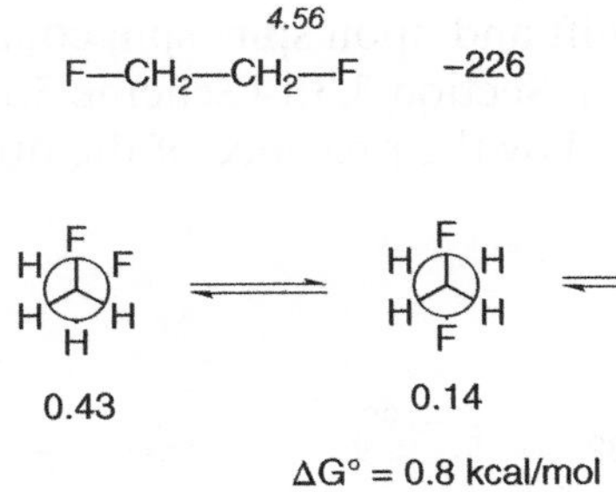

4.56

F—CH$_2$—CH$_2$—F   −226

0.43   0.14   0.43

$\Delta G° = 0.8$ kcal/mol

3.3.1.3. ^1H and ^{13}C NMR Spectra of Vicinal Difluoro Systems.

Some typical examples of proton and carbon NMR data for vicinal difluoro-substituted systems are given in Scheme 3.25.

Scheme 3.25

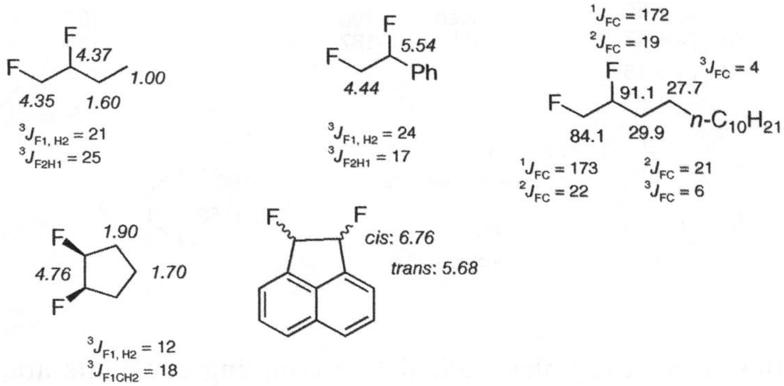

3.3.1.4. More Heavily Fluorinated Compounds.

The series of fluorinated ethanes indicates that the fluorine nucleus of a CH$_2$F group is increasingly shielded as the number of β-fluorines increases (Scheme 3.26), unlike the situation observed for increasing number of β-chlorine substituents (Scheme 3.16). Note that as one accumulates fluorines at the β-position, the three-bond H–F coupling constant becomes progressively smaller.

Replacing the hydrogens of CF$_3$CH$_2$F with CF$_3$ groups gives the secondary and tertiary fluoride compounds below, which absorb at

Scheme 3.26

	4.56	5.92 4.46	4.65
CH_3CH_2–F	FCH_2CH_2–F	F_2CH–CH_2–F	CF_3CH_2–F
–212	–226	–131 –239	–241
$^2J_{FH} = 48.5$	$^2J_{FH} = 45$	$^2J_{FH} = 46$	$^2J_{FH} = 46$
$^3J_{FH} = 27$	$^3J_{FH} = 17$	$^3J_{FH} = 18$	$^3J_{FF} = 16$
		$^3J_{FH} = 14, 7.0$	$^3J_{HF} = 8.2$
	5.07	$^3J_{HH} = 3.5$	
$(CH_3)_2CHF$	$(CF_3)_2CHF$		
–165	–214		
		$(CH_3)_3CF$	$(CF_3)_3CF$
$^2J_{FH} = 47$	$^2J_{FH} = 44$	–131	–188
	$^3J_{FF} = 11.5$		
	$^3J_{FH} = 5.5$	$^3J_{FH} = 21$	$^3J_{FF} = 6.1$

progressively lower field (Scheme 3.26). However, these secondary and tertiary fluorine substituents are the *most shielded* of any secondary and tertiary fluorines.

A few pertinent proton chemical shifts are also provided in this scheme.

3.3.2. Alcohol, Ether, Epoxide, Ester, Sulfide, Sulfone, Sulfonate, and Sulfonic Acid Groups

Compounds with fluorines bound directly to a carbon bearing a hydroxy group are generally very unstable, although there are exceptions. Hexafluoroacetone and hexafluorocyclobutanone both add HF to form stable α-fluoroalcohols, which release HF quickly in water to form the respective hydrates. The stability of these alcohols derives simply from the relative *instability* of the respective perfluoroketones. Fluorine NMR data for the one example available are provided in Scheme 3.27. Its chemical shift is obviously also influenced significantly by the six β-fluorines.

An ether oxygen bound directly to a CH_2F group, as in a fluoromethyl ether, deshields the fluorine more than does a chlorine substituent (Scheme 3.27).

Similar sulfur substitution, as in a fluoromethyl sulfide, also leads to deshielding, but somewhat less than for the analogous ether.

Again, secondary fluorides are deshielded considerably compared to the primary systems, with tertiary fluorides even more so.

Scheme 3.27

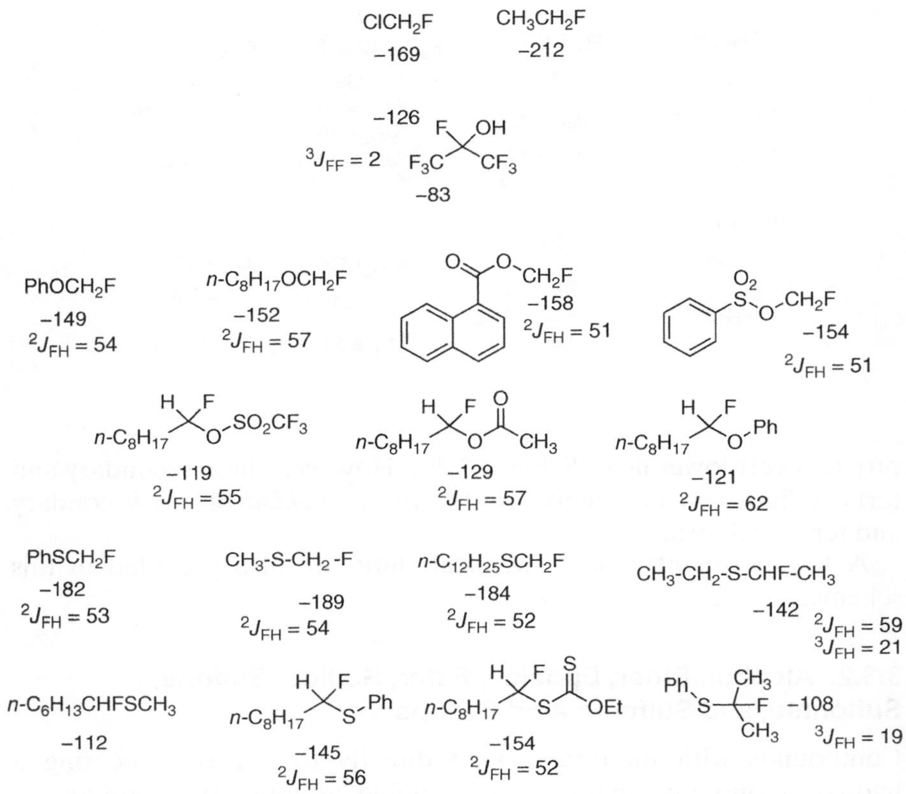

$$\text{ClCH}_2\text{F} \qquad \text{CH}_3\text{CH}_2\text{F}$$
$$-169 \qquad\qquad -212$$

$$-126$$
$$^3J_{FF} = 2 \quad \begin{array}{c} \text{F} \quad \text{OH} \\ \text{F}_3\text{C} \diagdown \diagup \text{CF}_3 \end{array}$$
$$-83$$

PhOCH$_2$F

-149
$^2J_{FH} = 54$

n-C$_8$H$_{17}$OCH$_2$F

-152
$^2J_{FH} = 57$

(naphthalene-CO-O-CH$_2$F)
-158
$^2J_{FH} = 51$

(Ph-SO$_2$-O-CH$_2$F)
-154

$^2J_{FH} = 51$

n-C$_8$H$_{17}$ (CHF-O-SO$_2$CF$_3$)

-119
$^2J_{FH} = 55$

n-C$_8$H$_{17}$ (CHF-O-CO-CH$_3$)

-129
$^2J_{FH} = 57$

n-C$_8$H$_{17}$ (CHF-O-Ph)

-121
$^2J_{FH} = 62$

PhSCH$_2$F
-182
$^2J_{FH} = 53$

CH$_3$-S-CH$_2$-F
-189
$^2J_{FH} = 54$

n-C$_{12}$H$_{25}$SCH$_2$F
-184
$^2J_{FH} = 52$

CH$_3$-CH$_2$-S-CHF-CH$_3$
-142 $^2J_{FH} = 59$
$^3J_{FH} = 21$

n-C$_6$H$_{13}$CHFSCH$_3$

-112

n-C$_8$H$_{17}$ (CHF-S-Ph)

-145
$^2J_{FH} = 56$

n-C$_8$H$_{17}$ (CHF-S-CO-OEt)

-154
$^2J_{FH} = 52$

Ph-S-C(CH$_3$)$_2$F -108

$^3J_{FH} = 19$

Two examples of epoxides bearing a fluorine substituent next to the epoxide oxygen are provided in Scheme 3.28. These fluorines are considerably shielded compared to what would be expected for an analogous straight chain ether system.

Scheme 3.28

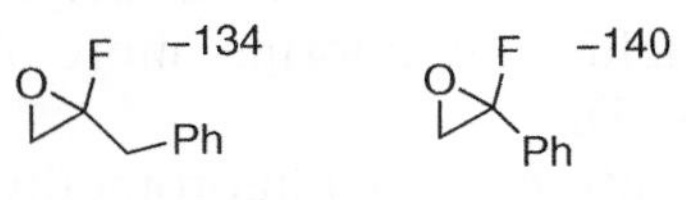

-134 (epoxide-F-CH$_2$Ph)

-140 (epoxide-F-Ph)

Sulfones and sulfoxides bearing a CH$_2$F group have very similar fluorine chemical shifts and are quite shielded relative to their unoxidized sulfide analogs. A CH$_2$F group attached to a sulfonium sulfur is slightly deshielded relative to sulfoxides and sulfones (Scheme 3.29).

Scheme 3.29

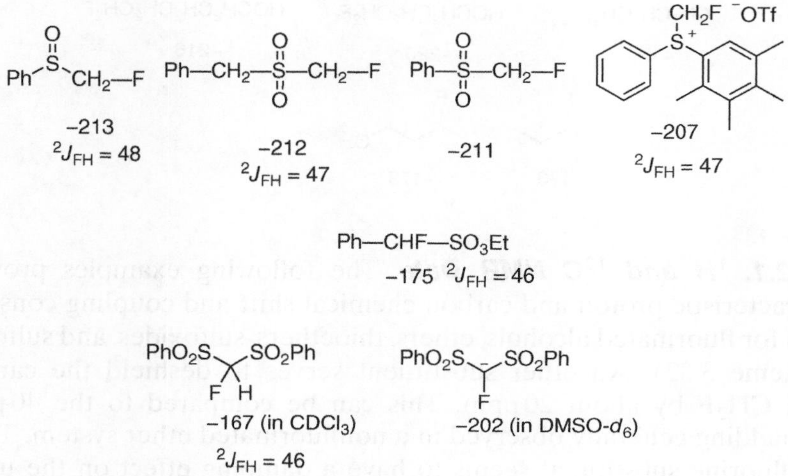

Ph—S(=O)—CH$_2$—F
−213
$^2J_{FH}$ = 48

Ph—CH$_2$—S(=O)$_2$—CH$_2$—F
−212
$^2J_{FH}$ = 47

Ph—S(=O)$_2$—CH$_2$—F
−211

CH$_2$F $^-$OTf
−207
$^2J_{FH}$ = 47

Ph—CHF—SO$_3$Et
−175 $^2J_{FH}$ = 46

PhO$_2$S   SO$_2$Ph
F H
−167 (in CDCl$_3$)
$^2J_{FH}$ = 46

PhO$_2$S   SO$_2$Ph
F
−202 (in DMSO-d_6)

An example of an α-fluorosulfonic acid ester is provided in the same scheme. Lastly, there is a comparison of a bis-sulfone and the carbanion formed by deprotonation. Note the significant shielding of the fluorine substituent in the carbanion.

As was the case when there was a β-fluorine substituent, placement of an ether or alcohol functionality β to a fluorine substituent leads to modest shielding (Scheme 3.30).

Scheme 3.30

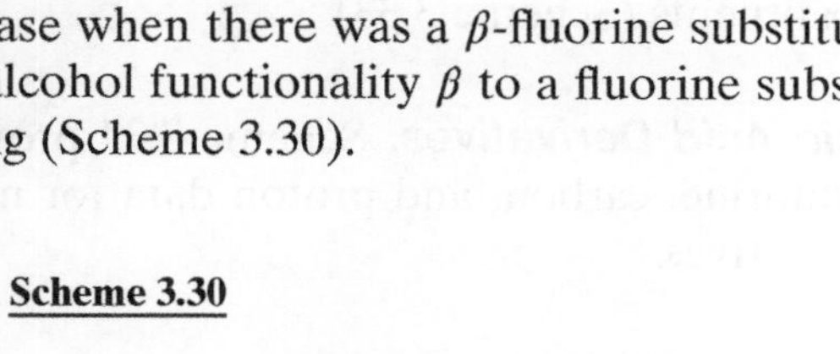

CH$_3$CH$_2$F
−212

CH$_3$OCH$_2$CH$_2$F
−223
$^2J_{FH}$ = 49

HOCH$_2$CH$_2$F
−227
$^2J_{FH}$ = 48
$^3J_{FH}$ = 32

−183

−191

A hydroxyl group one or two carbons further removed, γ or δ to the fluorine substituent, does not influence the fluorine chemical shift significantly (Scheme 3.31). In the case of a secondary system, the fluorine is also unaffected by a γ-hydroxy substituent.

Scheme 3.31

$CH_3CH_2CH_2F$ $HOCH_2CH_2CH_2F$ $HOCH_2CH_2CH_2CH_2F$

  –219   –220   –216

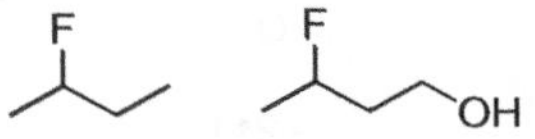

  –173   –173

3.3.2.1. ¹H and ¹³C NMR Data. The following examples provide characteristic proton and carbon chemical shift and coupling constant data for fluorinated alcohols, ethers, thioethers, sulfoxides, and sulfones (Scheme 3.32). An ether substituent serves to deshield the carbon of a CH_2F by about 20 ppm. This can be compared to the 40 ppm deshielding generally observed in a nonfluorinated ether system. Thus, the fluorine substituent seems to have a damping effect on the usual effects of other substituents.

3.3.2.2. Multiple α-Ether Substituents. There are few examples of such compounds. Three phenoxy substituents do not deshield a C–F fluorine as much as three chlorine substituents, but a little more than three additional F substituents (Scheme 3.33).

3.3.2.3. Sulfonic Acid Derivatives. Scheme 3.34 provides what data is available on fluorine, carbon, and proton data for monofluoroalkyl sulfonic acid derivatives.

3.3.3. Amino, Ammonium, Azide, and Nitro Groups

There are no α-amino fluorides because of the reactivity of this functional combination, and for the same reason, there are few β-amino fluorides. However, when attached to the less basic nitrogen of a benzimidazole or benzotriazole, the CH_2F or CHFR groups are much more stable (Scheme 3.35). Note that when attached to a partially positive nitrogen, as in an imidazolium compound, or a fully positive ammonium nitrogen, the respective fluorines become progressively more shielded. α-Fluoroazides can be prepared but are also not very stable.

Available *carbon, proton, and nitrogen* data are also provided for these compounds in the same scheme. Comparative data is provided in the benzotriazole case to show the effect of vicinal fluorine on its nitrogen chemical shifts.

Scheme 3.32

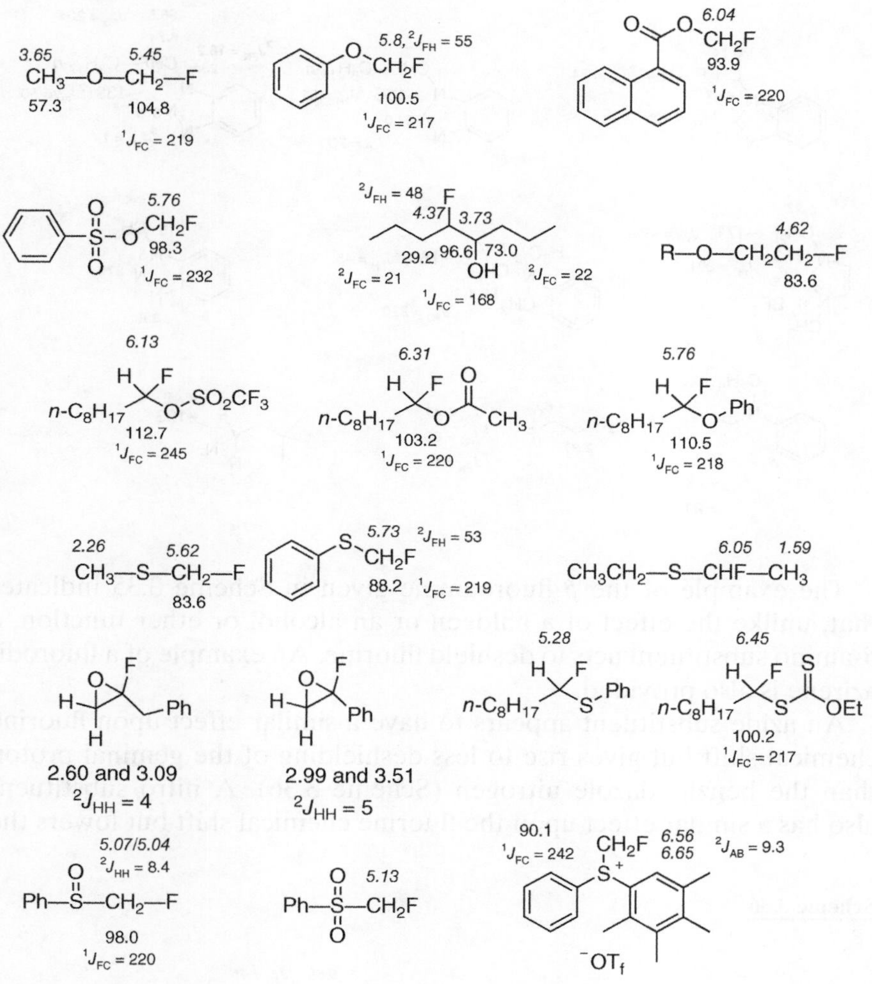

3.65 5.45
CH_3-O-CH_2-F
57.3 104.8
$^1J_{FC} = 219$

$5.8, ^2J_{FH} = 55$
$O-CH_2F$
100.5
$^1J_{FC} = 217$

6.04
$O-CH_2F$
93.9
$^1J_{FC} = 220$

5.76
CH_2F
98.3
$^1J_{FC} = 232$

$^2J_{FH} = 48$ F
4.37 3.73
29.2 96.6 73.0
$^2J_{FC} = 21$ OH $^2J_{FC} = 22$
$^1J_{FC} = 168$

4.62
$R-O-CH_2CH_2-F$
83.6

6.13
$n-C_8H_{17}$ H F
$O-SO_2CF_3$
112.7
$^1J_{FC} = 245$

6.31
$n-C_8H_{17}$ H F O
$O-C-CH_3$
103.2
$^1J_{FC} = 220$

5.76
$n-C_8H_{17}$ H F
$O-Ph$
110.5
$^1J_{FC} = 218$

2.26 5.62
CH_3-S-CH_2-F
83.6

$S-CH_2F$ 5.73 $^2J_{FH} = 53$
88.2 $^1J_{FC} = 219$

6.05 1.59
$CH_3CH_2-S-CHF-CH_3$

$n-C_8H_{17}$ H F
$S-Ph$
5.28

$n-C_8H_{17}$ H F S
$S-C-OEt$
6.45
100.2
$^1J_{FC} = 217$

O F
Ph H
2.60 and 3.09
$^2J_{HH} = 4$

O F
Ph H
2.99 and 3.51
$^2J_{HH} = 5$

$5.07/5.04$
O $^2J_{HH} = 8.4$
$Ph-S-CH_2-F$
98.0
$^1J_{FC} = 220$

O
$Ph-S-CH_2F$ 5.13
O

90.1
$^1J_{FC} = 242$ CH_2F 6.56 $^2J_{AB} = 9.3$
6.65
S^+
^-OT_f

Scheme 3.33

Cl_3C-F $(p-CF_3-C_6H_4O)_3C-F$ F_3C-F
$[0]$ -53.8 -64.6

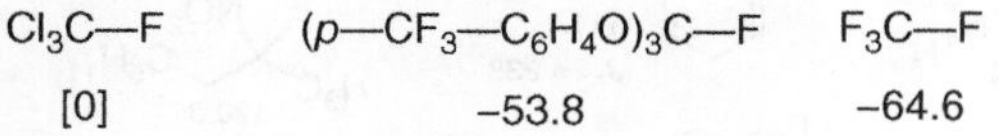

Scheme 3.34

H F
Ph SO_3Et
-175
$^2J_{FH} = 46$

5.45
$F-CH_2-SO_2Cl$
-200 96.6
$^2J_{FH} = 46$ $^1J_{FC} = 235$

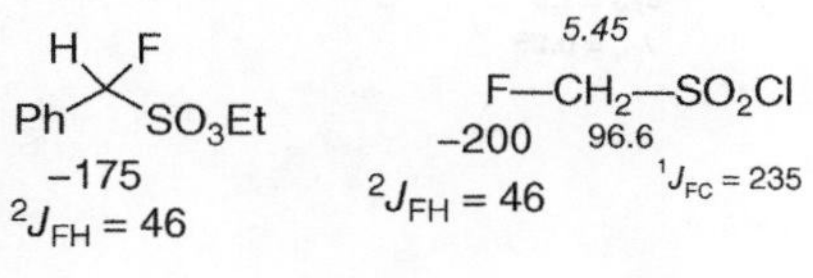

Scheme 3.35

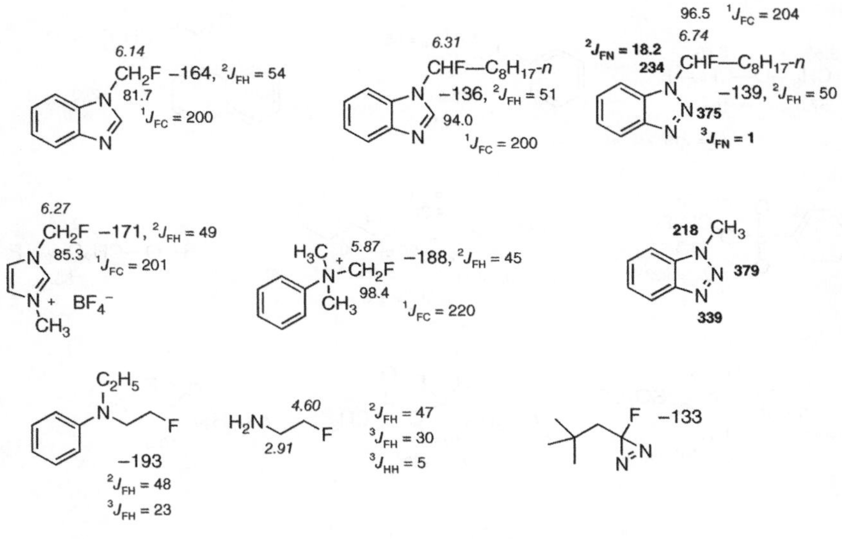

The example of the β-fluoroamine given in Scheme 3.35 indicates that, unlike the effect of a halogen or an alcohol or ether function, a β-amino substituent acts to deshield fluorine. An example of a fluorodiazirene is also provided.

An azide substituent appears to have a similar effect upon fluorine chemical shift but gives rise to less deshielding of the geminal proton than the benzimidazole nitrogen (Scheme 3.36). A nitro substituent also has a similar effect upon the fluorine chemical shift but lowers the

Scheme 3.36

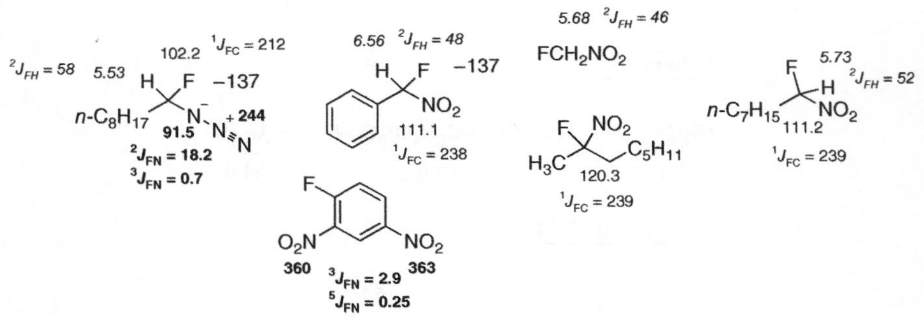

two-bond FH coupling constant and has a slightly greater deshielding influence upon the geminal proton than does the azide substituent.

3.3.4. Phosphorous Compounds

There do not appear to be any simple phosphines that bear a CH_2F group. However, fluorine NMR spectra of phosphonates, phosphane oxides and phosphonium compounds with CH_2F and $-CHF-$ bound to phosphorous have been reported. Examples are given in Scheme 3.37, including spectral data for the useful Horner–Wadsworth–Emmons reagent, triethyl 2-fluoro-2-phosphonoacetate. Spectral data for a phosphaalkene is also provided.

Scheme 3.37

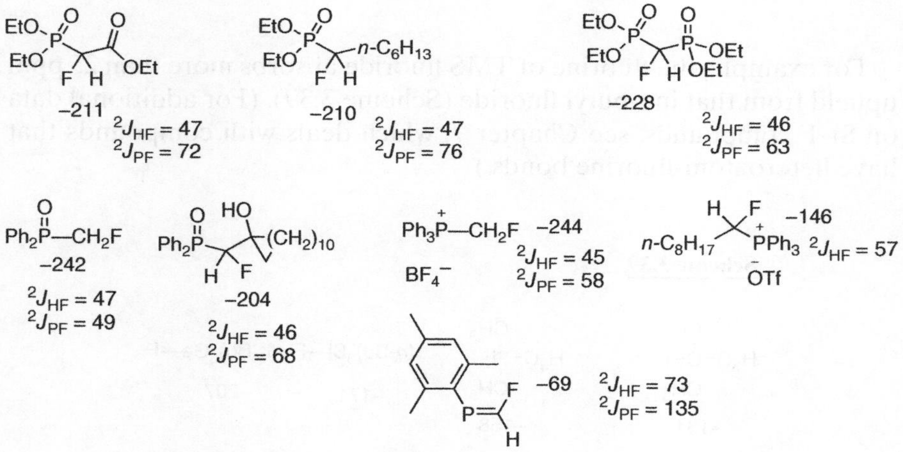

3.3.4.1. Proton, Carbon, and Phosphorous Spectra.
Proton and carbon NMR data, including ^{31}P chemical shift and P–C and P–F coupling constants for the above compounds are given in Scheme 3.38.

3.3.5. Silanes, Stannanes, and Germanes

Whether bound directly to the silicon or on a carbon bound to the silicon or germanium, a fluorine substituent is highly shielded compared to that in a hydrocarbon.

Scheme 3.38

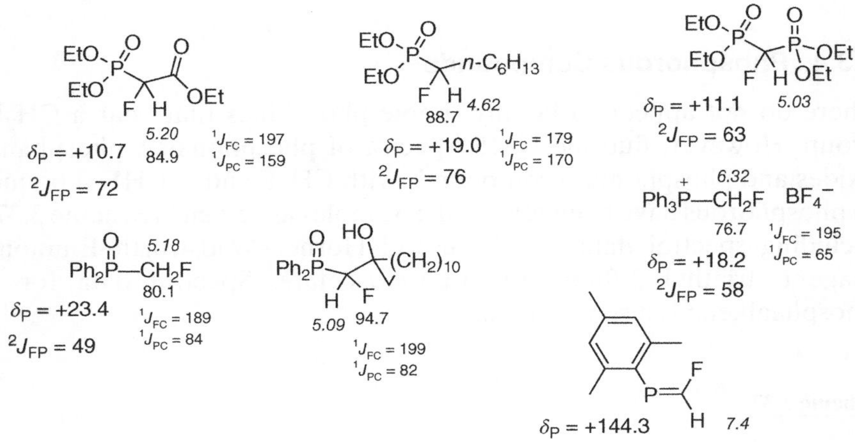

For the first structure (EtO)$_2$P(=O)—CHF—C(=O)OEt:
5.20, $^1J_{FC} = 197$
$\delta_P = +10.7$ 84.9 $^1J_{PC} = 159$
$^2J_{FP} = 72$

For the second structure (EtO)$_2$P(=O)—CHF—n-C$_6$H$_{13}$:
88.7   4.62
$\delta_P = +19.0$ $^1J_{FC} = 179$ $^1J_{PC} = 170$
$^2J_{FP} = 76$

For the third structure (EtO)$_2$P(=O)—CHF—P(=O)(OEt)$_2$:
$\delta_P = +11.1$ 5.03
$^2J_{FP} = 63$

Ph$_2$P(=O)—CH$_2$F:
5.18
$\delta_P = +23.4$ 80.1 $^1J_{FC} = 189$
$^2J_{FP} = 49$ $^1J_{PC} = 84$

Ph$_2$P(=O)—C(OH)(CH$_2$)$_{10}$—CHF:
5.09   94.7
$^1J_{FC} = 199$
$^1J_{PC} = 82$

Ph$_3$P$^+$—CH$_2$F   BF$_4^-$:
6.32
76.7 $^1J_{FC} = 195$
$\delta_P = +18.2$ $^1J_{PC} = 65$
$^2J_{FP} = 58$

$\delta_P = +144.3$ H   7.4

For example, the fluorine of TMS fluoride absorbs more than 25 ppm upfield from that in *t*-butyl fluoride (Scheme 3.39). (For additional data on Si–F compounds, see Chapter 7, which deals with compounds that have heteroatom-fluorine bonds.)

Scheme 3.39

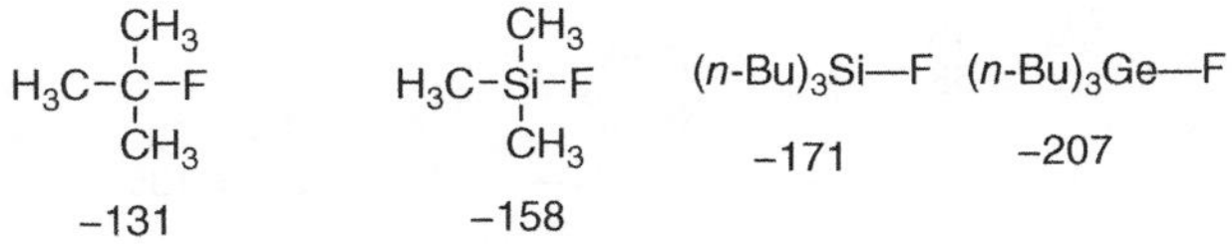

$$\begin{array}{cccc} H_3C-\underset{CH_3}{\overset{CH_3}{C}}-F & H_3C-\underset{CH_3}{\overset{CH_3}{Si}}-F & (n\text{-Bu})_3Si-F & (n\text{-Bu})_3Ge-F \\ -131 & -158 & -171 & -207 \end{array}$$

Similarly, a primary —CH$_2$F fluorine adjacent to silicon is shielded by about 50 ppm compared to the respective hydrocarbon, with similar tin or germanium compounds being shielded slightly less. The value of −277 ppm observed for tetrakis(fluoromethyl)silane has the largest chemical shift known for a single carbon-bound fluorine (Scheme 3.40).

3.4. CARBONYL FUNCTIONAL GROUPS

Carbonyl functional groups bound to a carbon bearing fluorine give rise to significant *shielding* of the fluorine.

Scheme 3.40

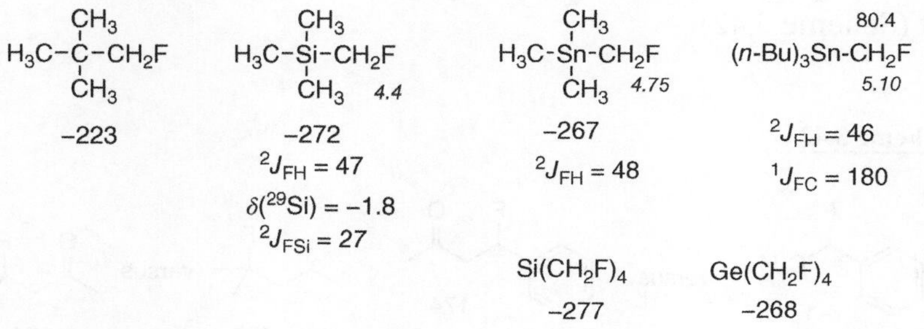

$$
\begin{array}{cccc}
\underset{\underset{CH_3}{|}}{\overset{\overset{CH_3}{|}}{H_3C-C-CH_2F}} &
\underset{\underset{CH_3}{|}}{\overset{\overset{CH_3}{|}}{H_3C-Si-CH_2F}} \;\; 4.4 &
\underset{\underset{CH_3}{|}}{\overset{\overset{CH_3}{|}}{H_3C-Sn-CH_2F}} \;\; 4.75 &
(n\text{-Bu})_3Sn-CH_2F \quad 80.4
\end{array}
$$

-223 -272 -267

$^2J_{FH} = 47$ (Si) $^2J_{FH} = 48$ (Sn) $^2J_{FH} = 46$

$\delta(^{29}Si) = -1.8$ $^1J_{FC} = 180$

$^2J_{FSi} = 27$

Si(CH$_2$F)$_4$ Ge(CH$_2$F)$_4$

-277 -268

3.4.1. Aldehydes and Ketones

Aldehydes have a greater impact on the chemical shifts of α-fluorine substituents than do ketones, and the chemical shifts of secondary fluorides are affected somewhat more than those of primary systems (Scheme 3.41).

Scheme 3.41

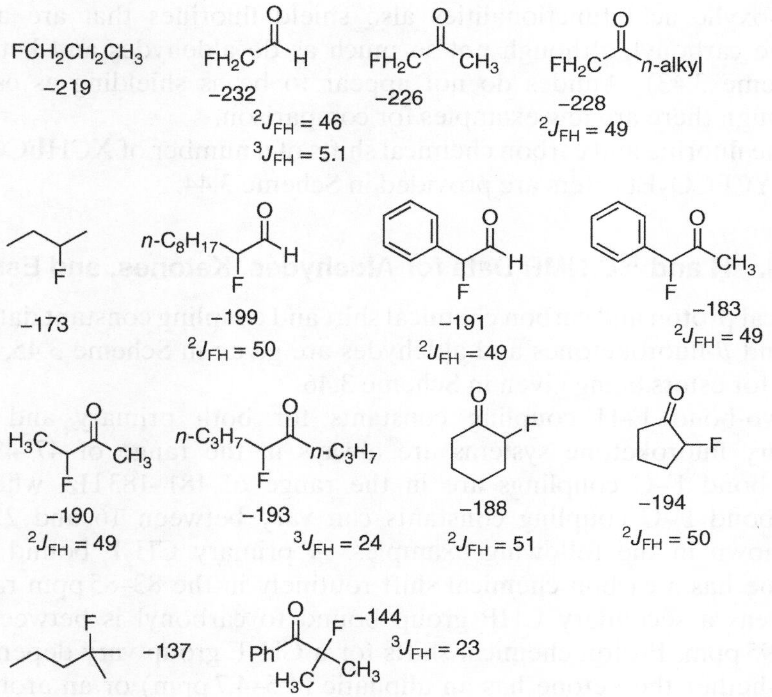

FCH$_2$CH$_2$CH$_3$ FH$_2$C–C(O)–H   FH$_2$C–C(O)–CH$_3$ FH$_2$C–C(O)–n-alkyl

-219 -232 -226 -228

$^2J_{FH} = 46$ $^2J_{FH} = 49$

$^3J_{FH} = 5.1$

(CH$_3$)$_2$CHF-type n-C$_8$H$_{17}$–CHF–CHO   Ph–CHF–CHO   Ph–CHF–CO–CH$_3$

-173 -199 -191 -183

$^2J_{FH} = 50$ $^2J_{FH} = 49$ $^2J_{FH} = 49$

H$_3$C–CO–CHF–CH$_3$ n-C$_3$H$_7$–CHF–CO–n-C$_3$H$_7$ cyclohexanone-F   cyclopentanone-F

-190 -193 -188 -194

$^2J_{FH} = 49$ $^3J_{FH} = 24$ $^2J_{FH} = 51$ $^2J_{FH} = 50$

(CH$_3$)$_2$CF–CH$_2$– (neopentyl type)   Ph–CO–CF(CH$_3$)$_2$

-137 -144

$^3J_{FH} = 23$

When the carbonyl function is one carbon further removed from the fluorine substituent, it has little if any influence on the fluorine chemical shift (Scheme 3.42).

Scheme 3.42

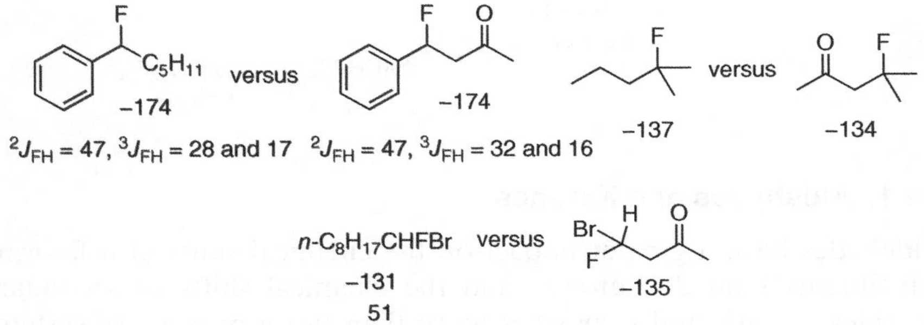

$^2J_{FH} = 47, {}^3J_{FH} = 28$ and 17 $^2J_{FH} = 47, {}^3J_{FH} = 32$ and 16

3.4.2. Carboxylic Acid Derivatives

Carboxylic acid functionalities also shield fluorines that are alpha to the carbonyl, although not so much as do aldehydes and ketones (Scheme 3.43). Amides do not appear to be as shielding as esters, although there are few examples for comparison.

The fluorine and carbon chemical shifts of a number of XCHFCO$_2$Et or XYCFCO$_2$Et esters are provided in Scheme 3.44.

3.4.3. ^{1}H and ^{13}C NMR Data for Aldehydes, Ketones, and Esters

Typical proton and carbon chemical shift and coupling constant data for α-, and β-fluoroketones and aldehydes are given in Scheme 3.45, with data for esters being given in Scheme 3.46.

Two-bond F–H coupling constants for both primary and sec-ondary fluoroketone systems are always in the range of 47–49 Hz. One-bond F–C couplings are in the range of 181–183 Hz, whereas two-bond F–C coupling constants can vary between 16 and 25 Hz, as shown in the following examples. A primary CH$_2$F bound to a ketone has a carbon chemical shift routinely in the 83–85 ppm range, whereas a secondary CHF group bound to carbonyl is between 92 and 95 ppm. Proton chemical shifts for a CH$_2$F group vary depending on whether the ketone has an aliphatic (4.5–4.7 ppm) or an aromatic

Scheme 3.43

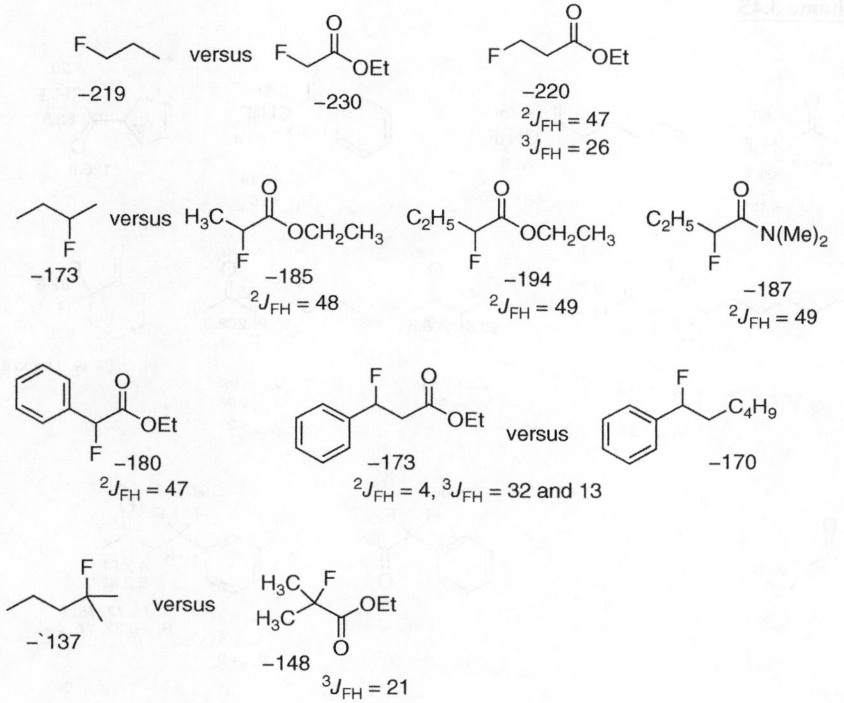

F⌒⌒   versus   F⌒C(O)OEt
−219   −230

F⌒⌒C(O)OEt
−220
$^2J_{FH} = 47$
$^3J_{FH} = 26$

⌒⌒F   versus   H₃C⌒C(O)OCH₂CH₃
−173   −185
                      $^2J_{FH} = 48$

C₂H₅⌒C(O)OCH₂CH₃
−194
$^2J_{FH} = 49$

C₂H₅⌒C(O)N(Me)₂
−187
$^2J_{FH} = 49$

Ph⌒C(O)OEt
F −180
$^2J_{FH} = 47$

Ph⌒F⌒C(O)OEt
−173
$^2J_{FH} = 4$, $^3J_{FH} = 32$ and 13

versus

Ph⌒F⌒C₄H₉
−170

⌒⌒F
−`137

versus

H₃C, H₃C, F⌒C(O)OEt
−148
$^3J_{FH} = 21$

Scheme 3.44

−151 F,H Br⌒CO₂Et 80.8

−133 F,H EtO⌒CO₂Et 104.5

−130 F,H PhO⌒CO₂Et 102.6

−163 F,H EtS⌒CO₂Et 92.3

−159 F,H PhS⌒CO₂Et 94.2

−65 Br,F Br⌒CO₂Et 81.2

−58 Br,F PhO⌒CO₂Et 106.4

−81 Br,F EtS⌒CO₂Et 99.6

−84 Br,F PhS⌒CO₂Et 98.1

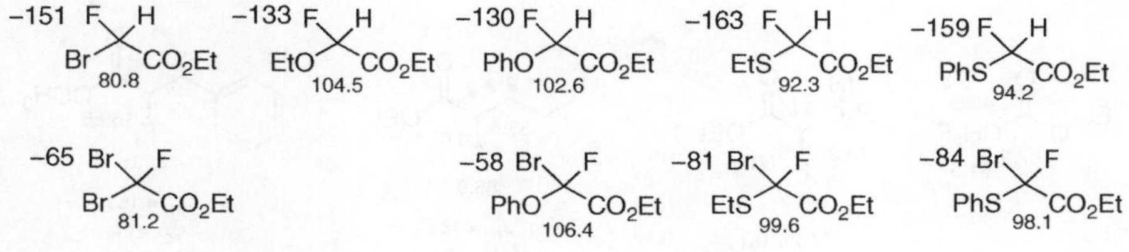

(5.3–5.6 ppm) substituent. Proton chemical shifts for a secondary CHF group adjacent to a ketone are around 4.7–4.8 ppm for aliphatic systems.

Both for primary and secondary α-fluoroesters, the α-protons absorb slightly downfield of analogous aliphatic ketones and somewhat upfield of analogous aromatic ketones.

Scheme 3.45

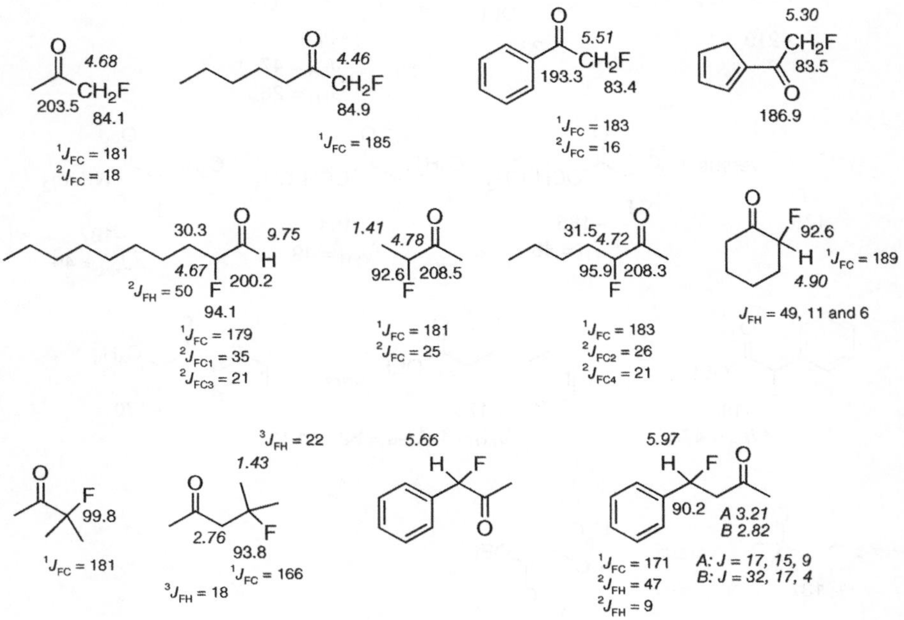

Scheme 3.46

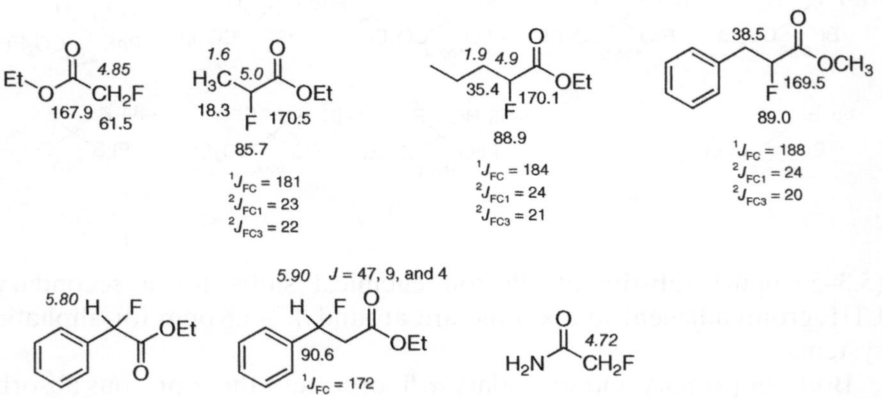

3.4.4. β-Ketoesters, Diesters, and Nitroesters

The fluorine chemical shifts of 3-fluoro-β-ketoesters and diesters exhibit little change from those of the monoesters (Scheme 3.47). The lone nitroester's proton experiences greater deshielding.

Scheme 3.47

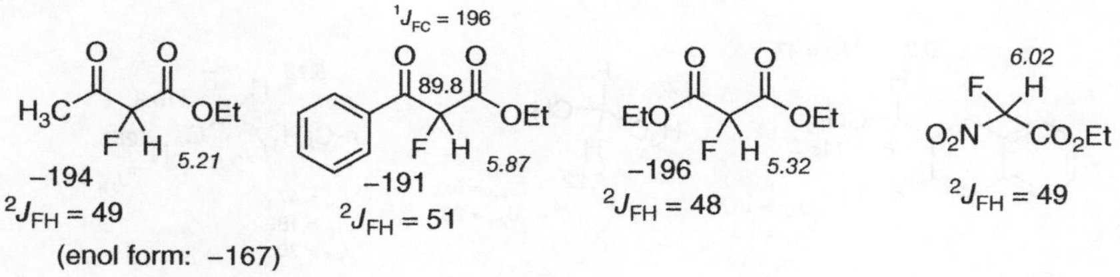

3.5. NITRILES

A nitrile function behaves much like a carbonyl functionality with respect to its influence upon an alkyl fluorine's chemical shift, acting to shield the fluorine modestly (Scheme 3.48).

Scheme 3.48

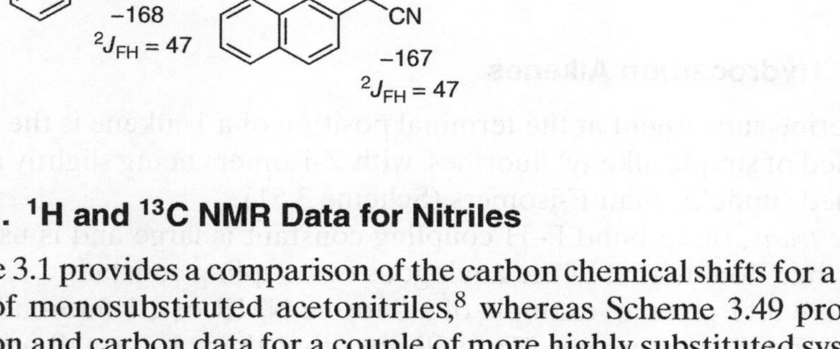

3.5.1. ¹H and ¹³C NMR Data for Nitriles

Table 3.1 provides a comparison of the carbon chemical shifts for a number of monosubstituted acetonitriles,[8] whereas Scheme 3.49 provides proton and carbon data for a couple of more highly substituted systems.

TABLE 3.1. ^{13}C Chemical Shifts for α-Substituted Nitriles

Compound	CH_3CN	$ClCH_2CN$	$MeOCH_2CN$	FCH_2CN
δ, CH_2X	1.5	24.6	59.0	66.7
δ, CN	116.3	114.5	115.6	113.7

Scheme 3.49

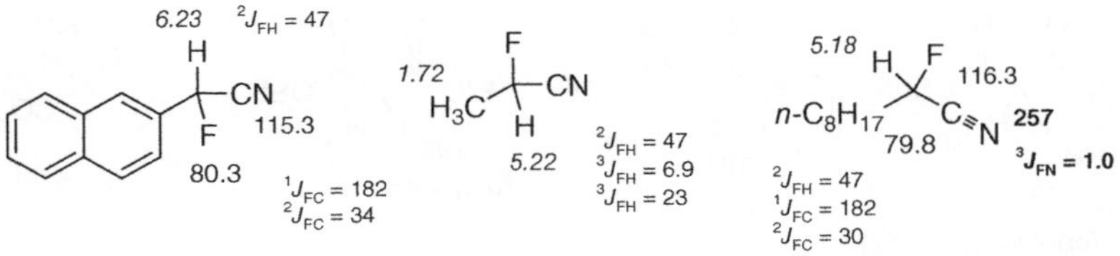

3.6. ALKENES WITH A SINGLE FLUORINE SUBSTITUENT

Single vinylic fluorine substituents absorb over quite a wide range of chemical shifts, with fluoroallene at the high field end (−169 ppm) and β-fluoroacrylate derivatives at the low field end (−75 ppm) (Scheme 3.50).

Scheme 3.50

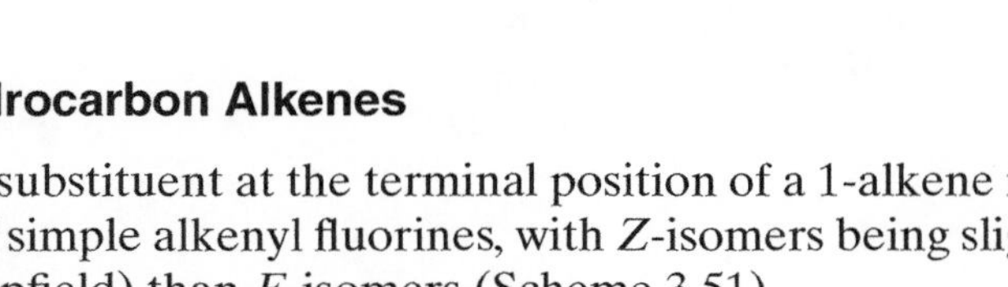

3.6.1. Hydrocarbon Alkenes

A fluorine substituent at the terminal position of a 1-alkene is the most shielded of simple alkenyl fluorines, with Z-isomers being slightly more shielded (upfield) than E-isomers (Scheme 3.51).

The *trans*, three-bond F–H coupling constant is large and is usually more than double that of the analogous *cis* coupling constant.

Figure 3.11 gives an example of a fluorine NMR of a 1-fluoroalkene, that of Z-1-fluoropentene. You will note that there is a small amount

Scheme 3.51

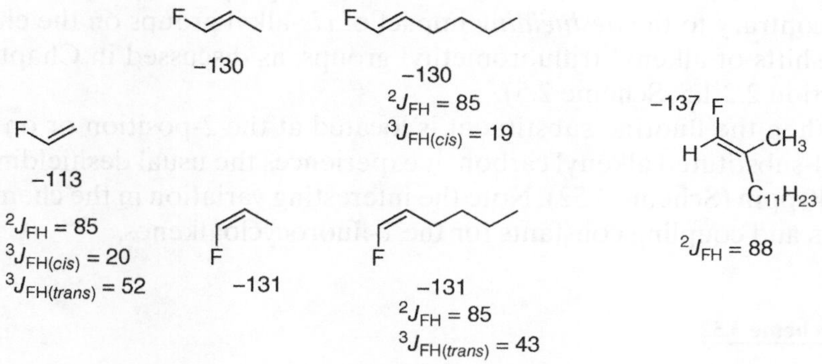

-130

-130
$^2J_{FH} = 85$
$^3J_{FH(cis)} = 19$

-137

-113
$^2J_{FH} = 85$
$^3J_{FH(cis)} = 20$
$^3J_{FH(trans)} = 52$

-131

-131
$^2J_{FH} = 85$
$^3J_{FH(trans)} = 43$

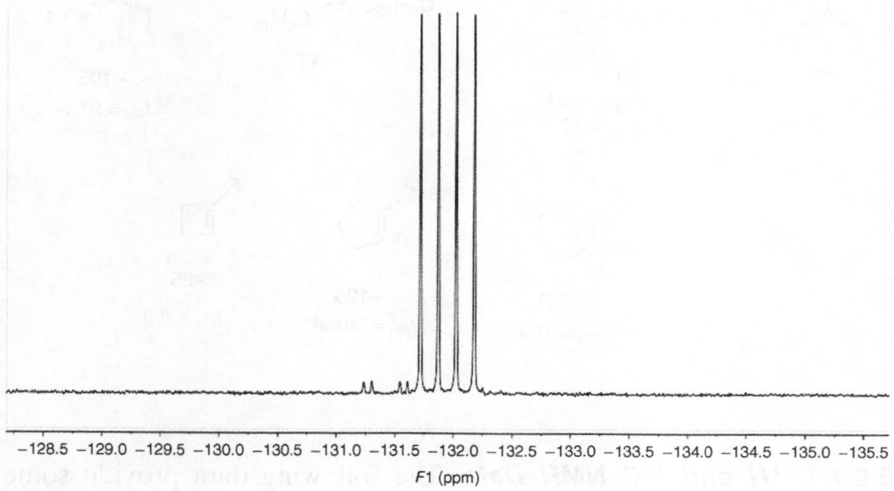

$^2J_{FH} = 88$

-128.5 -129.0 -129.5 -130.0 -130.5 -131.0 -131.5 -132.0 -132.5 -133.0 -133.5 -134.0 -134.5 -135.0 -135.5
F1 (ppm)

FIGURE 3.11. ^{19}F NMR spectrum of (Z)-1-fluoropentene, with a trace of the (E)-isomer evident

of the E-isomer also present in this sample, which exemplifies the significant difference between the *cis* and *trans*, three-bond, H–F coupling constants. The chemical shifts of the Z- and E-isomers are -131.9 and -131.4, respectively, the two-bond, H–F coupling constants of both isomers being 87 Hz, the *trans*, three-bond H–F coupling constant of the Z-isomer being 44 Hz, with the *cis*, three-bond H–F coupling constant of the E-isomer being 18.6 Hz.

The slight shielding of (Z)-1-fluoroalkenes versus (E)-1-fluoroalkenes is again worth reflecting on since, although only a small effect, it is consistent with the effect mentioned earlier with respect to chemical shifts of *cis*- and *trans*-2-methyl-1-fluorocyclopropanes (Section 3.2.4), but contrary to the *deshielding* impact of *cis*-alkyl groups on the chemical shifts of alkenyl trifluoromethyl groups, as discussed in Chapter 2 (Section 2.2.1.5, Scheme 2.5).

When the fluorine substituent is located at the 2-position or on any alkyl-substituted alkenyl carbon, it experiences the usual deshielding of 30–40 ppm (Scheme 3.52). Note the interesting variation in the chemical shifts and coupling constants for the 1-fluorocycloalkenes.

Scheme 3.52

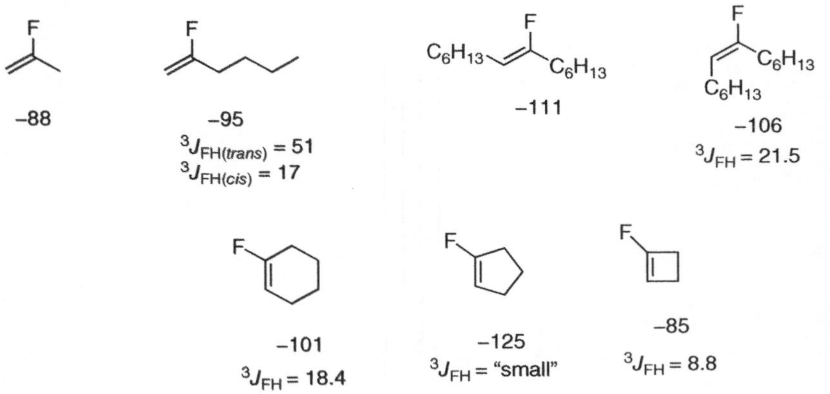

3.6.1.1. ¹H and ¹³C NMR Data.

3.6.1.1. ^1H and ^{13}C NMR Data. The following data provide some guidelines for proton and carbon NMR chemical shift and coupling constant data for fluoroalkenes (Scheme 3.53). Note that in all cases, hydrogens that are *cis* to the fluorine substituent are deshielded relative to those that are *trans*.

Figure 3.12a and b shows the proton NMR spectrum of (Z)-1-fluoropentene with the following assignments of shift and coupling constant data: δ 0.92 (t, $^3J_{HH} = 7.2$, 3H), 1.40 (sextet, $^3J_{HH} = 7.3$, 2H), 2.09 (qt, $^3J_{HH} = 7.5$, $^4J_{HH\&F} = 1.5$, 2H), 4.72 (ddt, $^3J_{HF(trans)} = 44$, $^4J_{HH(cis)} = 4.8$, $^3J_{HCH2} = 7.5$, 1H), 6.45 (ddt, $^2J_{FH} = 86$, $^3J_{HH(cis)} = 4.8$, $^4J_{HCH2} = 1.5$, 1H).

Figure 3.13 gives the ^{13}C NMR spectrum of (Z)-1-fluoropentene, with the following chemical shift assignments and F–C coupling constants: δ 147.9 (d, $^1J_{FC} = 255$), 111.0 (d, $^2J_{FC} = 5.1$), 24.9 (d, $^3J_{FC} = 5.0$), 22.6 (d, $^4J_{FC} = 2.0$), 13.7 (s).

Scheme 3.53

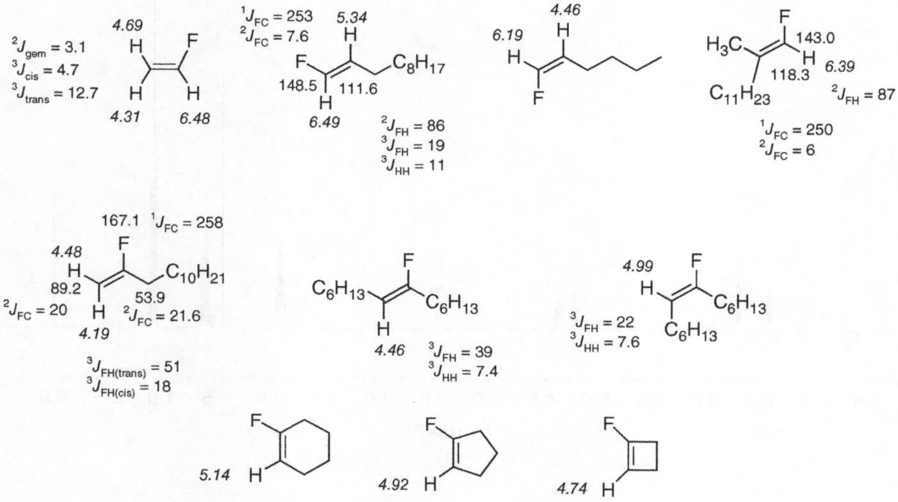

One can also detect the presence of the (E)-isomer in the above spectrum, with the (E)-1-fluoropentene having the following assignments: δ 148.8 (d, $^1J_{FC} = 252$), 111.6 (d, $^2J_{FC} = 5$), 27.2 (d, $^3J_{FC} = 5$), 23.0 (d, $^4J_{FC} = 2$), 13.6 (s).

3.6.2. Conjugated Alkenyl Systems

The chemical shifts of terminal (1°) vinylic fluorines are not affected significantly by conjugation of the fluorine bearing double bond with either another C=C double bond or a benzene ring (Scheme 3.54). In this case, however, the fluorines of the (Z)-isomers are slightly deshielded relative to those of (E)-isomers.

Unlike the terminal fluorines, fluorines placed at the internal (2°) position of conjugated systems are significantly shielded compared to their nonconjugated counterparts (Scheme 3.55).

3.6.2.1. *¹H and ¹³C NMR Data.* Some proton and carbon chemical shift and coupling constant data for conjugated CHF systems are presented as follows (Scheme 3.56). The fluorine-bearing carbons of the conjugated systems are deshielded relative to those of the nonconjugated systems. Similarly, the protons of the terminal CHF group of conjugated systems are also deshielded relative to those of the analogous nonconjugated systems.

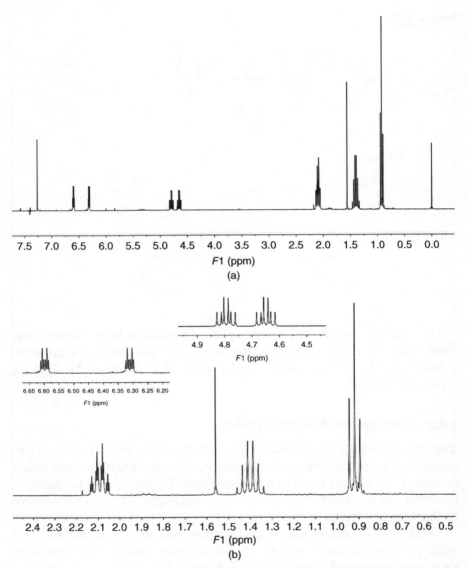

FIGURE 3.12. (a) Full ^{1}H NMR spectrum of (Z)-1-fluoropentene. (b) Details of ^{1}H NMR spectrum of (Z)-1-fluoropentene

3.6.3. Allylic Alcohols, Ethers, and Halides

Oxygen functionalities, such as alcohols, ethers, and acetate groups, and halogens at the allylic position, deshield when the fluorine is at the terminal position and shield when it is at the internal 2-position (Scheme 3.57).

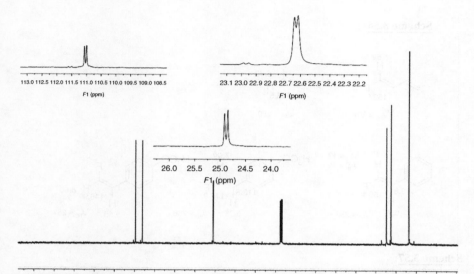

FIGURE 3.13. ^{13}C NMR spectrum of (Z)-1-fluoropentene

Scheme 3.54

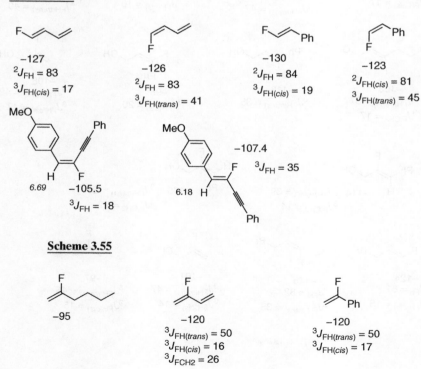

Scheme 3.56

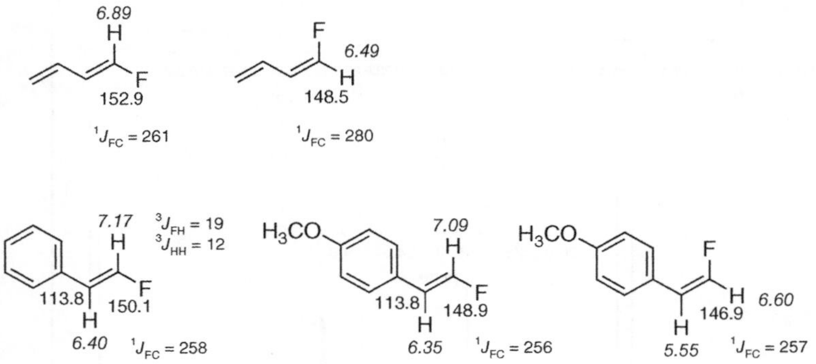

6.89
H
F
152.9
$^1J_{FC} = 261$

F
6.49
H
148.5
$^1J_{FC} = 280$

7.17 $^3J_{FH} = 19$
H $^3J_{HH} = 12$
113.8 150.1
F
H
6.40 $^1J_{FC} = 258$

H₃CO
7.09
H
113.8 148.9
F
H
6.35 $^1J_{FC} = 256$

H₃CO
F
H 6.60
146.9
H
5.55 $^1J_{FC} = 257$

Scheme 3.57

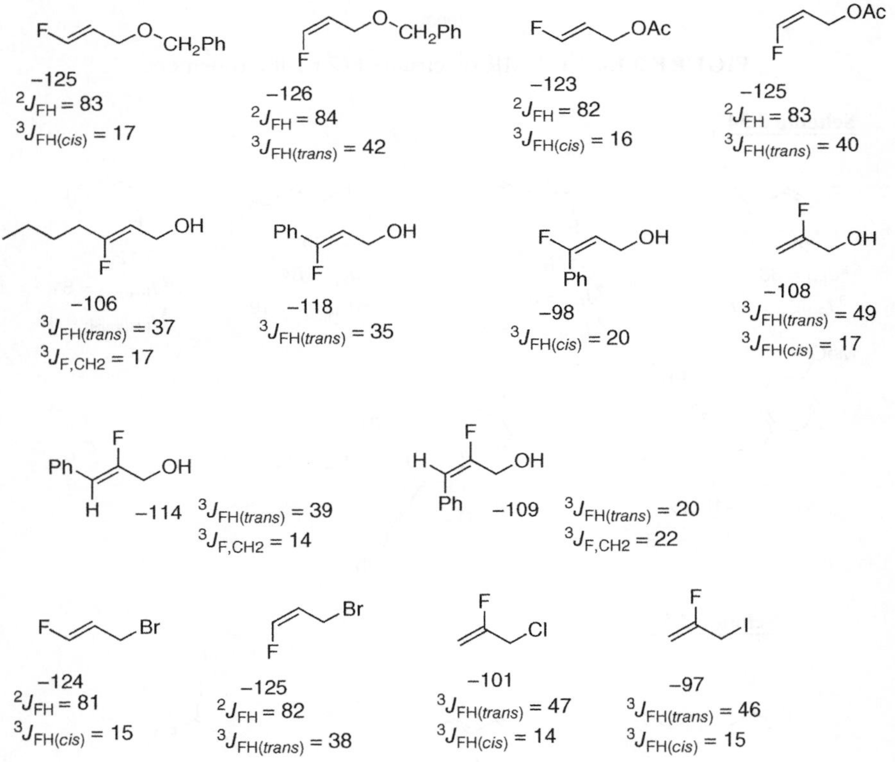

F
O-CH₂Ph

−125
$^2J_{FH} = 83$
$^3J_{FH(cis)} = 17$

F
O-CH₂Ph

−126
$^2J_{FH} = 84$
$^3J_{FH(trans)} = 42$

F
OAc

−123
$^2J_{FH} = 82$
$^3J_{FH(cis)} = 16$

F
OAc

−125
$^2J_{FH} = 83$
$^3J_{FH(trans)} = 40$

OH
F

−106
$^3J_{FH(trans)} = 37$
$^3J_{F,CH2} = 17$

Ph
OH
F

−118
$^3J_{FH(trans)} = 35$

F
OH
Ph

−98
$^3J_{FH(cis)} = 20$

OH

−108
$^3J_{FH(trans)} = 49$
$^3J_{FH(cis)} = 17$

F
Ph
OH
H   −114 $^3J_{FH(trans)} = 39$
$^3J_{F,CH2} = 14$

F
H
OH
Ph   −109 $^3J_{FH(trans)} = 20$
$^3J_{F,CH2} = 22$

F
Br

−124
$^2J_{FH} = 81$
$^3J_{FH(cis)} = 15$

Br
F

−125
$^2J_{FH} = 82$
$^3J_{FH(trans)} = 38$

F
Cl

−101
$^3J_{FH(trans)} = 47$
$^3J_{FH(cis)} = 14$

F
I

−97
$^3J_{FH(trans)} = 46$
$^3J_{FH(cis)} = 15$

The F–H spin–spin coupling constants of these compounds remain much the same as those of the simple alkenes.

3.6.3.1. Proton and Carbon NMR Data. Some characteristic ^{13}C and ^{1}H NMR data for fluorinated allylic alcohols and a bromide are provided in Scheme 3.58.

Scheme 3.58

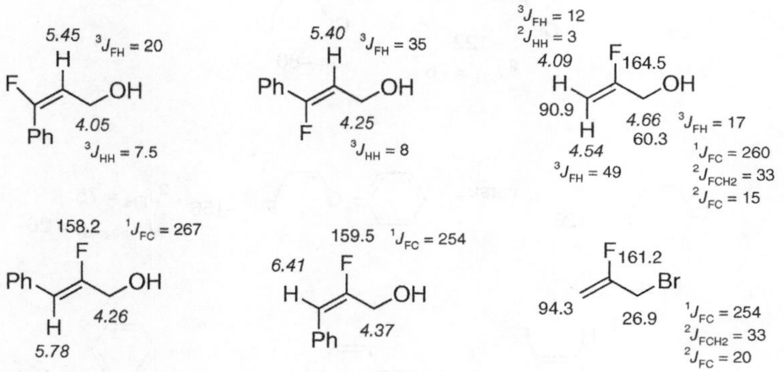

3.6.4. Halofluoroalkenes and Fluorovinyl Ethers

Geminal chlorine or bromine substituents deshield vinylic fluorine significantly, whereas a vicinal chlorine substituent shields the fluorine much as was the case for the saturated systems. Again similarly, a second vicinal chlorine substituent reverses the trend and shifts the fluorine signal downfield (Scheme 3.59).

Few monofluoro vinyl ethers have been reported in the literature. The NMR data for one example is given. It can be seen that the β-ether substituent shields the fluorine much more than does a β-chlorine substituent.

Note the impact of the geminal chlorine or bromine substituent to diminish the *cis* and *trans* F–H coupling constants in these systems.

3.6.4.1. Proton and Carbon NMR Data. Some selected chemical shift and coupling constant data from proton and carbon spectra of chloro- and bromofluoroethylenes are presented in Scheme 3.60.

Scheme 3.59

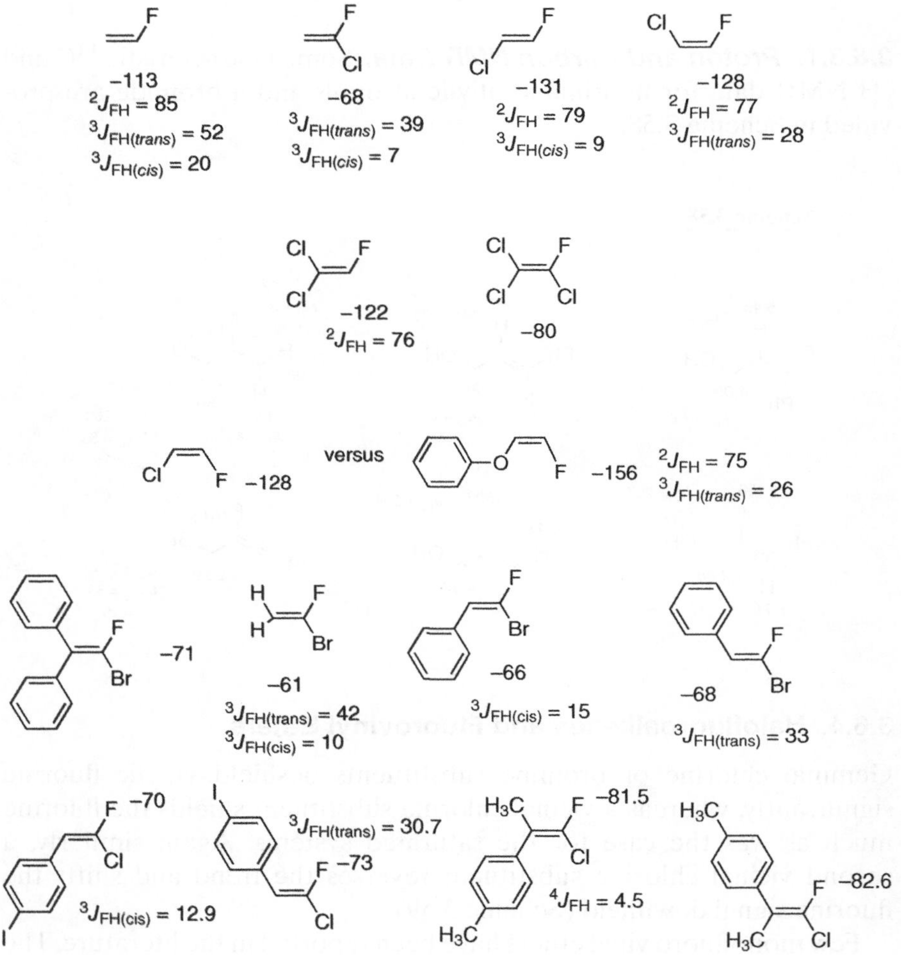
F
−113
$^2J_{FH}$ = 85
$^3J_{FH(trans)}$ = 52
$^3J_{FH(cis)}$ = 20
F
Cl
−68
$^3J_{FH(trans)}$ = 39
$^3J_{FH(cis)}$ = 7
F
Cl
−131
$^2J_{FH}$ = 79
$^3J_{FH(cis)}$ = 9
Cl
F
−128
$^2J_{FH}$ = 77
$^3J_{FH(trans)}$ = 28
Cl
F
Cl
−122
$^2J_{FH}$ = 76
Cl
F
Cl
Cl
−80
Cl
F
−128
versus
O
F
−156
$^2J_{FH}$ = 75
$^3J_{FH(trans)}$ = 26
F
Br
−71
H
F
H
Br
−61
$^3J_{FH(trans)}$ = 42
$^3J_{FH(cis)}$ = 10
F
Br
−66
$^3J_{FH(cis)}$ = 15
F
Br
−68
$^3J_{FH(trans)}$ = 33
F −70
Cl
I
$^3J_{FH(cis)}$ = 12.9
I
$^3J_{FH(trans)}$ = 30.7
F −73
Cl
H₃C
F −81.5
Cl
$^4J_{FH}$ = 4.5
H₃C
H₃C
F −82.6
H₃C
Cl

3.6.5. Geminal Fluoro, Hetero Alkenes

A few examples of a vinylic fluorine with geminal hetero-substituents are given in Scheme 3.61.

3.6.6. Multifluoroalkenes

3.6.6.1. Vicinal Difluoroalkenes. Each of the two vicinal fluorine substituents is significantly shielded by the presence of the other

Scheme 3.60

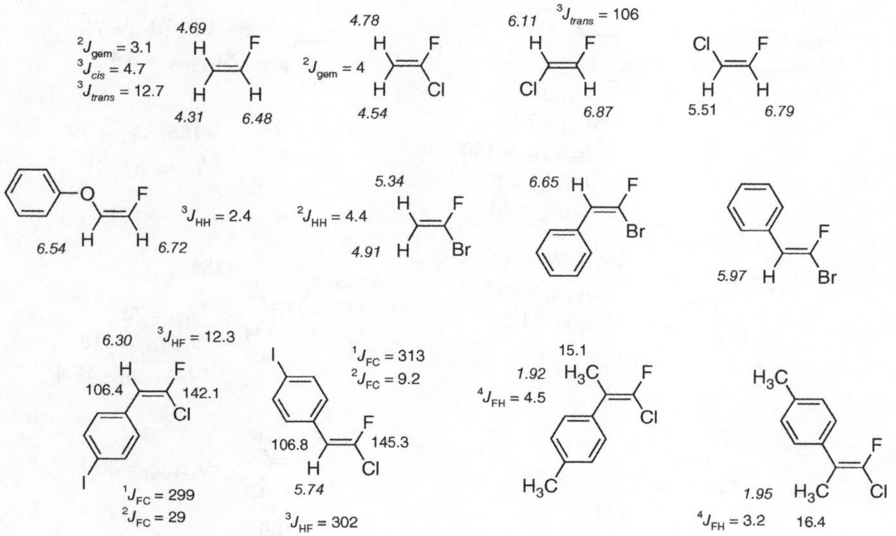

$^2J_{gem} = 3.1$
$^3J_{cis} = 4.7$
$^3J_{trans} = 12.7$

4.69   H   F
4.31   H   H   6.48

4.78   H   F
4.54   H   Cl
$^2J_{gem} = 4$

6.11 $^3J_{trans} = 106$ H   F
Cl   H   6.87

Cl   F
5.51   H   H   6.79

Ph—O   F
6.54   H   H   6.72
$^3J_{HH} = 2.4$

5.34   H   F
4.91   H   Br
$^2J_{HH} = 4.4$

6.65   H   F
Ph   Br

Ph   F
5.97   H   Br

6.30 $^3J_{HF} = 12.3$
106.4   H   F   142.1
(4-I-C6H4)   Cl
$^1J_{FC} = 299$
$^2J_{FC} = 29$

I—C6H4
$^1J_{FC} = 313$
$^2J_{FC} = 9.2$
106.8   H   F   145.3
Cl
5.74
$^3J_{HF} = 302$

15.1
1.92   H3C   F
(4-CH3-C6H4)   Cl
$^4J_{FH} = 4.5$
H3C

H3C—C6H4
F
1.95   H3C   Cl   16.4
$^4J_{FH} = 3.2$

Scheme 3.61

−132   4.59
F   H
−62   F3C—O   F
−188
$^2J_{FH} = 72$
$^3J_{FF(trans)} = 123$
$^3J_{FH(cis)} = 4.3$

−105   −180
F   F
−63   F3C—O   H   4.86
$^2J_{FH} = 70$
$^3J_{FH(trans)} = 11.7$
$^3J_{FH(cis)} = 242$

MeO   Ph
−88   F   H
$^2J_{FH} = 8$

−127   Bn2N
F   CH3
(EtO)2OP   150.6
$\delta_P = 4.86$
$^3J_{FH} = 41$
$^2J_{FP} = 105$
$^1J_{FC, PC} = 276$ and 233

−130
F
154.1
Ph—SO2   H   6.23
$^2J_{FH} = 33$
$^1J_{FC} = 293$

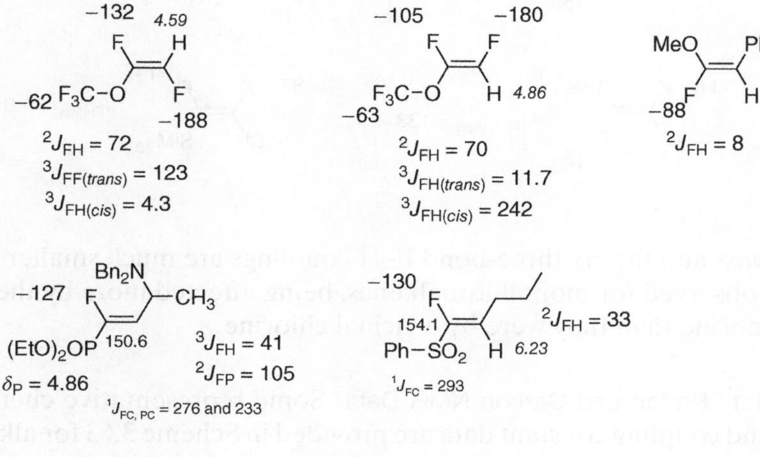

(Scheme 3.62). However, it can be seen that when the vicinal fluorines are *cis* to each other both of the fluorines appear at much lower field than when they are *trans* to each other.

The observed *trans* F–F coupling constants are very large (>130 Hz), whereas the analogous *cis* couplings are much smaller (<15 Hz). Both

Scheme 3.62

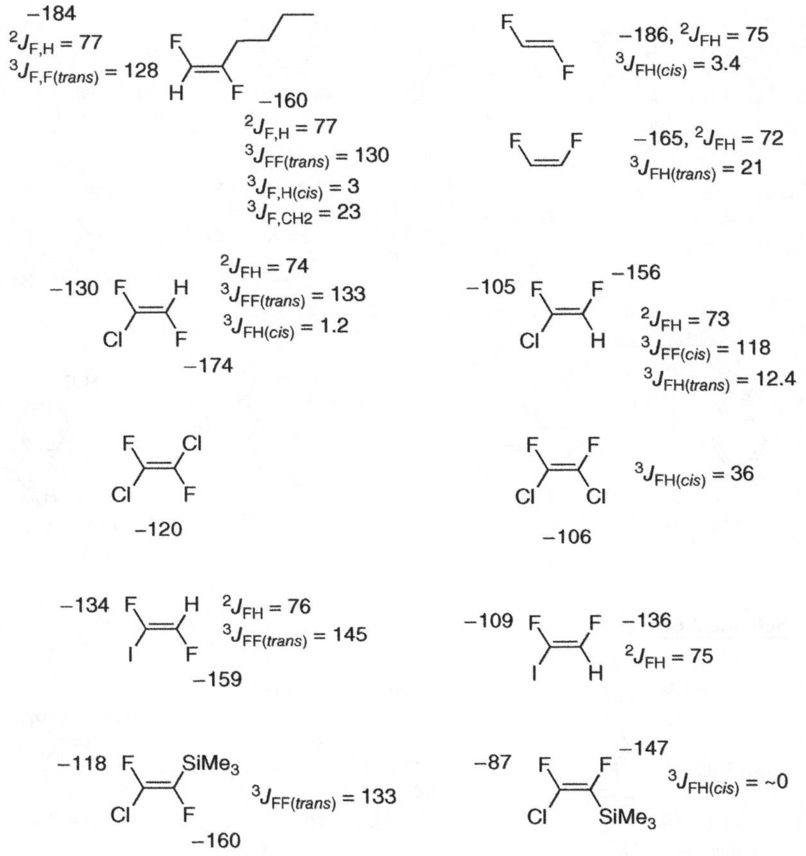

−184
$^2J_{F,H} = 77$
$^3J_{F,F(trans)} = 128$

−160
$^2J_{F,H} = 77$
$^3J_{FF(trans)} = 130$
$^3J_{F,H(cis)} = 3$
$^3J_{F,CH2} = 23$

−186, $^2J_{FH} = 75$
$^3J_{FH(cis)} = 3.4$

−165, $^2J_{FH} = 72$
$^3J_{FH(trans)} = 21$

−130
$^2J_{FH} = 74$
$^3J_{FF(trans)} = 133$
$^3J_{FH(cis)} = 1.2$
−174

−105
−156
$^2J_{FH} = 73$
$^3J_{FF(cis)} = 118$
$^3J_{FH(trans)} = 12.4$

−120

−106
$^3J_{FH(cis)} = 36$

−134
$^2J_{FH} = 76$
$^3J_{FF(trans)} = 145$
−159

−109
−136
$^2J_{FH} = 75$

−118
$^3J_{FF(trans)} = 133$
−160

−87
−147
$^3J_{FH(cis)} = ~0$

the *trans-* and the *cis*-three-bond F–H couplings are much smaller than those observed for monofluoroalkenes, being affected more by the vicinal fluorine than they were by a vicinal chlorine.

3.6.6.1.1. Proton and Carbon NMR Data. Some representative chemical shift and coupling constant data are provided in Scheme 3.63 for alkenes with vicinal fluorines.

Comparing geometric isomers of the type CHF=CFR, a proton *cis* to fluorine is more deshielded than one that is *trans*.

3.6.6.2. Trifluorovinyl Groups. Trifluorovinyl groups have characteristic chemical shifts and coupling constants that are exemplified with

Scheme 3.62 (*Continued*)

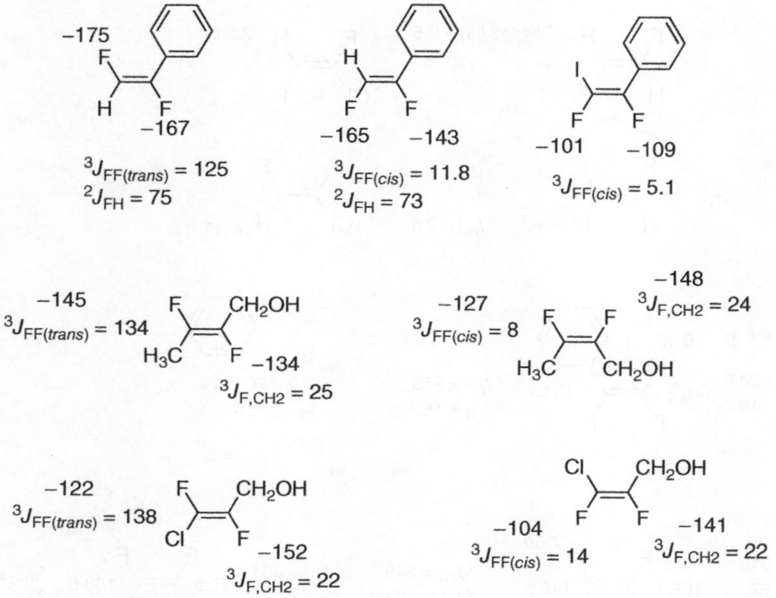

−175
F
H
F
−167
$^3J_{FF(trans)}$ = 125
$^2J_{FH}$ = 75
H
F
F
−165
−143
$^3J_{FF(cis)}$ = 11.8
$^2J_{FH}$ = 73
I
F
F
−101
−109
$^3J_{FF(cis)}$ = 5.1
−145
$^3J_{FF(trans)}$ = 134
F
CH₂OH
H₃C
F
−134
$^3J_{F,CH2}$ = 25
−127
$^3J_{FF(cis)}$ = 8
F
F
H₃C
CH₂OH
−148
$^3J_{F,CH2}$ = 24
−122
$^3J_{FF(trans)}$ = 138
F
CH₂OH
Cl
F
−152
$^3J_{F,CH2}$ = 22
Cl
CH₂OH
F
F
−104
$^3J_{FF(cis)}$ = 14
−141
$^3J_{F,CH2}$ = 22

the example given in Scheme 3.64. See Chapter 6 for more details and examples.

3.6.7. *α,β*-Unsaturated Carbonyl Compounds

The usual deshielding that is observed at the *β*-position of *α,β*-unsaturated carbonyl compounds in both proton and carbon NMR is also observed for a fluorine substituted at this position (Scheme 3.65).

Fluorines in the *β*-position are deshielded by as much as 20 ppm relative to a simple fluoroalkene, whereas those at the *α*-position are shielded by about 20 ppm, similar to fluorines at the 2-position of a 1,3-diene (Scheme 3.66). Generally, in pairs of geometric isomers, fluorines that are *cis* to the carbonyl function appear at higher fields than those that are *trans* to the carbonyl function.

3.6.7.1. ^{1}H and ^{13}C NMR Data. The carbons at the *β*-position of *α,β*-unsaturated carbonyl compounds are also deshielded relative to ordinary terminal fluoroalkenes (Scheme 3.67).

Scheme 3.63

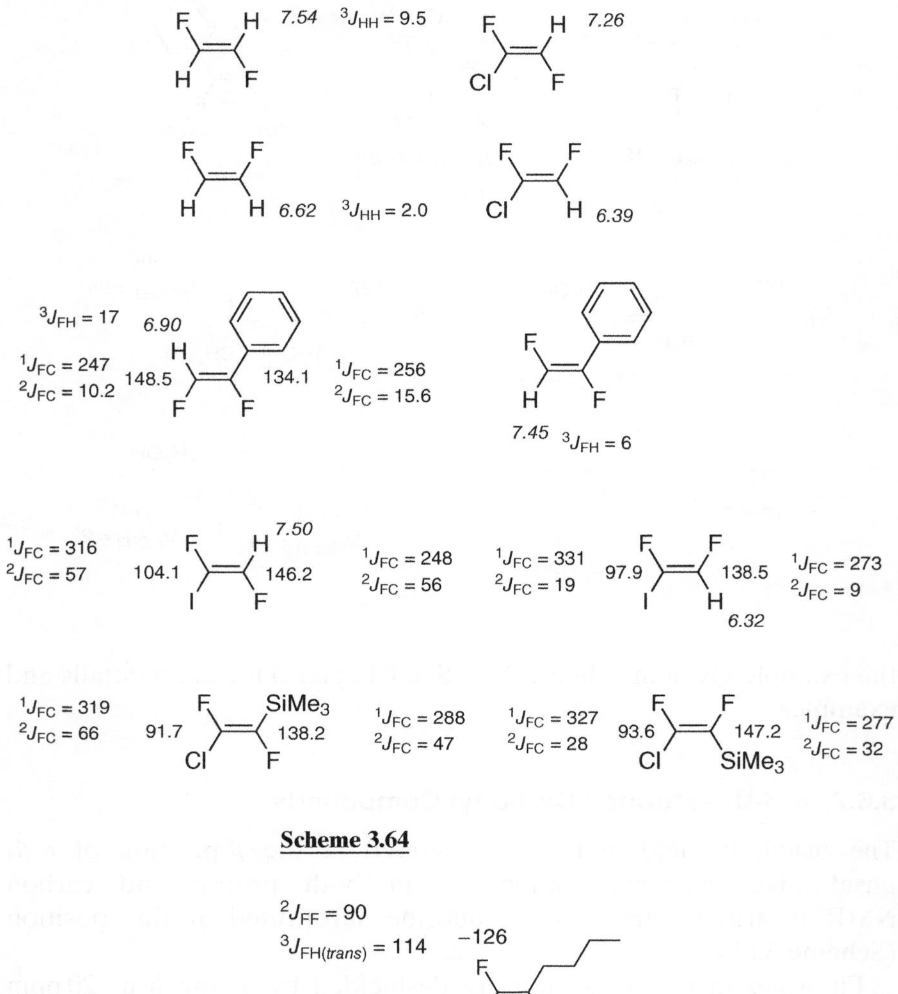

$^3J_{HH} = 9.5$ (structure: F, H, H, F) 7.54

(structure: F, H, Cl, F) 7.26

(structure: F, F, H, H) 6.62 $^3J_{HH} = 2.0$

(structure: F, F, Cl, H) 6.39

$^3J_{FH} = 17$ 6.90
$^1J_{FC} = 247$ 148.5   (phenyl vinyl structure)   134.1 $^1J_{FC} = 256$
$^2J_{FC} = 10.2$ (F, F) $^2J_{FC} = 15.6$

(phenyl vinyl structure: F, H, F)
7.45 $^3J_{FH} = 6$

$^1J_{FC} = 316$ 104.1   (F, H, I, F) 7.50   146.2 $^1J_{FC} = 248$ $^1J_{FC} = 331$ 97.9   (F, F, I, H) 138.5 $^1J_{FC} = 273$
$^2J_{FC} = 57$ $^2J_{FC} = 56$ $^2J_{FC} = 19$ 6.32 $^2J_{FC} = 9$

$^1J_{FC} = 319$ 91.7   (F, SiMe$_3$, Cl, F) 138.2 $^1J_{FC} = 288$ $^1J_{FC} = 327$ 93.6   (F, F, Cl, SiMe$_3$) 147.2 $^1J_{FC} = 277$
$^2J_{FC} = 66$ $^2J_{FC} = 47$ $^2J_{FC} = 28$ $^2J_{FC} = 32$

Scheme 3.64

$^2J_{FF} = 90$
$^3J_{FH(trans)} = 114$ −126
(F)

$^2J_{FF} = 90$ (F, F)
$^3J_{FF(cis)} = 32$ −104   −174

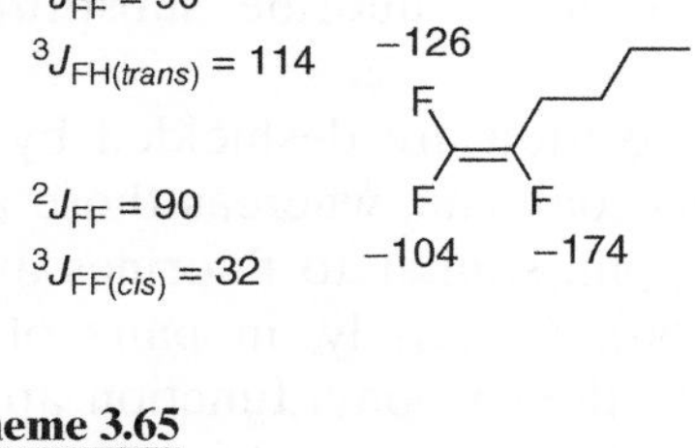

Scheme 3.65

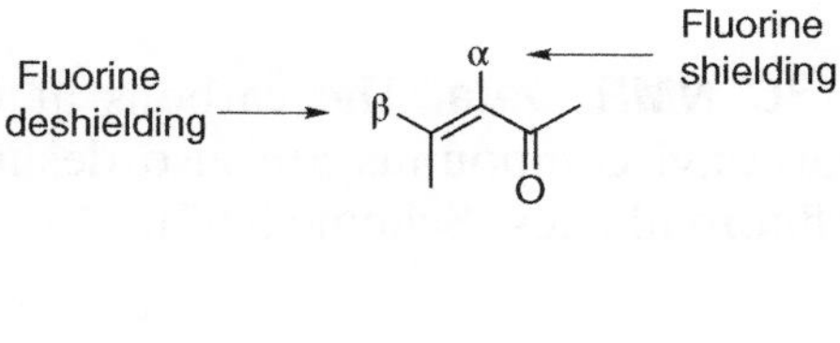

Fluorine deshielding →   β (structure with α, O)   ← Fluorine shielding

Scheme 3.66

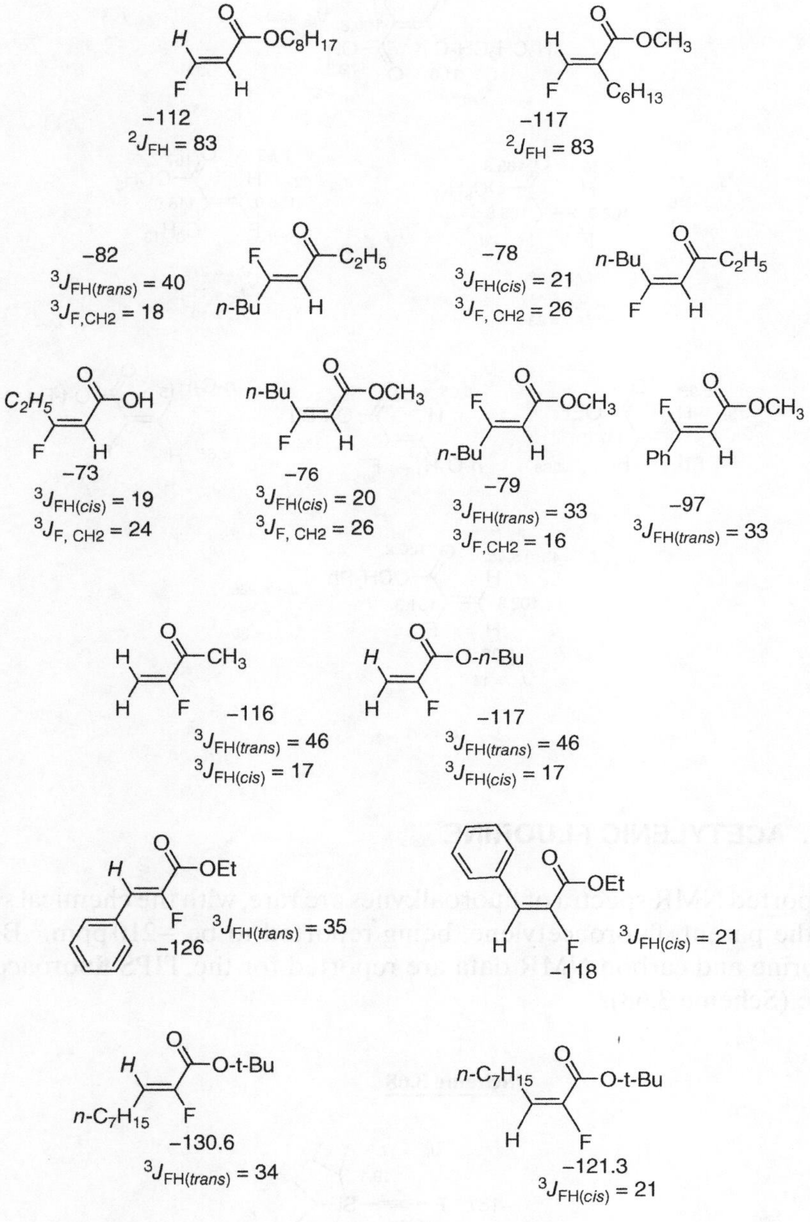

−112
$^2J_{FH} = 83$

−117
$^2J_{FH} = 83$

−82
$^3J_{FH(trans)} = 40$
$^3J_{F,CH2} = 18$

−78
$^3J_{FH(cis)} = 21$
$^3J_{F, CH2} = 26$

−73
$^3J_{FH(cis)} = 19$
$^3J_{F, CH2} = 24$

−76
$^3J_{FH(cis)} = 20$
$^3J_{F, CH2} = 26$

−79
$^3J_{FH(trans)} = 33$
$^3J_{F,CH2} = 16$

−97
$^3J_{FH(trans)} = 33$

−116
$^3J_{FH(trans)} = 46$
$^3J_{FH(cis)} = 17$

−117
$^3J_{FH(trans)} = 46$
$^3J_{FH(cis)} = 17$

−126
$^3J_{FH(trans)} = 35$

−118
$^3J_{FH(cis)} = 21$

−130.6
$^3J_{FH(trans)} = 34$

−121.3
$^3J_{FH(cis)} = 21$

Scheme 3.67

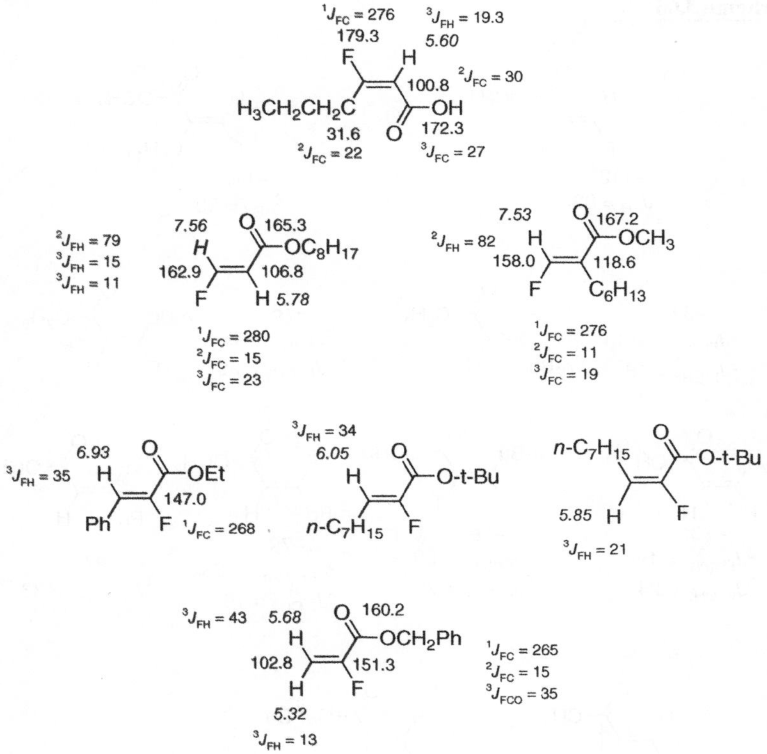

3.7. ACETYLENIC FLUORINE

Reported NMR spectra of fluoroalkynes are rare, with the chemical shift of the parent, fluoroacetylene, being reported to be −210 ppm.[9] Both fluorine and carbon NMR data are reported for the TIPS fluoroacetylene (Scheme 3.68).

Scheme 3.68

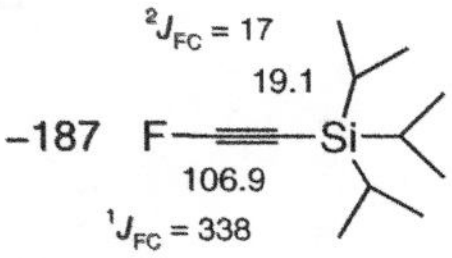

3.8. ALLYLIC AND PROPARGYLIC FLUORIDES

The proximity of carbon–carbon double or triple bonds, as in allylic and propargylic systems has little impact upon a fluorine substituent's chemical shift (Scheme 3.69). Note that one would not intuitively expect allyl fluoride and *n*-butyl fluoride to have such similar chemical shifts.

Scheme 3.69

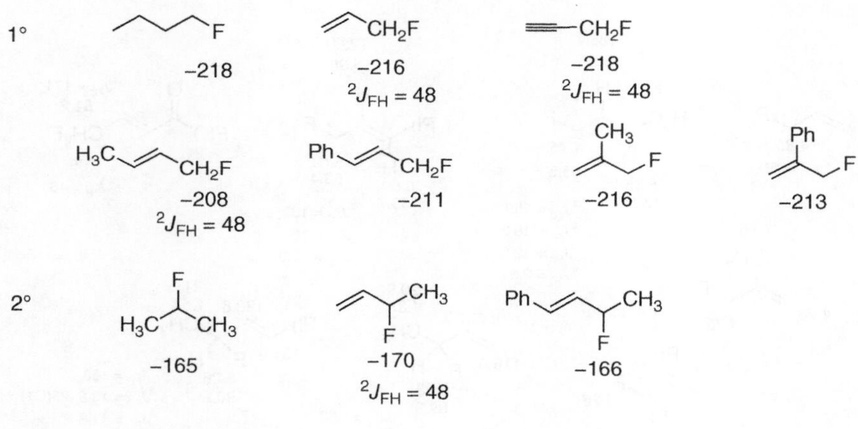

1°

-218

-216
$^2J_{FH} = 48$

-218
$^2J_{FH} = 48$

-208
$^2J_{FH} = 48$

-211

-216

-213

2°

-165

-170
$^2J_{FH} = 48$

-166

It is worth noting that the same kind of dramatic hyperconjugative $\pi-\sigma_{CF}{}^*$ interactions that were observed for substituted benzyl fluorides (see Section 2.2.1) are also apparently in play here. The observed trends for the two series of allyl fluorides in Scheme 3.70 show this to be clearly the case.

Scheme 3.70

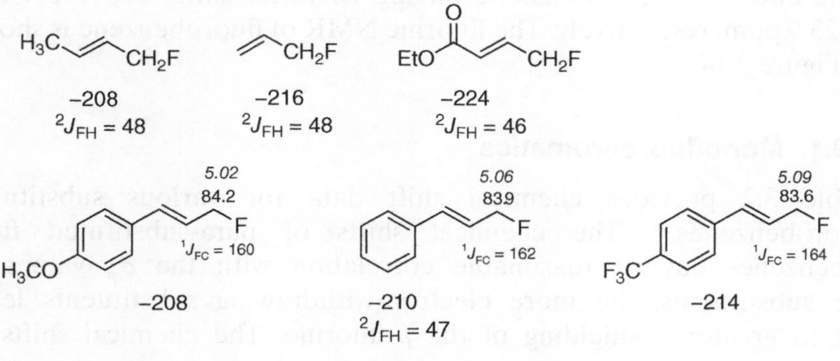

-208
$^2J_{FH} = 48$

-216
$^2J_{FH} = 48$

-224
$^2J_{FH} = 46$

5.02
84.2
$^1J_{FC} = 160$

5.06
83.9
$^1J_{FC} = 162$

5.09
83.6
$^1J_{FC} = 164$

-208

-210
$^2J_{FH} = 47$

-214

3.8.1. ^{1}H and ^{13}C NMR Data

Some typical proton and carbon chemical shift and coupling constant data for allylic systems are given in Scheme 3.71. An alkenyl substituent on either a CH$_2$F or a —CHF— group has virtually no effect upon that carbon's chemical shift, and they also only affect the proton chemical shift by about 0.5 ppm.

Scheme 3.71

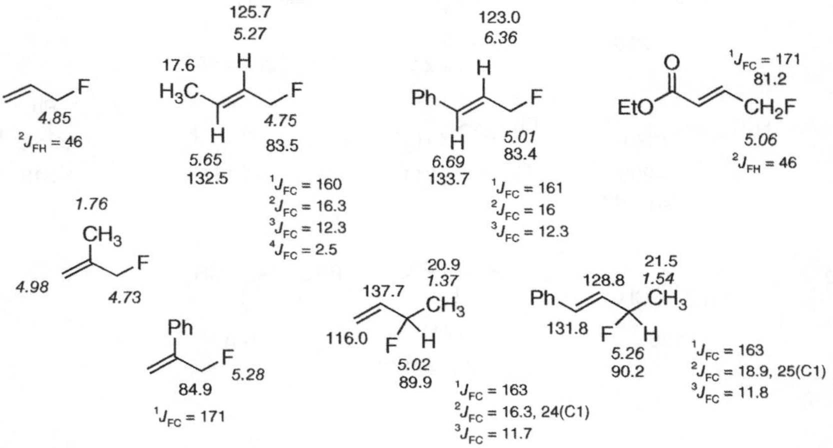

3.9. FLUOROAROMATICS

Ring current (anisotropic) effects do not play a significant role in fluorine NMR. Therefore, fluorine substituents on a benzene ring absorb in the general region of fluoroalkenes, with fluorobenzene and 1-fluoronaphthalene having chemical shifts of −113.4 and −123.9 ppm, respectively. The fluorine NMR of fluorobenzene is shown in Figure 3.14.

3.9.1. Monofluoroaromatics

Table 3.2 provides chemical shift data for various substituted fluorobenzenes.[10] The chemical shifts of para-substituted fluorobenzenes have a reasonable correlation with the σ_p values of the substituents, the more electron-withdrawing substituents leading to greater deshielding of the p-fluorine. The chemical shifts of

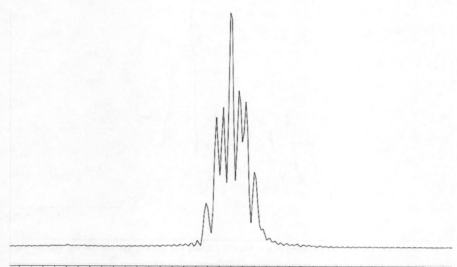

-112.95 -113.00 -113.05 -113.10 -113.15 -113.20 -113.25 -113.30 -113.35 -113.40 -113.45 -113.50 -113.55 -113.60 -113.65 -113.70 -113.75 -113.80 -113.85

*F*1 (ppm)

FIGURE 3.14. ^{19}F NMR of fluorobenzene

TABLE 3.2. ^{19}F Chemical Shifts for Fluorobenzenes[10]

Substituent	Ortho		Meta		Para		σ_p Value
	Acetone-d_6	DMSO	Acetone-d_6	DMSO	Acetone-d_6	DMSO	
COCl	−109.5		−113.6		−101.8		0.61
NO$_2$	−119.7	−119.0	−110.0	−109.5	−103.0	−102.4	0.78
CN	−108.6	−107.9	−110.9	−110.0	−104.0	−102.8	0.66
CHO	−122.4	−120.7	−112.6	−111.6	−104.3	−103.2	0.42
COCH$_3$	−110.6	−110.0	−113.1	−112.0	−107.1	−105.9	0.50
CO$_2$H	−110.0	−110.1	−113.3	−112.0	−107.2	−106.5	0.45
CF$_3$	−115.8	−115.4	−111.4	−110.3	−108.0	−106.8	0.54
CONH$_2$	−113.6	−113.3	−113.4	−112.6	−109.8	−109.2	0.36
H					−113.8	−112.6	0
I	−94.4	−106.2	−110.9	−110.3	−114.8	−114.2	0.18
Br	−108.1	−107.7	−110.8	−110.0	−115.6	−114.7	0.23
Cl	−116.3	−115.9	−111.2	−110.3	−116.7	−115.2	0.23
F	−139.7	−138.8	−110.6	−109.5	−120.0	−119.4	0.06
CH$_3$	−118.4	−117.3	−114.9	−113.7	−119.2	−118.0	−0.17
NHAc	−125.6	−124.6	−112.8	−111.8	−120.3	−119.4	0.0
OCH$_3$	−136.1	−135.3	−112.6	−111.4	−125.2	−124.0	−0.27
OH	−138.0	−136.3	−113.2	−112.1	−126.8	−125.0	−0.37
NH$_2$	−136.3	−134.9	−115.6	−113.5	−129.7	−129.4	−0.66

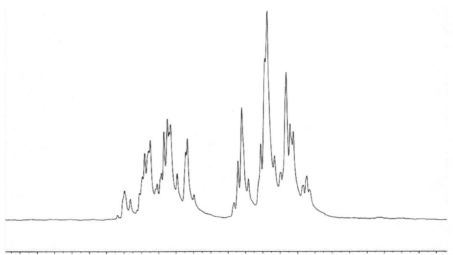

7.04 7.02 7.00 6.98 6.96 6.94 6.92 6.90 6.88 6.86 6.84 6.82 6.80 6.78 6.76 6.74 6.72 6.70 6.68 6.66 6.64 6.62 6.60 6.58 6.56

*F*1 (ppm)

FIGURE 3.15. ^{1}H NMR spectrum of fluorobenzene (benzene-d_6)

ortho-substituted fluorobenzenes also exhibit a rough correlation, but there are some significant aberrations. The chemical shifts of meta-substituted fluorobenzenes exhibit no correlation and vary over a much smaller range.

It should be noted (and can be seen from Table 3.2) that there can be significant solvent effects upon the chemical shifts of fluorobenzenes.

3.9.1.1. Interplay of ^{19}F, ^{13}C, and ^{1}H NMR Spectra for

Fluoroaromatics. The second order character of the ^{1}H NMR spectrum (Figure 3.15) makes analysis by examination impossible.

A fluorine substituent on benzene has a characteristic effect upon the ^{13}C spectrum of benzene and it couples in a distinctive and highly consistent manner with the ipso, ortho, meta, and para carbons (Scheme 3.72).

The ^{13}C NMR of fluorobenzene itself, shown in Figure 3.16, exemplifies this nicely with four doublets being clearly observable. The chemical shifts seen in this spectrum are slightly different from those given in Scheme 3.72 because of the choice of solvent (C_6D_6).

3.9.1.2. Complete NMR Analysis of o-, m-, and p-Nitrofluoro-

benzenes. The complete set of NMR data for one series of *o*-, *m*-, and *p*-disubstituted fluorobenzene compounds, that of the nitrofluorobenzenes, will serve to further exemplify the interplay of ^{19}F, ^{13}C, and ^{1}H chemical shifts and coupling constants that provide unique

Scheme 3.72

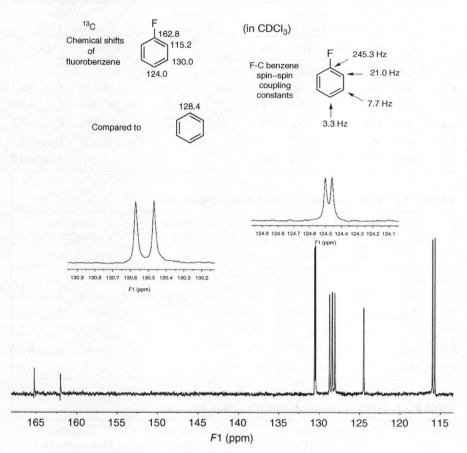

FIGURE 3.16. ^{13}C NMR of fluorobenzene (benzene-d_6)

insight into the structures of disubstituted fluorobenzenes. These data are given in Tables 3.3–3.5.

3.9.1.3. *Coupling Constants.* The usual three-bond H–H coupling constant in fluorobenzenes is about 8 Hz, whereas the four-bond coupling constant is between 1 and 3 Hz. Five-bond coupling is usually not observed. Similarly, the three-bond F–H coupling constant is about 8 Hz, the four-bond value 5–6 Hz, and the five-bond coupling constant about 1 Hz.

The F coupling to carbon can vary considerably for the carbon directly substituted (ipso), depending on its substitution environment,

TABLE 3.3. NMR Analysis of *ortho*-Nitrofluorobenzene

	Coupling Constants to Carbon (Hz)				Chemical Shifts	Coupling Constants (Hz)	
	F	H3	H4	H5	H6		F/H	H/H
C1	−262.6	−8.1	−1.9	−11.5	−5.0		$^4J_{FH3}=7.9$	$^3J_{H3H4}=8.1$
C2	8.8	—	—	—	—		$^5J_{FH4}=-0.9$	$^3J_{H4H5}=7.5$
C3	−2.6	169.4	2.6	9.0	1.0		$^4J_{FH5}=4.6$	$^3J_{H5H6}=8.5$
C4	4.0	0.9	167.8	0.9	8.7		$^3J_{FH6}=11.5$	$^4J_{H3H5}=1.7$
C5	8.8	9.2	1.5	165.5	0.5			$^4J_{H4H6}=1.2$
C6	20.7	1.3	8.3	1.3	167.2			

Chemical Shifts diagram: −119.6 F; 8.14 H; 156.1 NO$_2$; 119.1; 138.2; 136.9; 126.8; H 7.48; 125.8; 7.50 H; 7.83

TABLE 3.4. NMR Analysis of *meta*-Nitrofluorobenzene

	Coupling Constants to Carbon (Hz)				Chemical Shifts	Coupling Constants (Hz)	
	F	H2	H4	H5	H6		F/H	H/H
C1	−249.6	−6.0	−1.4	−11.9	−4.5		$^3J_{FH2}=8.9$	$^3J_{H4H5}=8.3$
C2	26.5	171.1	−5.2	−1.5	−4.3		$^5J_{FH4}=-1.0$	$^3J_{H5H6}=8.3$
C3	8.8	−3.9	−1.5	−11.4	−2.6		$^4J_{FH5}=5.7$	$^4J_{H2H4}=2.2$
C4	3.1	−4.2	171.3	−1.7	−8.0		$^3J_{FH6}=8.3$	$^4J_{H2H6}=2.6$
C5	8.8	0.0	0.0	167.9	0.0			$^4J_{H4H6}=0.9$
C6	21.6	4.0	8.0	−2.4	168.1			

Chemical Shifts diagram: −110.2 F; 7.65 H; 163.2 H 8.00; 122.8; 111.7; 132.2; 150.0; H 7.76; 120.3 NO$_2$; H 8.12

TABLE 3.5. NMR Analysis of *para*-Nitrofluorobenzene

	Coupling Constants to Carbon (Hz)				Chemical Shifts	Coupling Constants (Hz)	
	F	H2	H4	H5	H6		F/H	H/H
C1	−255.7	4.6	10.9	−10.9	−4.6		$^3J_{FH2}=8.2$	$^3J_{H2H3}=9.2$
C2	24.1	168.9	0.0	0.05	−4.5		$^5J_{FH3}=-4.8$	
C3	10.2	0.0	171.4	5.4	0.0			
C4	4.5	10.2	5.8	5.8	10.2			

Chemical Shifts diagram: −103.5 F; H; 167.2 H 7.43; 117.3; 127.2; H; 145.5 H 8.35; NO$_2$

but it is always very large, 250 Hz or larger. F coupling to the ortho-position is usually about 20–26 Hz, to the meta-position about 8–10 Hz, and to the para-position about 4 Hz.

Ipso (one-bond) coupling of H to C is consistently between 165 and 172 Hz, whereas its two-bond coupling constants (0–5 Hz) are usually much smaller than three-bond H–C couplings (4–10 Hz).

Usually, a careful analysis of the combination of fluorine, proton, and carbon NMR chemical shifts and spin–spin coupling constants provides definitive information regarding the structure of disubstituted fluoroaromatics.

3.9.2. Fluoropolycyclic Aromatics: Fluoronaphthalenes

The isomeric 1- and 2-fluoronaphthalenes have fluorine chemical shifts of −124 and −116 ppm, respectively. A full analysis of the proton and carbon spectra of 1-fluoronaphthalene is given in Scheme 3.73. NMR data for a number of other fluoropolycyclic aromatic compounds are available.[11]

3.9.2.1. Steric Deshielding of Fluorine Nucleus. As can be seen from the data in Table 3.2 and for 4-methyl-1-fluoronaphthalene (Scheme 3.74), ordinarily a methyl group on a fluoroaromatic gives rise to shielding of the fluorine nucleus. However, a methyl group in the *peri*-8-position of 1-fluoronaphthalene provides a rare example of *steric deshielding* of a fluorine atom.[12] Other groups, such as ethyl and acetyl, in this position give rise to similar deshielding effects, with the

Scheme 3.73

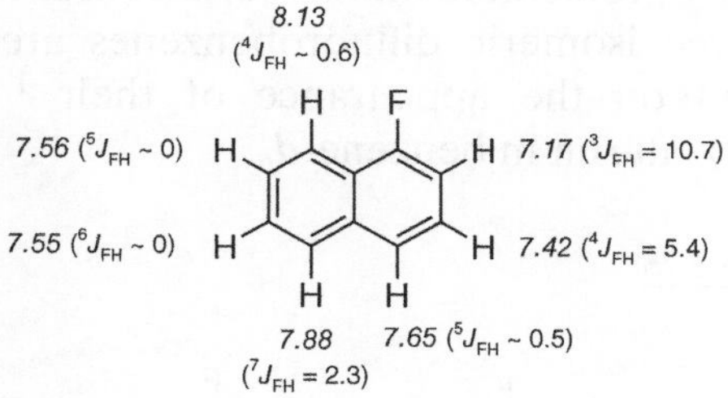

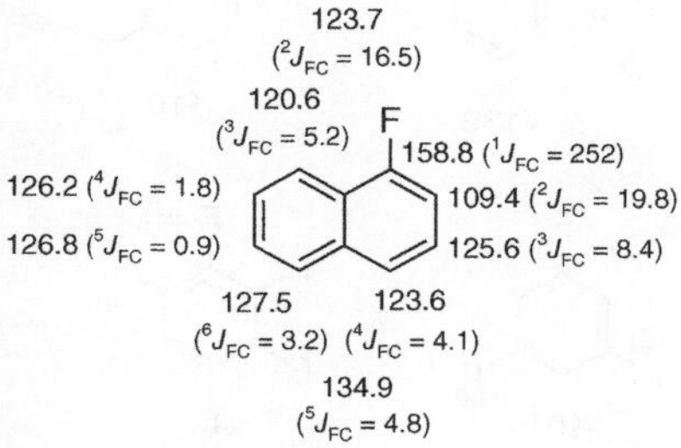

$$8.13$$
$$(^4J_{FH} \sim 0.6)$$

$$7.56 \, (^5J_{FH} \sim 0)$$

$$7.55 \, (^6J_{FH} \sim 0)$$

$$7.17 \, (^3J_{FH} = 10.7)$$

$$7.42 \, (^4J_{FH} = 5.4)$$

$$7.88 \quad 7.65 \, (^5J_{FH} \sim 0.5)$$
$$(^7J_{FH} = 2.3)$$

$$123.7$$
$$(^2J_{FC} = 16.5)$$
$$120.6$$
$$(^3J_{FC} = 5.2)$$

$$158.8 \, (^1J_{FC} = 252)$$
$$109.4 \, (^2J_{FC} = 19.8)$$
$$125.6 \, (^3J_{FC} = 8.4)$$

$$126.2 \, (^4J_{FC} = 1.8)$$
$$126.8 \, (^5J_{FC} = 0.9)$$

$$127.5 \quad 123.6$$
$$(^6J_{FC} = 3.2) \quad (^4J_{FC} = 4.1)$$
$$134.9$$
$$(^5J_{FC} = 4.8)$$

Scheme 3.74

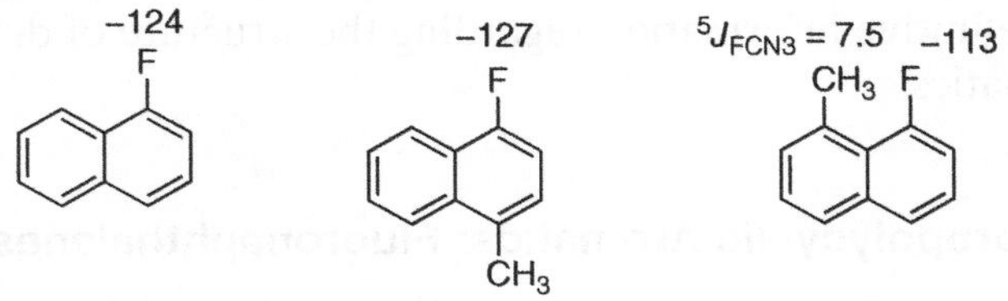

t-butyl group providing the largest observed effect, deshielding the fluorine by approximately 27 ppm (to −96 ppm). Further discussion of such deshielding effects has been provided in Section 2.2.2.

One should also note the significant 7.5 Hz F–H coupling constant between the methyl hydrogens of the 8-methyl-1-fluoronaphthalene and its fluorine substituent. This likely derives, at least in part, from through-space F–H coupling (see Section 2.3.2).

3.9.3. Polyfluoroaromatics

A second fluorine substituent shields in the ortho- and especially in the para-position, but one in the meta-position deshields, with 1,3-5-trifluorobenzene having the most deshielded fluorines in a polyfluoroaromatic system (Scheme 3.75). On the other hand, hexafluorobenzene has highly shielded fluorines. The fluorine spectra of these multifluorobenzenes are second order in nature and their appearance is thus not generally predictable on the basis of first-order logic.

Indeed, the three isomeric difluorobenzenes are not readily distinguishable based on the appearance of their ^{19}F NMR spectra (Figures 3.17–3.19), all run in benzene-d_6.

Scheme 3.75

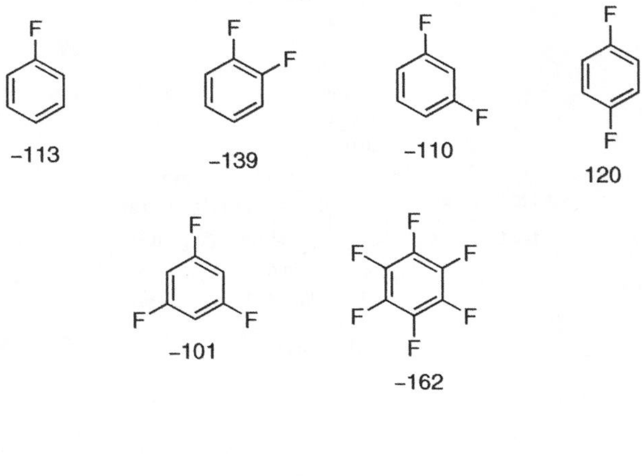

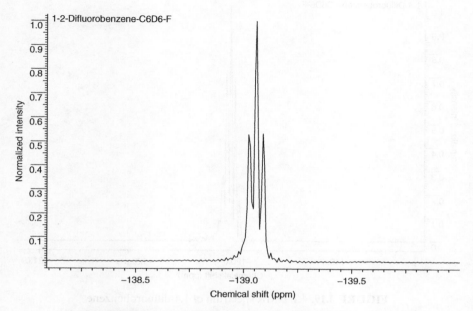

FIGURE 3.17. ^{19}F NMR spectrum of 1,2-difluorobenzene

The spectra, although all are not readily interpretable upon a cursory review, the pattern in spectra of the three isomeric difluorobenzenes are distinctly distinct (Figures 3.17–3.19) in this region.

The ^{19}F NMR spectra of the three isomeric difluorobenzenes are quite different and are readily interpretable about peak positions, with the 1,2-isomer exhibiting more significant by a resonance host, and the 1,4-isomer at different (delta) 3.19. The signal observed in each of these (despite a small 0.5-ppm interval) on the actual spectrum is—

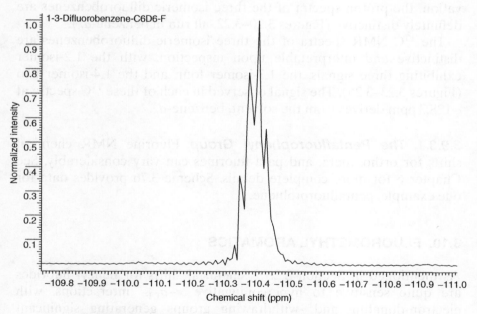

FIGURE 3.18. ^{19}F NMR spectrum of 1,3-difluorobenzene

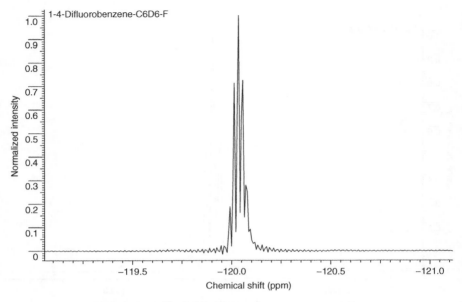

FIGURE 3.19. ¹⁹F NMR spectrum of 1,4-difluorobenzene

In contrast, although all are not readily interpretable upon observation, the proton spectra of the three isomeric difluorobenzenes are definitely distinctive (Figures 3.20–3.22, all run in benzene-d_6).

The ^{13}C NMR spectra of the three isomeric difluorobenzenes are distinctive and interpretable upon inspection, with the 1,2-isomer exhibiting three signals, the 1,3-isomer four, and the 1,4-isomer two (Figures 3.23–3.25). The signal observed in each of these ^{13}C spectra at ~128.3 ppm derives from the solvent, benzene-d_6.

3.9.3.1. The Pentafluorophenyl Group. Fluorine NMR chemical shifts for ortho, meta, and para fluorines can vary considerably. See Chapter 6 for more complete details. Scheme 3.76 provides data for one example, pentafluorotoluene.

3.10. FLUOROMETHYL AROMATICS

As mentioned in Section 2.2.1.2, chemical shifts for benzyl fluorides are quite sensitive to hyperconjugative π–σ_{CF}^* interactions, with electron-donating and -withdrawing groups generating significant deshielding and shielding effects, respectively. Benzyl fluoride itself

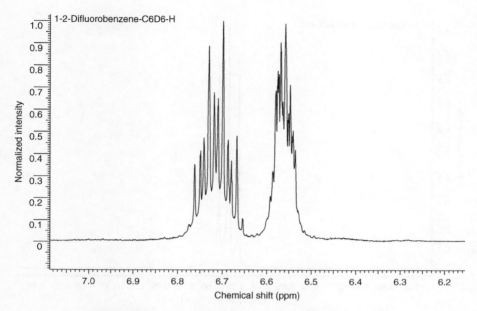

FIGURE 3.20. ^{1}H NMR spectrum of 1,2-difluorobenzene

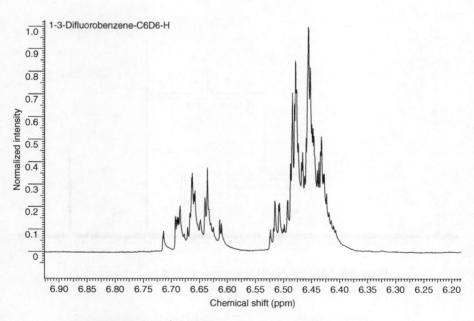

FIGURE 3.21. ^{1}H NMR spectrum of 1,3-difluorobenzene

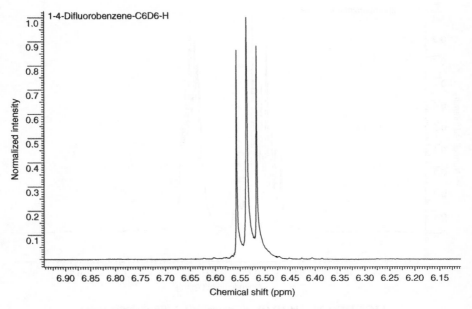

FIGURE 3.22. ^{1}H NMR spectrum of 1,4-difluorobenzene

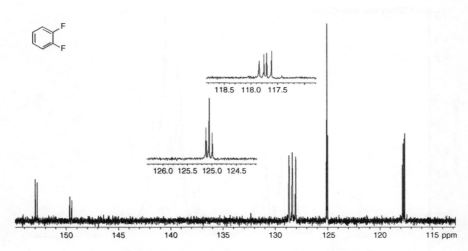

FIGURE 3.23. ^{13}C NMR spectrum of 1,2-difluorobenzene

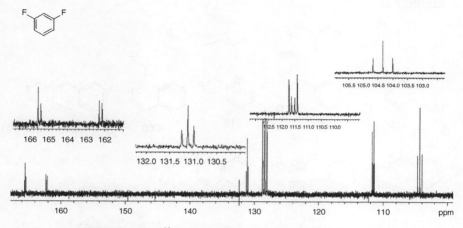

FIGURE 3.24. ^{13}C NMR spectrum of 1,3-difluorobenzene

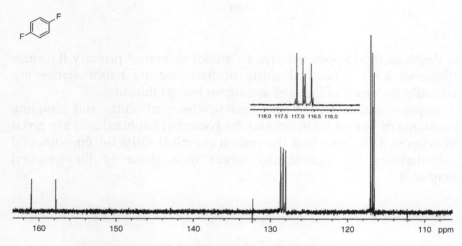

FIGURE 3.25. ^{13}C NMR spectrum of 1,4-difluorobenzene

Scheme 3.76

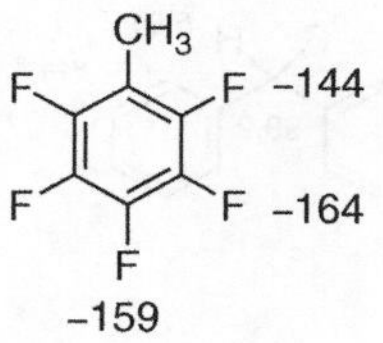

$^3J_{23} = 20.4$
$^3J_{34} = 18.9$
$^5J_{25} = 8.6$

–144
–164
–159

Scheme 3.77

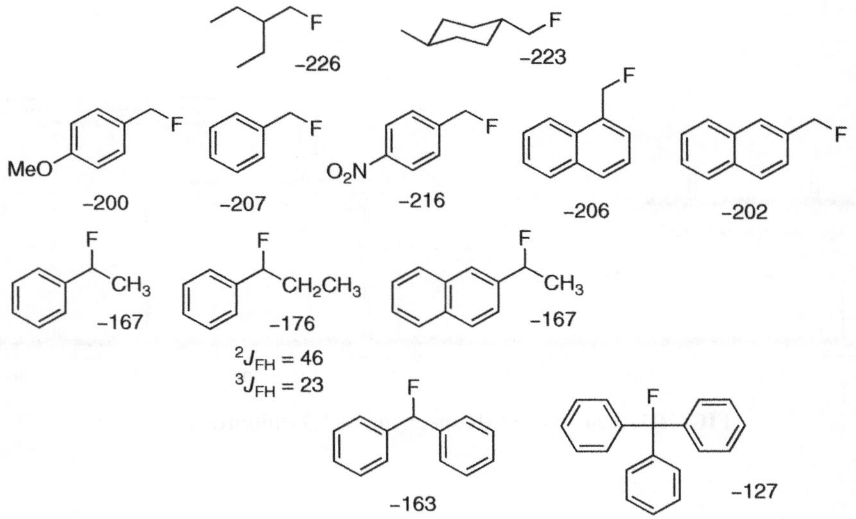

is deshielded >15 ppm, relative to model saturated primary fluorides (Scheme 3.77). Chemical shifts of fluoromethyl naphthalenes are virtually the same as those of analogous benzyl fluorides.

Representative proton and carbon chemical shifts and coupling constants of benzyl fluorides and fluoromethyl naphthalenes are given in Scheme 3.78. Note that the proton chemical shifts for fluoromethyl naphthalenes are significantly larger than those of fluoromethyl benzenes.

Scheme 3.78

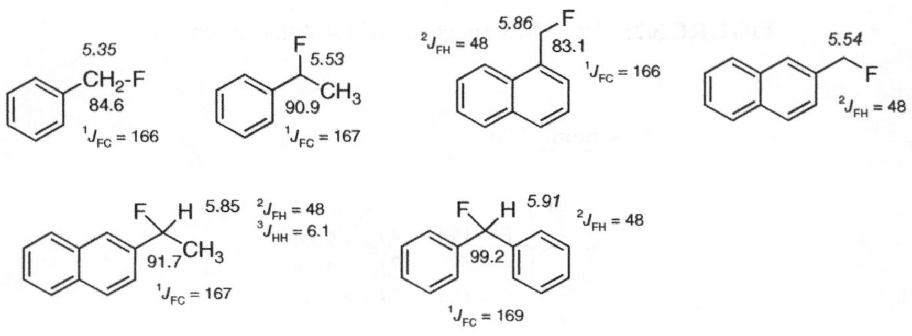

3.11. FLUOROHETEROCYCLES

The increasing amounts of data available for simple fluoroheterocycles allow one to understand how the position of the fluorine substituent on a heterocycle can significantly affect its chemical shift. For all nitrogen and oxygen heterocycles, a fluorine substituent on a carbon bound to the nitrogen or oxygen will be deshielded compared to a fluorine at any other position. The situation is reversed for sulfur heterocycles, although the observed differences are small.

3.11.1. Fluoropyridines, Quinolines, and Isoquinolines

In the case of pyridine, large differences in chemical shift are observed for fluorines at the 2-, 3-, and 4-positions, with fluorines at the 2-position of pyridines and quinolines being the most deshielded and those at the 3-position being the most shielded (Scheme 3.79). The fluorine NMR spectrum of 2-fluoropyridine is given in Figure 3.26. The various small H couplings to the fluorine at 282 MHz appear only to combine to give rise to a broadening of the fluorine signal.

3.11.1.1. Carbon, Proton, and Nitrogen NMR Data. Detailed proton and carbon NMR data are available for the three isomers of fluoropyridine,[13] as are nitrogen data. These data are provided in Scheme 3.80. The ^{1}H and ^{13}C NMR spectra of 2-fluoropyridine are given in Figures 3.27 and 3.28, respectively.

In the proton spectrum, the C-6 proton is most deshielded (at 8.2 ppm), the proton at C-3, vicinal to fluorine is the most shielded. The fluorine's largest F–H coupling constant is its 8.3 Hz, five-bond coupling constant to the C4 hydrogen. In the ^{13}C spectrum, note that the three-bond F–C coupling with C-6, through the nitrogen, is significantly larger (14.5 Hz) than that to the C-4 carbon (7.7 Hz), in contrast to the relative magnitude of the four-bond F–H coupling constants to the hydrogen at C-6 (1.3 Hz), which is smaller than that to the hydrogen at C-4 (5.3 Hz).[7]

Note that fluorine substitution at the 2-position has a significant shielding effect upon the ^{15}N chemical shift of pyridine, considerably less at the 3-position, and fluorine at the 4-position actually has a *deshielding* influence on the nitrogen.

Proton and carbon NMR data for fluoroquinolines and isoquinolines are given in Scheme 3.81.

Scheme 3.79

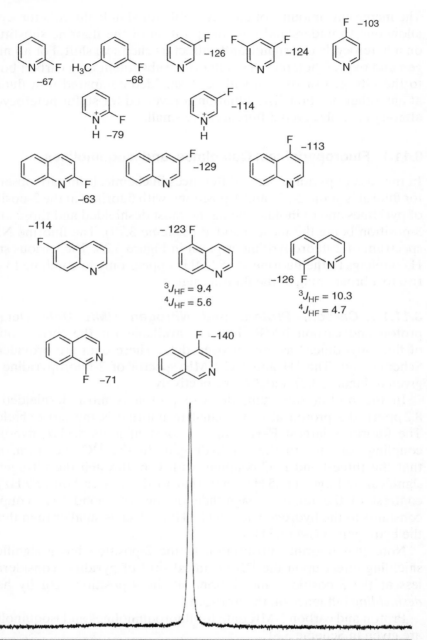

FIGURE 3.26. ^{19}F NMR spectrum of 2-fluoropyridine

Scheme 3.80

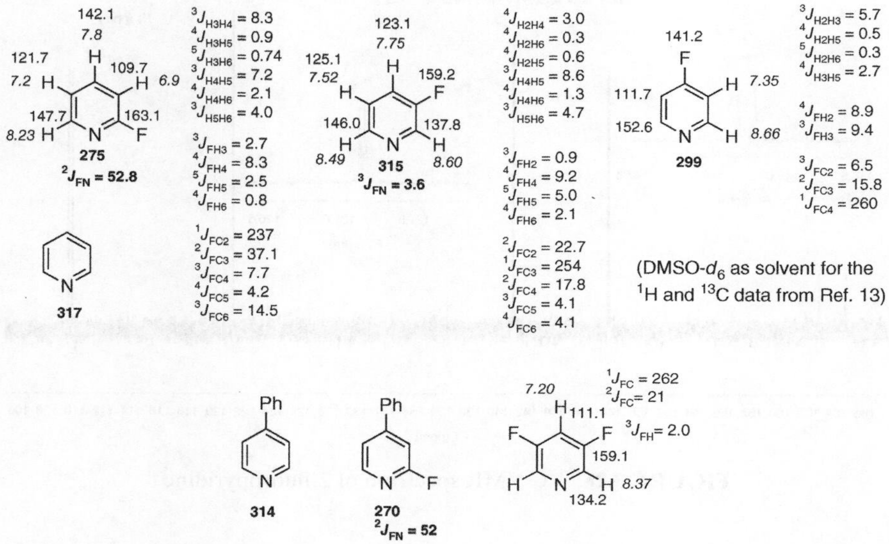

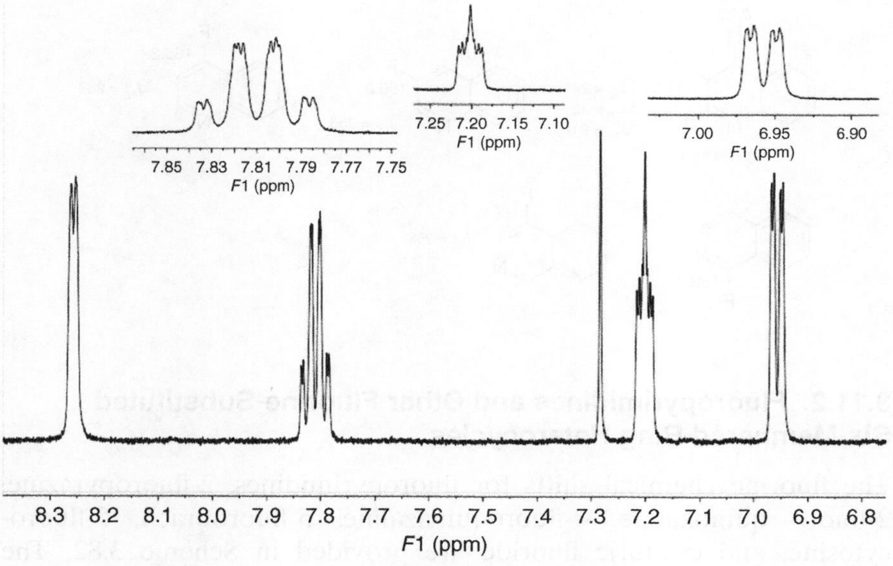

275

142.1
7.8
121.7
7.2 H
109.7
H 6.9
147.7
8.23 H
163.1
N   F
$^2J_{FN}$ = 52.8

$^3J_{H3H4}$ = 8.3
$^4J_{H3H5}$ = 0.9
$^5J_{H3H6}$ = 0.74
$^3J_{H4H5}$ = 7.2
$^4J_{H4H6}$ = 2.1
$^3J_{H5H6}$ = 4.0

$^3J_{FH3}$ = 2.7
$^4J_{FH4}$ = 8.3
$^5J_{FH5}$ = 2.5
$^4J_{FH6}$ = 0.8

315

123.1
7.75
125.1
7.52 H
159.2
F
146.0
H
137.8
8.49 H   N   H 8.60
$^3J_{FN}$ = 3.6

$^4J_{H2H4}$ = 3.0
$^4J_{H2H6}$ = 0.3
$^4J_{H2H5}$ = 0.6
$^3J_{H4H5}$ = 8.6
$^4J_{H4H6}$ = 1.3
$^3J_{H5H6}$ = 4.7

$^3J_{FH2}$ = 0.9
$^4J_{FH4}$ = 9.2
$^5J_{FH5}$ = 5.0
$^4J_{FH6}$ = 2.1

299

141.2
F
111.7
H 7.35
152.6
N
H 8.66

$^3J_{H2H3}$ = 5.7
$^4J_{H2H5}$ = 0.5
$^5J_{H2H6}$ = 0.3
$^4J_{H3H5}$ = 2.7

$^4J_{FH2}$ = 8.9
$^3J_{FH3}$ = 9.4

$^3J_{FC2}$ = 6.5
$^2J_{FC3}$ = 15.8
$^1J_{FC4}$ = 260

317

$^1J_{FC2}$ = 237
$^2J_{FC3}$ = 37.1
$^3J_{FC4}$ = 7.7
$^4J_{FC5}$ = 4.2
$^3J_{FC6}$ = 14.5

$^2J_{FC2}$ = 22.7
$^1J_{FC3}$ = 254
$^2J_{FC4}$ = 17.8
$^3J_{FC5}$ = 4.1
$^4J_{FC6}$ = 4.1

(DMSO-d_6 as solvent for the ^{1}H and ^{13}C data from Ref. 13)

314
Ph

270
Ph
F
N
$^2J_{FN}$ = 52

7.20
H
111.1
F
$^1J_{FC}$ = 262
$^2J_{FC}$ = 21
$^3J_{FH}$ = 2.0
F
159.1
H   N   H 8.37
134.2

FIGURE 3.27. ^{1}H NMR spectrum of 2-fluoropyridine

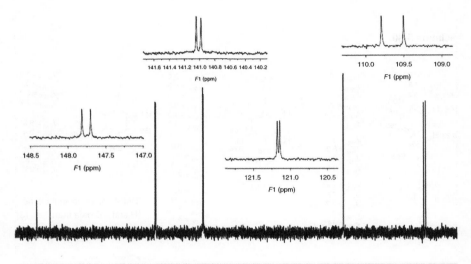

FIGURE 3.28. ^{13}C NMR spectrum of 2-fluoropyridine

Scheme 3.81

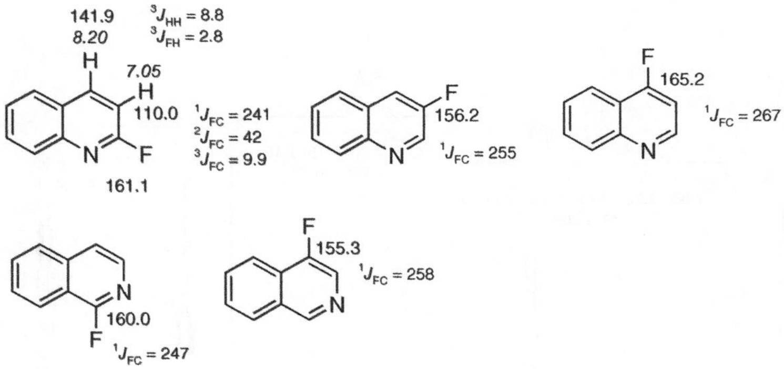

141.9 $^3J_{HH}$ = 8.8
8.20 $^3J_{FH}$ = 2.8

$^1J_{FC}$ = 241
$^2J_{FC}$ = 42
$^3J_{FC}$ = 9.9

161.1   110.0   7.05   141.9

156.2 $^1J_{FC}$ = 255

165.2 $^1J_{FC}$ = 267

160.0 $^1J_{FC}$ = 247

155.3 $^1J_{FC}$ = 258

3.11.2. Fluoropyrimidines and Other Fluorine-Substituted Six-Membered Ring Heterocycles

The fluorine chemical shifts for fluoropyrimidines, 2-fluoropyrazine, 2-fluoro quinoxaline, 4-fluoroquinazaline, 5-fluorouracil, 5-fluoro-cytosine, and cyanuric fluoride are provided in Scheme 3.82. The fluorine of 2-fluoropyrimidine is considerably deshielded relative to

Scheme 3.82

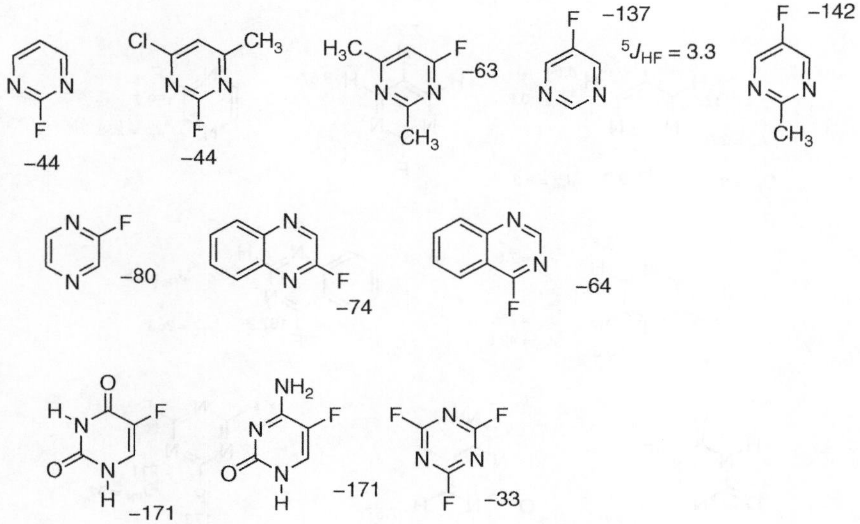

that of 2-fluoropyridine, whereas the 4-fluoro isomer is only slightly so. On the other hand, the fluorine of 2-fluoropyrazine is actually shielded relative to 2-fluoropyridine.

Proton and carbon NMR data for these compounds are provided in Scheme 3.83.

3.11.3. Fluoromethyl Pyridines and Quinolines

The fluorine chemical shifts of a few 2- and 4-fluoromethyl pyridines and quinolines are available. The 4-substituted isomers appear to exhibit the shielding (relative to the phenyl and naphthyl analogs) that would be expected based on pyridine's electron-withdrawing hyperconjugative $\pi-\sigma_{CF}^{*}$ impact (Scheme 3.84) (see Section 2.2.1.1).

3.11.4. Fluoropyrroles and Indoles

Being electron-rich aromatics, fluorine nuclei on fluoropyrroles are generally somewhat shielded relative to those of fluorobenzenes, but the chemical shifts can be significantly affected by further substitution (Scheme 3.85). Also, fluorines at the 2-position are considerably deshielded relative to those at the 3-position.

Scheme 3.83

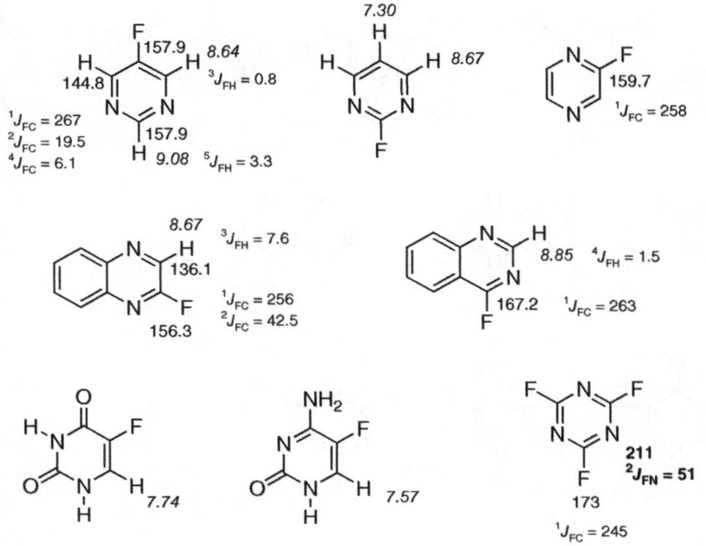

Scheme 3.84

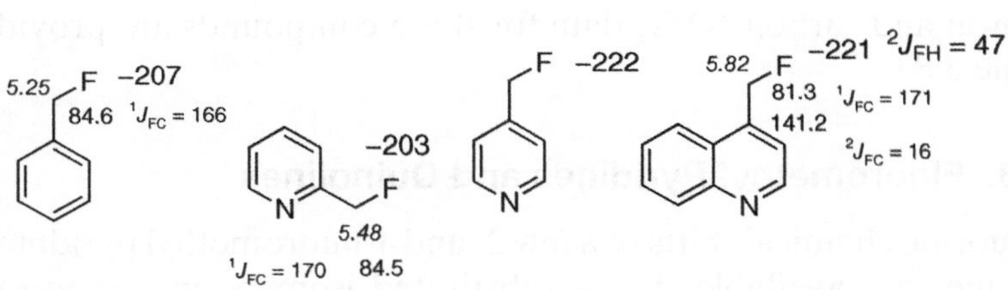

Scheme 3.85

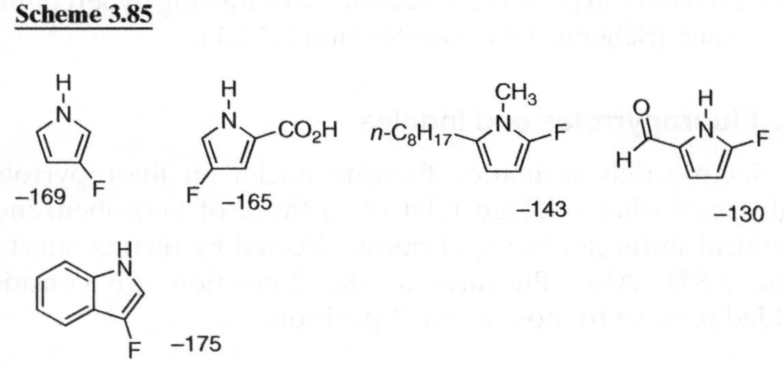

3.11.4.1. Carbon and Proton NMR Data for Fluoropyrroles and Indoles.
Some typical proton and carbon chemical shift and coupling constant data are provided in Scheme 3.86.

Scheme 3.86

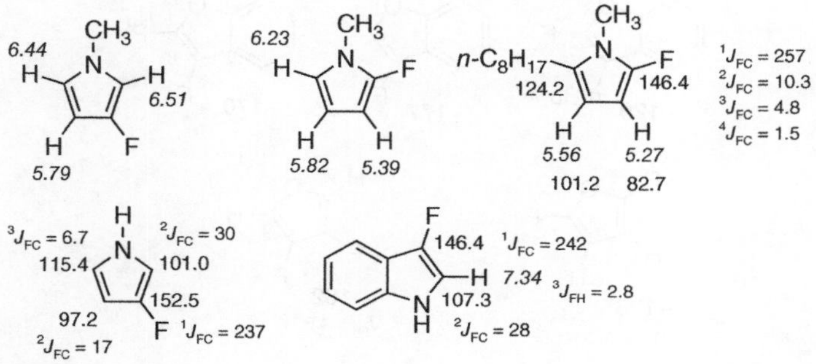

3.11.5. Fluoromethyl Pyrroles and Indoles

The only such a compound for which NMR data are reported is the indole given in Scheme 3.87.

Scheme 3.87

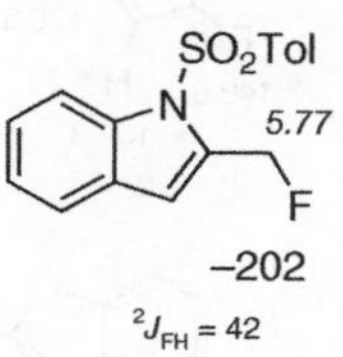

3.11.6. Fluorofurans and Benzofurans

As was the case for pyrroles, fluorine substituents at the 2-position are greatly deshielded relative to those at the 3-position of furans and benzofurans (Scheme 3.88).

The pair of difluorofurans given at the bottom of the same scheme also exhibit the significant difference in fluorine chemical shift between the 2- and the 3-positions of furan.

Scheme 3.88

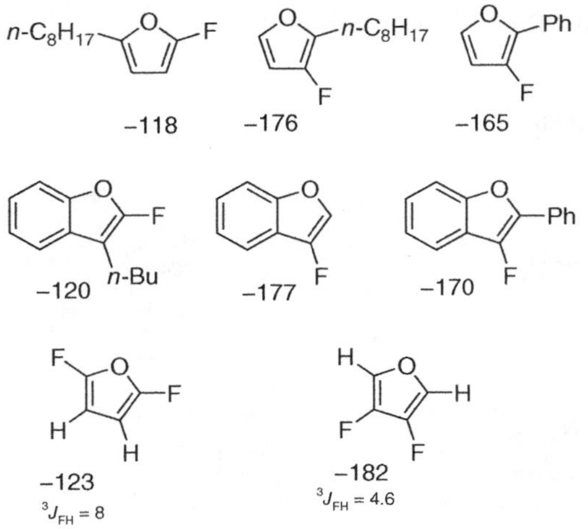

-118   -176   -165

-120   -177   -170

-123   -182
$^3J_{FH} = 8$ $^3J_{FH} = 4.6$

3.11.6.1. Carbon and Proton NMR Data.
Some typical proton and carbon chemical shift and coupling constant data for fluorofurans are provided in Scheme 3.89.

Scheme 3.89

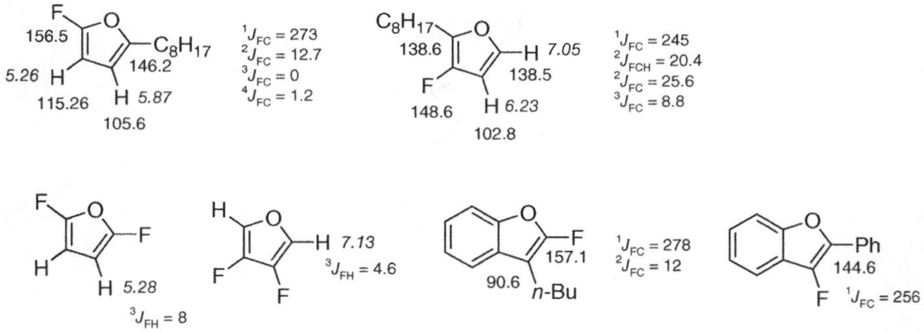

$^1J_{FC} = 273$
$^2J_{FC} = 12.7$
$^3J_{FC} = 0$
$^4J_{FC} = 1.2$

$^1J_{FC} = 245$
$^2J_{FCH} = 20.4$
$^2J_{FC} = 25.6$
$^3J_{FC} = 8.8$

$^3J_{FH} = 8$

$^3J_{FH} = 4.6$

$^1J_{FC} = 278$
$^2J_{FC} = 12$

$^1J_{FC} = 256$

3.11.6.2. Fluorodibenzofurans.
Three fluorodibenzofurans have been reported. Their fluorine NMR data are given in Scheme 3.90.

3.11.7. Fluoromethyl Furans and Benzofurans

The difference between these two similar compounds (Scheme 3.91) regarding their reported fluorine and proton chemical shifts is not readily explainable, even considering the different solvents.

Scheme 3.90

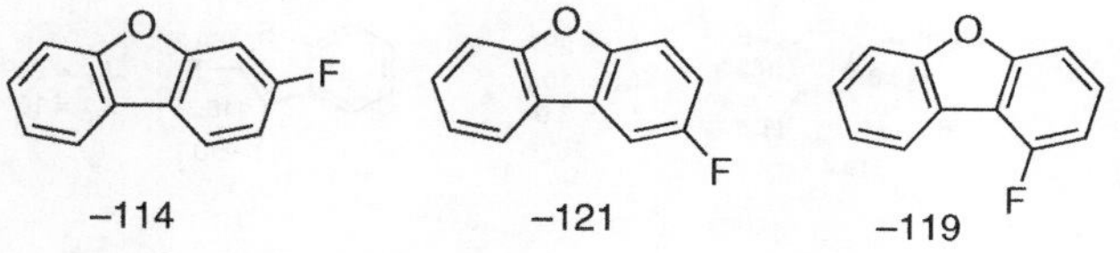
O
F
−114
−121
−119

Scheme 3.91

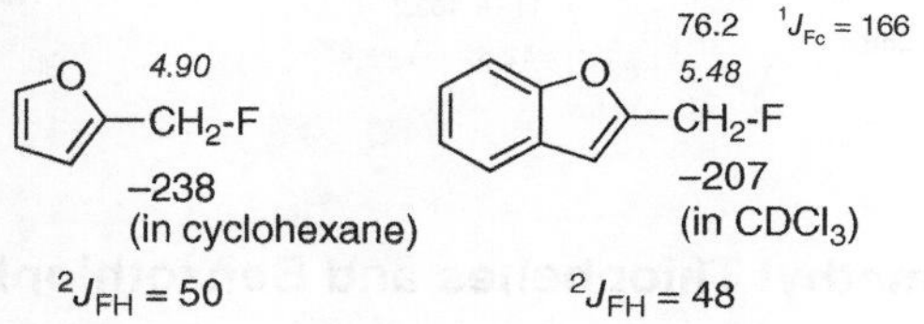
O
4.90
CH2-F
−238
(in cyclohexane)
2JFH = 50
76.2
1JFC = 166
5.48
CH2-F
−207
(in CDCl3)
2JFH = 48

3.11.8. Fluorothiophene and Benzothiophene

Unlike the nitrogen of pyrroles and the oxygen of furans, the sulfur of thiophenes does not significantly affect fluorine chemical shifts either inductively or as an electron donor. Thus, the fluorine chemical shifts of fluorothiophenes are generally in the region of electron-rich fluorobenzenes (Scheme 3.92). Moreover, in this heterocycle, fluorines at the 2-position are slightly shielded compared to those at the 3-position.

Scheme 3.92

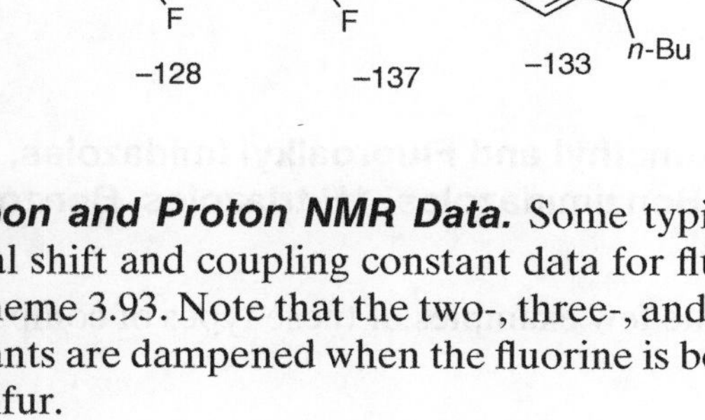
S
F
H3C
−135
−134
−131
C8H17
n-Bu
−128
−137
−133

3.11.8.1. Carbon and Proton NMR Data. Some typical carbon and proton chemical shift and coupling constant data for fluorothiophenes are given in Scheme 3.93. Note that the two-, three-, and four-bond F–C coupling constants are dampened when the fluorine is bound to the carbon bearing sulfur.

Scheme 3.93

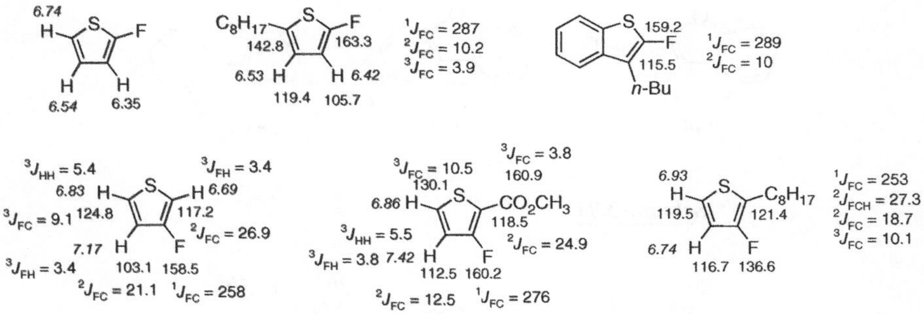

3.11.9. Fluoromethyl Thiophenes and Benzothiophenes

The only example of such a compound is given in Scheme 3.94.

Scheme 3.94

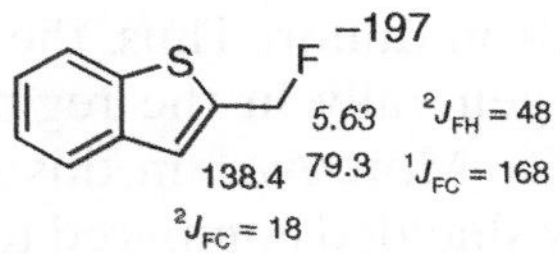

3.11.10. Fluoroimidazoles and Pyrazoles

Fluorine, proton, and carbon NMR spectra for 4-fluoro- and 4,5-di-fluoroimidazole, for 4-fluoropyrazole and substituted 3-fluoropyrazoles have been reported (Scheme 3.95).

Comparing the monofluoropyrazoles with the difluoropyrazoles, note the significant effect of the *second* fluorine on the chemical shifts of both fluorines!

3.11.11. Fluoromethyl and Fluoroalkyl Imidazoles, 1*H*-pyrazoles, Benzimidazoles, 1*H*-triazoles, Benzotriazoles, and Sydnones

NMR data for the few examples of these types of compounds are given in Scheme 3.96.

Scheme 3.95

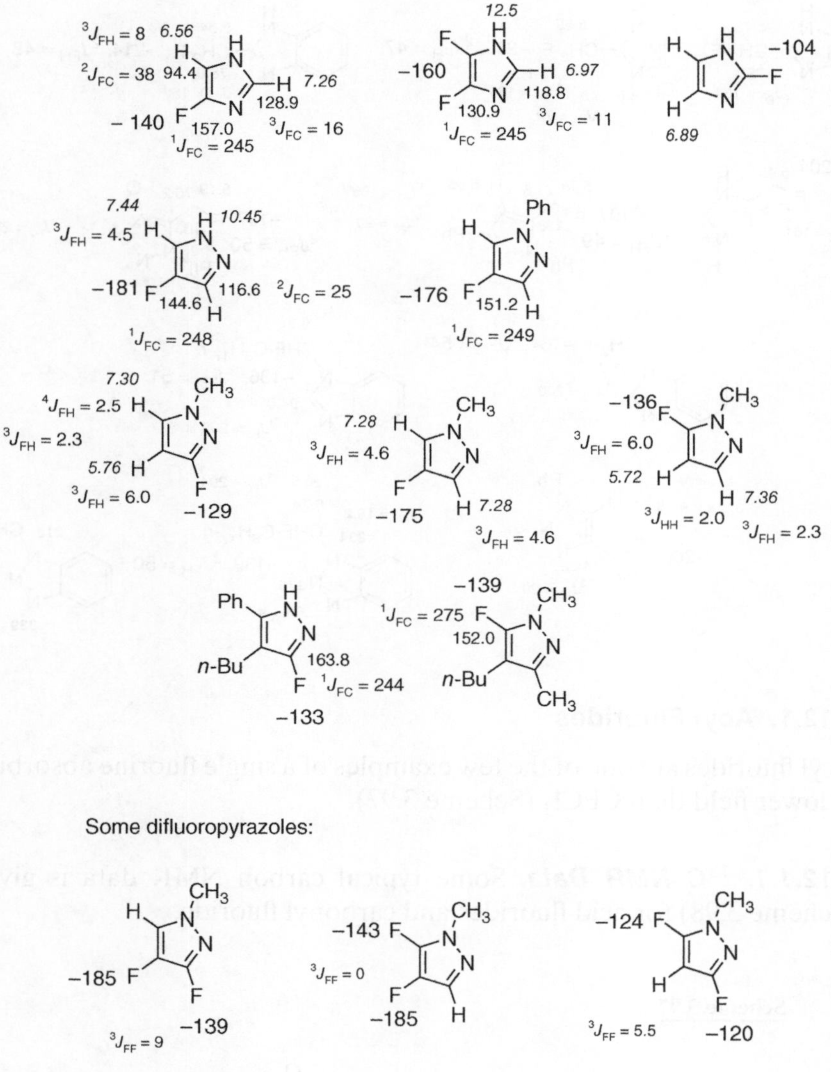

Some difluoropyrazoles:

3.12. OTHER COMMON GROUPS WITH A SINGLE FLUORINE SUBSTITUENT

Two other functional groups that contain a single fluorine substituent are acyl fluorides and sulfonyl fluorides. Such fluorines are among the rare few that absorb in the highly deshielded region downfield of $CFCl_3$.

Scheme 3.96

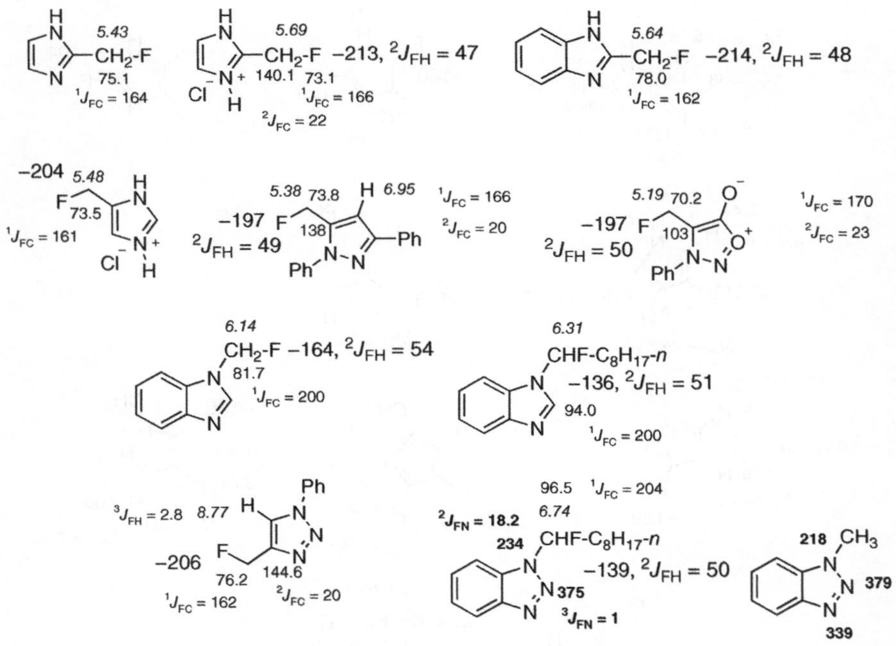

3.12.1. Acyl Fluorides

Acyl fluorides are one of the few examples of a single fluorine absorbing at lower field than $CFCl_3$ (Scheme 3.97).

3.12.1.1. ^{13}C NMR Data. Some typical carbon NMR data is given (Scheme 3.98) for acid fluorides and carbonyl fluoride.

Scheme 3.97

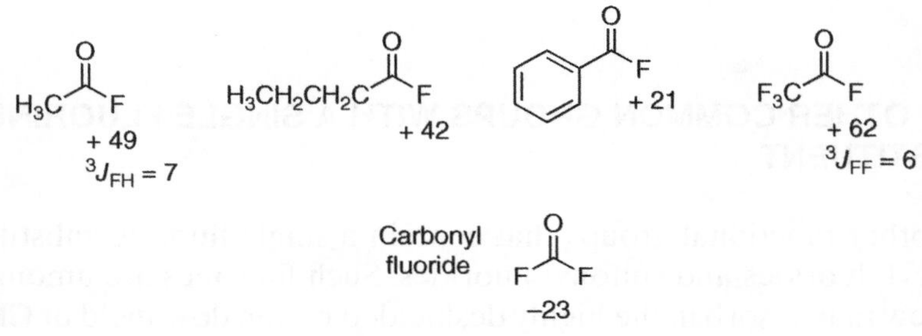

Scheme 3.98

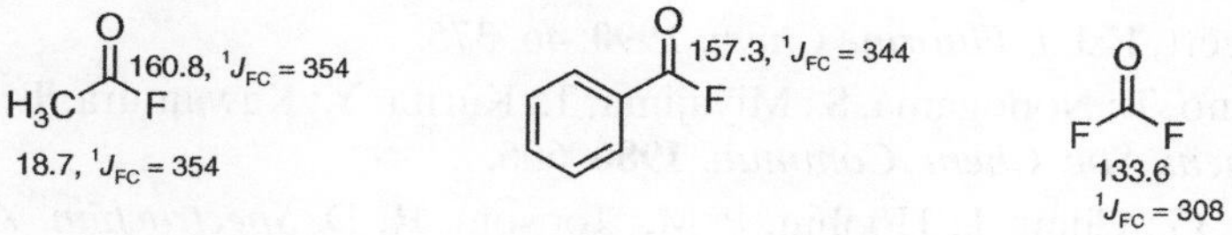

3.12.2. Fluoroformates

This class of compounds is exemplified by the data for methyl fluoroformate (Scheme 3.99).

Scheme 3.99

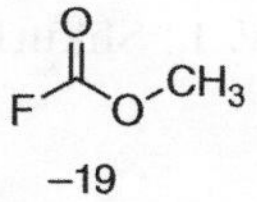

3.12.3. Sulfinyl and Sulfonyl Fluorides

Chapter 6 provides a more thorough coverage of compounds with fluorine bound directly to sulfur, but typical examples of sulfinyl and sulfonyl fluorides are given in Scheme 3.100.

Scheme 3.100

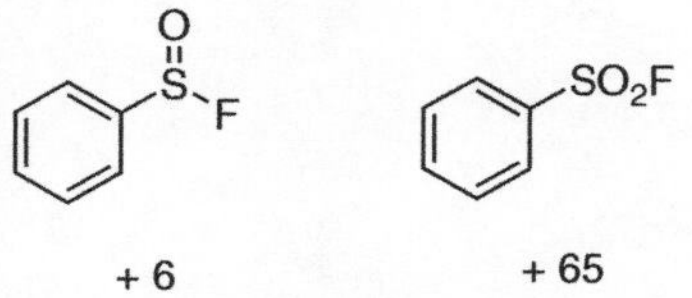

REFERENCES

1. Wiberg, K. B.; Zilm, K. W. *J. Org. Chem.* **2001**, *66*, 2809.
2. Durie, A. J.; Slawin, A. M. Z.; Lebl, T.; Kirsch, P.; O'Hagan, D. *Chem. Commun.* **2011**, *47*, 8265.
3. Giuffredi, O. T.; Jennings, L. E.; Bernet, B.; Gouverneur, V. *J. Fluorine Chem.* **2011**, *132*, 772.

4. Bradshaw, T. K.; Hine, P. T.; Della, E. W. *Org. Magn. Reson.* **1981**, *16*, 26.

5. Brey, W. S. *Magn. Res. Chem.* **2008**, *46*, 480.

6. Weigert, F. J. *J. Fluorine Chem.* **1990**, *46*, 375.

7. Hirano, T.; Nonogama, S.; Miyajima, T.; Kurita, Y.; Kawamura, T.; Sato, H. *J. Chem. Soc. Chem. Commun.* **1986**, 606.

8. Butt, G.; Cilmi, J.; Hoobin, P. M.; Topsom, R. D. *Spectrochim. Acta Part* **A1980**, *36A*, 521.

9. Simonnin, M.-P. *J. Organometal. Chem.* **1966**, *5*, 155.

10. Fifolt, M. J.; Sojka, S. A.; Wolfe, R. A.; Hojnicki, D. S.; Bieron, J. F.; Dinon, F. J. *J. Org. Chem.* **1989**, *54*, 3019.

11. Lutnaes, B. F.; Luthe, G.; Brinkman, U. A. T.; Johansen, J. E.; Krane, J. *Magn. Reson. Chem.* **2005**, *43*, 588.

12. Gribble, G. W.; Keavy, D. J.; Olson, E. R.; Rae, I. D.; Staffa, A.; Herr, T. E.; Ferrara, M. B.; Contreras, R. H. *Magn. Reson. Chem.* **1991**, *29*, 422.

13. Denisov, A. Y.; Mamatyuk, V. I.; Shkurko, O. P. *Magn. Res. Chem.* **1985**, *23*, 482.

CHAPTER 4

THE CF$_2$ GROUP

4.1. INTRODUCTION

The difluoromethylene (CF$_2$) group is encountered in a large variety of environments within molecules of pharmaceutical and agrochemical interest. It imparts significant effects upon the acidity and basicity of proximate OH and NH functions, and it has significant effects upon the reactivity of most organic functional groups because of fluorine's potent inductive, electron-withdrawing power. Thus, the CF$_2$ group can function as an important structural component within a molecule, having a significant impact upon that molecule's chemical and biological activity.[1] A few illustrative examples of bioactive compounds that contain a difluoromethylene group are (a) the difluoro analog of thromboxane A$_2$ (**4-1**), which enhances the bioactivity of the powerful vasoconstrictor and platelet-aggregating nonfluorinated compound, while greatly enhancing its hydrolytic stability; (b) the fluorinated prostaglandin antifertility drug 16,16-difluoro PGE$_1$ (**4-2**), in which the CF$_2$ group enhances the acidity and inhibits the metabolic oxidation of its neighboring OH group, while enhancing the activity of the compound, (c) the fungicide, fludioxonil (**4-3**), (d) the herbicide, primisulfuron-methyl (**4-4**), the CF$_2$ groups of which impart multiple

Guide to Fluorine NMR for Organic Chemists, Second Edition. William R. Dolbier, Jr.
© 2016 John Wiley & Sons, Inc. Published 2016 by John Wiley & Sons, Inc.

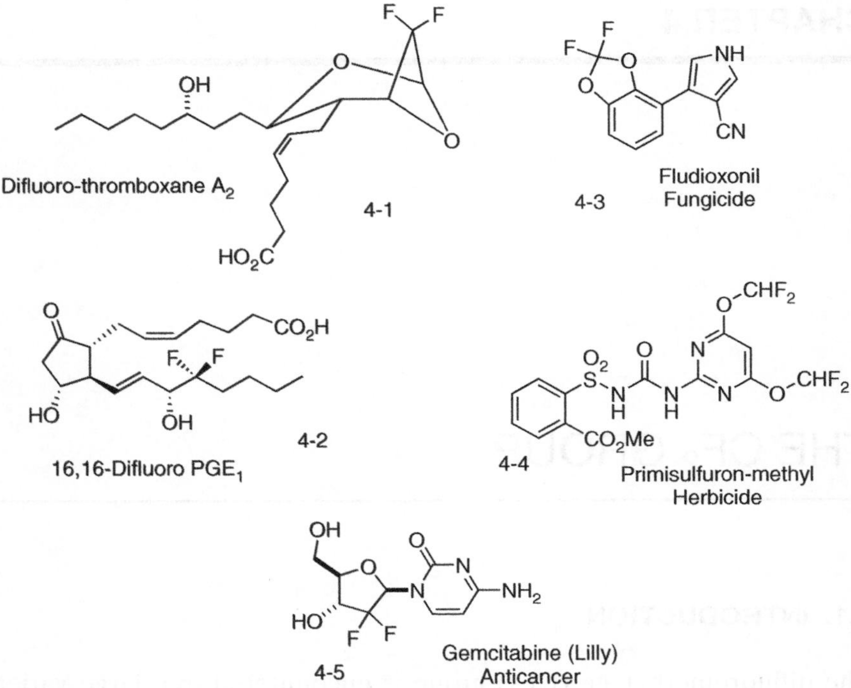

FIGURE 4.1. Examples of bioactive compounds containing a CF$_2$ group

beneficial effects upon their efficacy as agrochemicals, and (e) the anticancer drug Gemcitabine (**4-5**) (Figure 4.1).

Therefore, it is quite common for organic chemists with an interest in designing and synthesizing novel bioactive compounds to consider how a well-placed CF$_2$ group might help them attain their goals. This will require that they be able (a) to synthesize and (b) to characterize the structures of new compounds that they prepare, which contain CF$_2$ groups in specific structural environments. Explicit interpretation of the ^{19}F NMR spectra of such compounds usually will provide definitive structural characterization, especially when combined with ^{1}H and ^{13}C NMR spectra. The chemical shift and coupling constant data contained in this chapter should be all that one needs to derive the most information possible from fluorine, proton, and carbon NMR spectra of compounds containing a CF$_2$ group.

4.1.1. Chemical Shifts – General Considerations

The CF$_2$ group exhibits a wide range of chemical shift possibilities, with, for example, difluoromethane compounds appearing at both extremes, upfield at -143.6 ppm for CF_2H_2 and downfield at $+18.6$ ppm for CF_2I_2.

However, realistically, most CF_2 groups that one would encounter in molecules of synthetic interest to organic chemists have ^{19}F chemical shifts that lie within a much narrower range of -80 to -120 ppm. In fact, the lack of sensitivity to structural environment of a CF_2 group can sometimes be quite surprising, a good example being the similarity of chemical shifts for the vinylic and alkyl CF_2 groups in $CF_2=CH_2$ (δ -81.8) and $CH_3CF_2CH_3$ (δ -84.5), respectively.

In the different classes of CF_2 compounds, those bound to saturated and unsaturated carbon, those bound to hydrogen, and those bound to heteroatoms and proximate to functional groups, there are predictable trends in chemical shifts.

4.1.2. Spin–Spin coupling Constants – General Considerations

With respect to *spin–spin coupling constants*, although not as large as those to a single fluorine substituent, the normal three-bond H–F spin–spin coupling constants between a CF_2 group and vicinal hydrogens remain quite large and consistent in magnitude (generally between 15 and 22 Hz).

Geminal, two-bond H–F coupling constants for CF_2H groups are *larger* than those seen for CH_2F groups and are usually in the range of 57 Hz.

One-bond F–C coupling constants for $-CF_2-$ or $-CF_2H$ groups are in the 234–250 Hz range, which is characteristically larger than the 160–170 Hz observed for $-CHF-$ or $-CH_2F$ groups, but much smaller than the 275–285 Hz observed for CF_3 groups.

Two-bond F–F coupling constants between diastereotopic fluorines in a CF_2 group can be quite variable. They can be as small as 14 Hz for some vinylic $C=CF_2$ groups (see Section 4.7.1), of moderate magnitude ($\sim$150 Hz) for cyclopropyl CF_2 groups, or as large as 240–285 Hz for diastereotopic, acyclic CF_2 groups.

4.2. SATURATED HYDROCARBONS CONTAINING A CF_2 GROUP[2]

The rules governing trends in fluorine chemical shifts for hydrocarbon CF_2 groups are virtually the same as those that governed monofluoroalkanes, with difluoromethane having the most shielded fluorines, *primary CF_2 groups* (that is *CF_2H groups*) being downfield approximately

Scheme 4.1

Primary CF$_2$H groups

CH$_3$CF$_2$H. −110   CH$_3$CH$_2$CF$_2$H, −120, $^2J_{HF}$ = 57, $^3J_{HF}$ = 17.5

n-C$_7$H$_{15}$CF$_2$H   −116, $^2J_{HF}$ = 57, $^3J_{HF}$ = 18

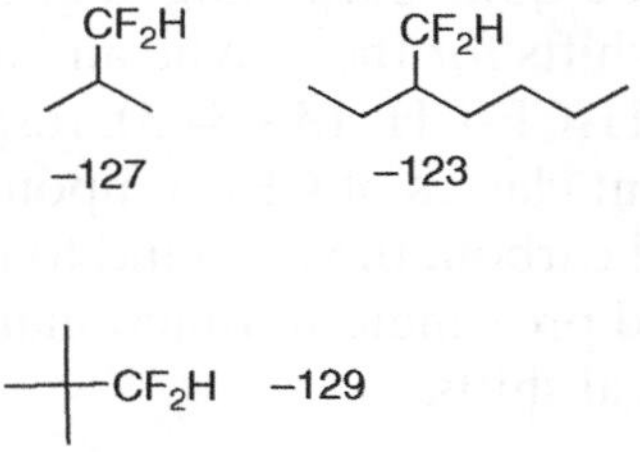

CF$_2$H   CF$_2$H

−127   −123

CF$_2$H   −129

25–30 ppm, and secondary CF$_2$ groups (i.e., those bound to two alkyl groups) being further deshielded by 20–30 ppm (Scheme 4.1). Again, branching of the chain near the CF$_2$ group increases the shielding of both 1° and 2° CF$_2$ groups (that is more negative chemical shifts).

4.2.1. Alkanes Bearing a Primary CF$_2$H Group

Typical ^{19}F chemical shifts for *n*-C$_n$H$_{2n+1}$CF$_2$H compounds are between −116 and −117 ppm, with the usual branching effects being observed. Thus, a CF$_2$H group attached to a secondary carbon is more shielded (7–10 ppm), and the one attached to a tertiary carbon is still further upfield.

The coupling constants given in Scheme 4.1 are typical two- and three-bond F–H values for such systems, and as mentioned earlier, such two-bond F–H coupling constants within CF$_2$H groups are almost invariably around 57 Hz regardless of the environment. Three-bond F–H coupling constants between CF$_2$H groups and vicinal hydrogens are usually in the 17–20 Hz range.

When an ^{19}F NMR spectrum is obtained for typical compounds containing a CF$_2$H group, scanning the usual huge range for fluorine NMR, from about 0 to −220 ppm, the signals will generally look like singlets (in spite of the typical FH coupling constant of 58 Hz), as can be seen in Figure 4.2, for the spectrum of 1,1-difluorobutane. However, when one expands the region of the signal, one can clearly see both the larger two-bond HF coupling and the smaller three-bond HF coupling, as depicted in the expanded inset in Figure 4.2. The chemical shift and

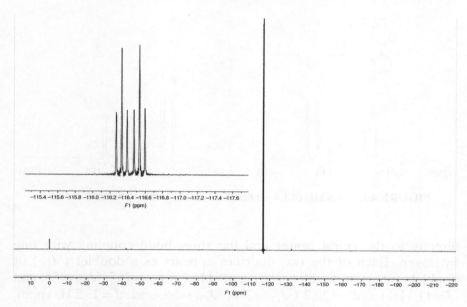

FIGURE 4.2. ^{19}F NMR spectrum of 1,1-difluorobutane

Scheme 4.2

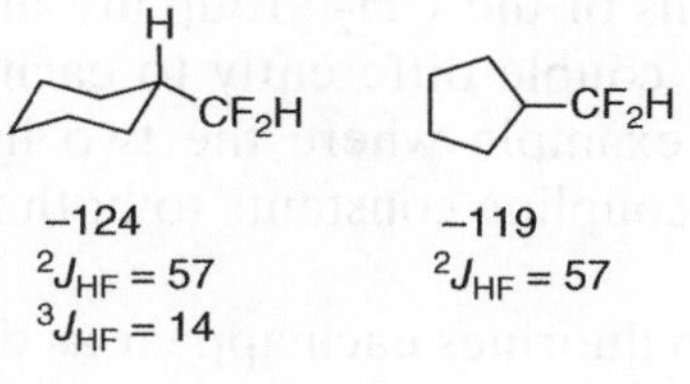

−124
$^2J_{HF} = 57$
$^3J_{HF} = 14$

−119
$^2J_{HF} = 57$

coupling constant data for this fluorine spectrum of 1,1-difluorobutane are as follows: δ −116.4 (d of t, $^2J_{FH} = 58\,Hz$, $^3J_{FH} = 16.6\,Hz$).

When the CF$_2$H is attached to a carbocyclic ring, its chemical shift is not significantly affected (Scheme 4.2) as compared to the analogous acyclic systems above.

4.2.1.1. Coupling for CF$_2$H Groups Within Chiral Compounds.
An interesting coupling situation arises when a CF$_2$H group is proximate to a chiral center in a molecule. In such cases, the two fluorines become diastereotopic and appear as two distinct signals, with a large two-bond coupling between them. A simple example of this is shown in Figure 4.3, the ^{19}F NMR spectrum of 1,4-diphenyl-2-(difluoromethyl)butan-1,4-dione, where the CF$_2$H group is attached

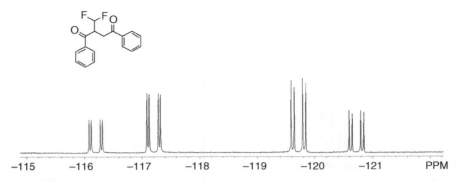

FIGURE 4.3. ^{19}F NMR of 1,4-diphenyl-2-(difluoromethyl)butan-1,4-dione

directly to the chiral center and has three-bond coupling with one hydrogen. Each of the two fluorines appears as a doublet ($^2J_{FF}$) of doublets ($^2J_{FH}$) of doublets ($^3J_{FH}$): δ −116.7 ($^2J_{FF} = 285$, $^2J_{FH} = 56$, and $^3J = 11.3$ Hz) and −120.2 ($^2J_{FF} = 285$, $^2J_{FH} = 56$, and $^3J = 15.2$ Hz) ppm. Note that the three-bond H–F coupling is different for each of the two fluorines of the CF$_2$H group.

When the CF$_2$H is more distant from the chiral center and is attached to a CH$_2$ group, the spectrum can become more complicated, because now the two hydrogens of the CH$_2$ group are also diastereotopic and they may or may not couple differently to each of the two fluorines. Figure 4.4 shows an example where the two hydrogens fortuitously couple with identical coupling constants to both fluorines of the CF$_2$H group.

In this case, the two fluorines each appear as doublets ($^2J_{FF}$) of doublets ($^2J_{FH}$) of *triplets* ($^3J_{FH}$) at −116.3 and −118.9, each with coupling constants of 289, 57, and 16 Hz.

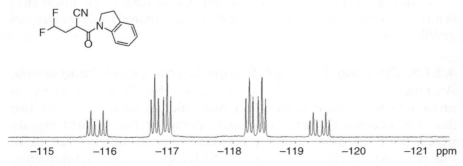

FIGURE 4.4. ^{19}F NMR spectrum of 2-cyano-4,4-difluoro-1-(1H-indol-1-yl)butan-1-one

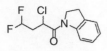

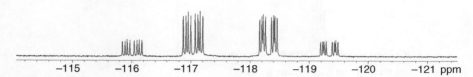

FIGURE 4.5. ^{19}F NMR spectrum of 2-chloro-4,4-difluoro-1-(1H-indol-1-yl)butan-1-one

If one simply changes the cyano group of this example to a chloro substituent, the spectrum becomes more complicated, with different coupling to each of the fluorines by each of the β-hydrogens, giving rise to the spectrum shown in Figure 4.5, where the two fluorines appear as doublets ($^2J_{FF}$) of doublets ($^2J_{FH}$) of doublets ($^3J_{FHA}$) of doublets ($^3J_{FHB}$): δ −116.5 ($^2J_{FF} = 290$, $^2J_{FH} = 57$, $^3J_{FHA} = 23$ and $^3J_{FHB} = 13$ Hz) and −118.8 ($^2J_{FF} = 290$, $^2J_{FH} = 57$, $^3J_{FHA} = 16$ and $^3J_{FHB} = 10$ Hz). Note that in this case both three-bond H–F coupling constants are different for each of the fluorines in the CF$_2$H group.

4.2.2. Secondary CF$_2$ Groups

Alkanes that contain an internal CF$_2$ group exhibit a significant down-field shift (15–20 ppm) in their ^{19}F NMR spectra compared to those compounds bearing a CF$_2$H group, typically absorbing at about −102 ppm. Again, there is a significant shielding impact due to branching as can be seen in Scheme 4.3.

Scheme 4.3

Secondary CF$_2$ groups

CH$_3$CF$_2$CH$_3$ −84.5, $^3J_{HF} = 18$

CH$_3$CH$_2$CF$_2$CH$_3$ −93.3

CH$_3$CH$_2$CF$_2$CH$_2$CH$_3$ −102.4

(CH$_3$)$_3$CCF$_2$CH$_3$ −102.2

(CH$_3$)$_2$CHCF$_2$CH(CH$_3$)$_2$ −120

A typical ^{19}F NMR spectrum of a compound with a secondary CF$_2$ group, that of 2,2-difluoropentane, is given in Figure 4.6. Its single signal appears as a sextet at −90.8 ppm with identical three-bond coupling of 18.6 Hz to both its vicinal CH$_2$ and CH$_3$ protons.

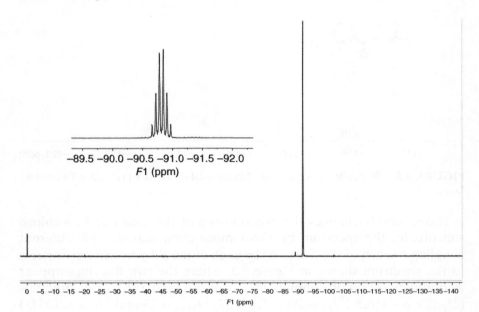

FIGURE 4.6. ^{19}F NMR spectrum of 2,2-difluoropentane

CF$_2$ groups within a *carbocyclic ring system* are unremarkable, generally absorbing slightly downfield from those contained in a straight chain acyclic system, with the remarkable exception of cyclopropane systems, the fluorines of which exhibit a characteristic (~40 ppm) upfield shift to absorb at about −139 ppm (Scheme 4.4).

Scheme 4.4

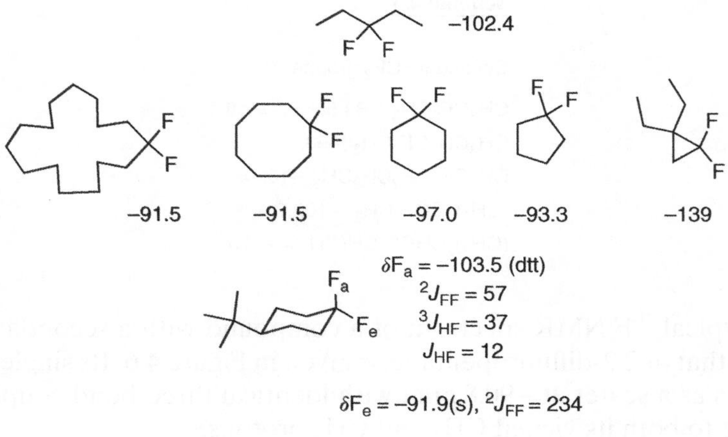

Note the very large difference in chemical shift between the diastereotopic axial and equatorial fluorines in the rigid 4-*t*-butyl-1, 1-difluorocyclohexane system (11.6 ppm), with axial fluorines being more shielded. Of course, because of the presence of the *t*-butyl group, the axial and equatorial fluorines cannot interchange via a ring flipping process. However, in 1,1-difluorocyclohexane itself, the interchange of axial and equatorial fluorines can be readily examined via *dynamic NMR*. In spite of the fact that the energy required for ring flip in fluorinated cyclohexanes is not much different from that of cyclohexane itself ($\Delta G^{\ddagger} \approx 10$ kcal/mol), the relatively huge difference in axial–equatorial fluorine chemical shifts causes the CF$_2$ group of 1,1-difluorocyclohexane to exhibit broadening, even at room temperature, with coalescence occurring at a much higher temperature than for nonfluorine-containing systems. Equation 4.1 defines the relationship between $\Delta G^{\ddagger}$, coalescence temperature, and chemical shift difference of the equilibrating nuclei.

$$\Delta G^{\ddagger}(\text{in kcal/mol}) = 4.57 \times 10^{-3} \cdot T_c(9.97 + \log T_c - \log|\nu_A - \nu_B|) \quad (4.1)$$

The temperature dependence of the ^{19}F NMR spectrum of 1,1-difluoro-cyclohexane is shown in Figure 4.7. With an observed $\Delta\nu$ of 15.3 ppm

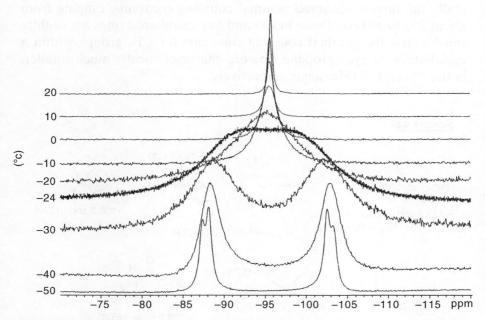

FIGURE 4.7. Temperature dependence of ^{19}F NMR spectrum of 1,1-difluoro-cyclohexane

(4315 Hz) and a coalescence temperature (T_c) of 249 K, the $\Delta G^{\ddagger}$ for ring flipping of 1,1-difluorocyclohexane is calculated to be 9.9 kcal/mol. One can see the axial–equatorial AB system emerging in the −50 °C spectrum.

4.2.3. Discussion of Coupling Constants Within CF$_2$ Groups

Three-bond F–H coupling constants between CF$_2$ fluorines and vicinal hydrogens are usually in the 18–20 Hz range (Scheme 4.5).

<u>**Scheme 4.5**</u>

$$CH_3CF_2CH_3 (^3J_{HF} = 17.8\,Hz)\quad CH_3CF_2CH_2CH_2CH_2CH_3 (^3J_{HF} = 18.6\,Hz)$$

4.2.3.1. Coupling Between Diastereotopic Fluorines in CF$_2$ Groups. The examples in Scheme 4.6 provide further insight into trends observed for the geminal coupling constants between CF$_2$ fluorine atoms when they are in a diastereotopic environment. Diastereotopic fluorines in an *acyclic* CF$_2$ group appear to have generally the largest observed geminal coupling constants, ranging from about 250 to 290 Hz. Those in six- and five-membered rings are slightly smaller, but the geminal coupling constants for CF$_2$ groups within a cyclobutane or cyclopropane ring are characteristically much smaller, in the 190 and 150 Hz range, respectively.

<u>**Scheme 4.6**</u>

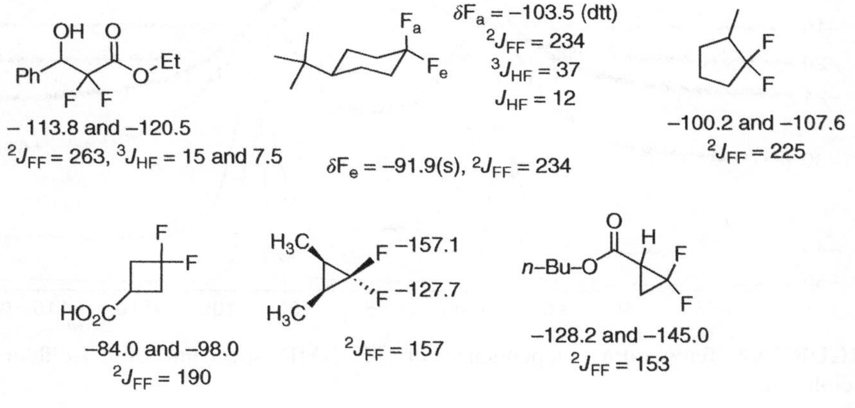

OH O
Ph⟍⟍⟍O⁻Et
F F
− 113.8 and −120.5
$^2J_{FF}$ = 263, $^3J_{HF}$ = 15 and 7.5

δF_a = −103.5 (dtt)
$^2J_{FF}$ = 234
$^3J_{HF}$ = 37
J_{HF} = 12
δF_e = −91.9(s), $^2J_{FF}$ = 234

−100.2 and −107.6
$^2J_{FF}$ = 225

−84.0 and −98.0
$^2J_{FF}$ = 190

F −157.1
F −127.7
$^2J_{FF}$ = 157

−128.2 and −145.0
$^2J_{FF}$ = 153

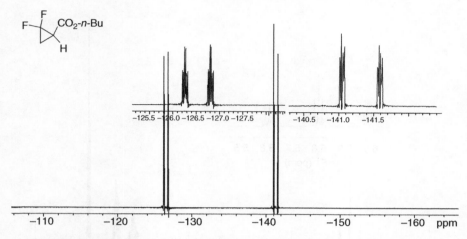

FIGURE 4.8. ^{19}F NMR spectrum of n-butyl 2,2-difluorocyclopropanecarboxylate

Although Figure 4.8 shows the AB system of 1,1-difluorocyclohexane emerging, Figure 4.5 provides a classic example of a CF_2 AB system, which derives from two diastereotopic fluorines in n-butyl 2,2-difluorocyclopropanecarboxylate. In this case, $^2J_{AB} = 153\,Hz$.

4.2.4. Pertinent ^1H Chemical Shift Data

The protons of CH_2F_2 appear at 4.70 ppm, whereas those of CF_2H groups at the terminus of a straight-chained saturated hydrocarbon appear almost invariably at about 5.79 ppm. When the CF_2H group is attached to a secondary carbon, it appears at a slightly higher field. $^2J_{FH}$ coupling constants for such systems are always about 56–58 Hz (Scheme 4.7).

Scheme 4.7

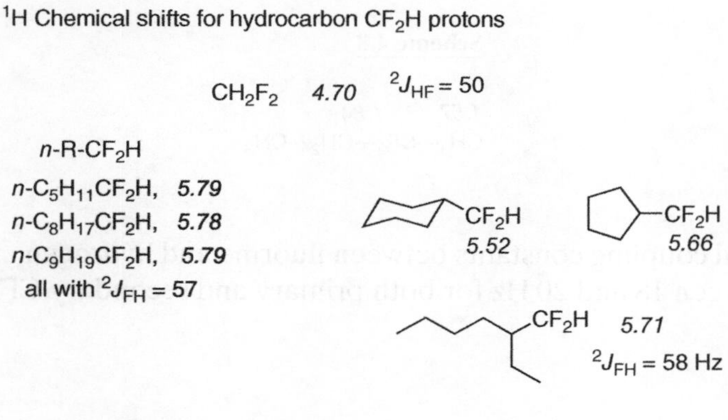

^1H Chemical shifts for hydrocarbon CF_2H protons

CH_2F_2 *4.70* $^2J_{HF} = 50$

n-R-CF_2H

n-$C_5H_{11}CF_2H$, *5.79*
n-$C_8H_{17}CF_2H$, *5.78*
n-$C_9H_{19}CF_2H$, *5.79*
all with $^2J_{FH} = 57$

CF_2H
5.52

CF_2H
5.66

CF_2H *5.71*
$^2J_{FH} = 58\,Hz$

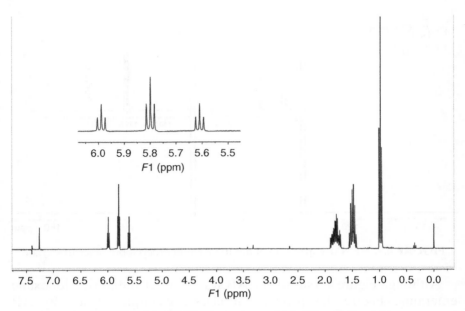

FIGURE 4.9. ^{1}H NMR spectrum of 1,1-difluorobutane

A typical proton NMR spectrum for a primary CF$_2$H system, that of 1,1-difluorobutane, is given in Figure 4.9. Note the characteristically large two-bond F–H coupling constant of 57 Hz, along with the small three-bond H–H coupling constant of 4.5 Hz depicted in the inserted expansion. Complete details as to chemical shift and coupling constant data for this proton spectrum of 1,1-difluorobutane are as follows: δ 0.98 (t, $^3J_{HH} = 7.2$ Hz, 3H), 1.49 (sextet, $^3J_{HH} = 7.5$ Hz, 2H), 1.80 (m, 2H), 5.80 (t of t, $^2J_{FH} = 57$ Hz, $^3J_{HH} = 4.5$ Hz, 1H).

Typical chemical shifts for CH$_3$ and CH$_2$ groups contiguous to a CF$_2$ group are given in Scheme 4.8.

Scheme 4.8

$$\begin{array}{cc} 1.57 & 1.84 \\ \end{array}$$
$$CH_3\text{—}CF_2\text{—}CH_2\text{—}CH_3$$

Vicinal coupling constants between fluorine and hydrogen are generally between 18 and 20 Hz for both primary and secondary CF$_2$ groups.

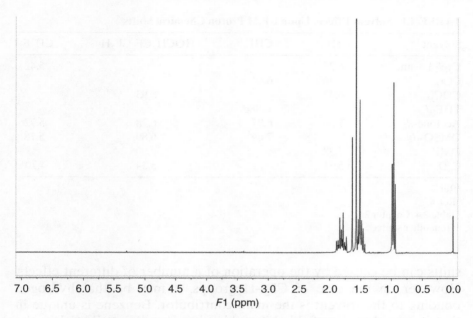

FIGURE 4.10. ^{1}H NMR spectrum of 2,2-difluoropentane

On the other hand, H–H coupling constants between vicinal hydrogens are much smaller in these compounds, between 4 and 8 Hz.

Figure 4.10 provides a typical ^{1}H NMR spectrum of a hydrocarbon containing a secondary CF$_2$ group. In this spectrum, one can distinguish the triplet at 1.58 ppm deriving from the C-1 methyl group, which is coupled to two adjacent fluorines of the CF$_2$ group with a characteristically large 18.6 Hz three-bond F–H coupling constant. Note the triplet at δ 0.96 deriving from the C-5 methyl group, which has the characteristically much smaller three-bond H–H coupling constant of 7.5 Hz. Complete details as to chemical shift and coupling constant data for this proton spectrum of 2,2-difluoropentane are as follows: δ 0.96 (t, $^3J_{HH} = 7.5$ Hz, 3H), 1.51 (sextet, $^3J_{HH} = 7.8$ Hz, 2H), 1.58 (t, $^3J_{FH} = 18.6$ Hz, 3H), 1.81 (m, 2H).

4.2.4.1. Solvent Effects on –CF$_2$H Protons.
Significant solvent effects on proton chemical shifts are observed for $-$CF$_2$H protons (even greater for CF$_3$H). As exemplified in Table 4.1, a significant downfield shift is observed for such protons in proceeding from non-polar to polar solvents. Such influence of solvent on proton chemical

TABLE 4.1. Solvent Effects Upon CF$_2$H Proton Chemical Shifts

Solvent	CHF$_3$[a]	CHF$_3$[b]	HOCH$_2$CF$_2$CF$_2$H[c]	CH$_2$F$_2$[b]
Cyclohexane	6.25			5.42
CCl$_4$	6.46	6.44		
CDCl$_3$	6.47		5.93	
THF-d_4		6.88e[d]		
Acetone-d_6	7.04	6.97	6.28	5.72
DMSO-d_6		7.09	6.89	5.78
DMF	7.32			
C$_6$D$_6$	5.31		5.34	4.70[d]

[a]Ref. 3.
[b]Ref. 4.
[c]Table 2.4, Chapter 2.
[d]From other sources.

shifts can be caused by the operation of a number of different effects. However, in the case of CF$_2$–H groups, it may be that hydrogen bonding to the solvent is the main contributor. Benzene is unique in that it provides an upfield shift, which is generally attributed to the formation of a 1:1 complex with the CF$_2$–H bond aligned with the sixfold symmetry axis of the benzene ring.

There are only small and for most purposes negligible solvent effects upon the CF$_2$H *fluorine* chemical shifts and the F–H coupling constants.

4.2.5. Pertinent ^{13}C NMR Data

Both the ^{13}C chemical shifts and the F–C coupling constants for the CF$_2$ carbons are quite characteristic in value, as can be seen from the following examples (Scheme 4.9). A review article on ^{13}C NMR spectra of fluorinated cyclopropanes is available.[5]

Typical examples of such spectra, those of 1,1-difluorobutane and 2,2-difluoropentane are given in Figures 4.11 and 4.12, respectively. The chemical shift and F–C coupling constant data for each are given below the respective spectra.

δ 117.59 (t, $^1J_{FC} = 239$ Hz), 36.23 (t, $^2J_{FC} = 20$ Hz), 15.84 (t, $^3J_{FC} = 6.0$ Hz), 13.79 (s)

δ 124.57 (t, $^1J_{FC} = 238$ Hz), 40.25 (t, $^2J_{FC} = 25$ Hz), 23.39 (t, $^2J_{FC} = 28$ Hz), 16.46 (t, $^3J_{FC} = 4.8$ Hz), 14.11 (s)

Scheme 4.9

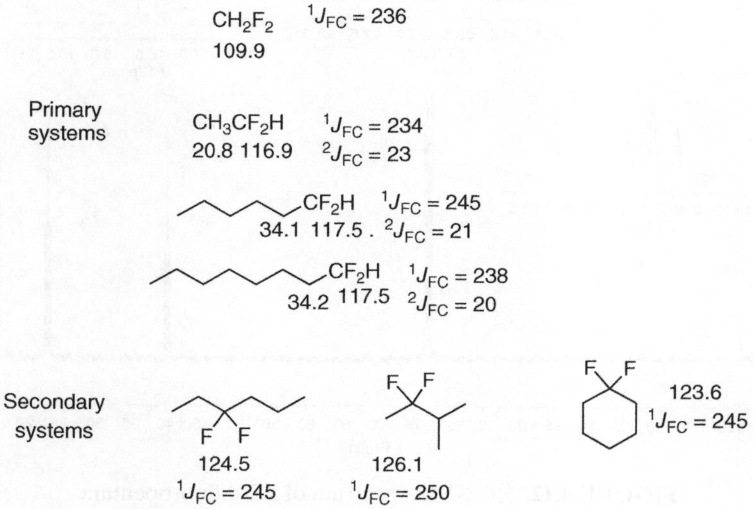

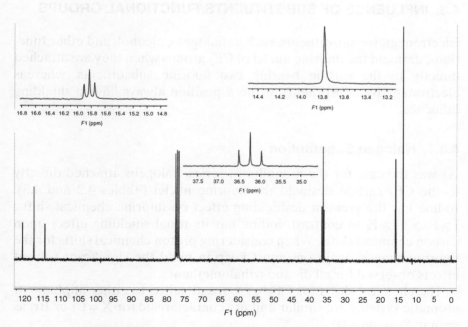

CH_2F_2 $^1J_{FC} = 236$
109.9

Primary systems CH_3CF_2H $^1J_{FC} = 234$
20.8 116.9 $^2J_{FC} = 23$

CF₂H $^1J_{FC} = 245$
34.1 117.5 . $^2J_{FC} = 21$

CF₂H $^1J_{FC} = 238$
34.2 117.5 $^2J_{FC} = 20$

Secondary systems

124.5
$^1J_{FC} = 245$

126.1
$^1J_{FC} = 250$

123.6
$^1J_{FC} = 245$

FIGURE 4.11. ¹³C NMR spectrum of 1,1-difluorobutane

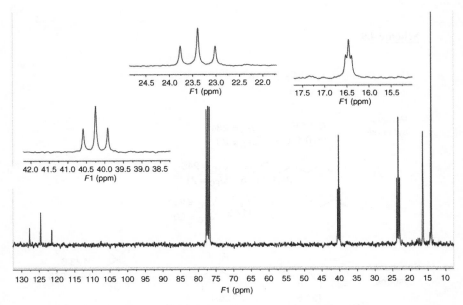

FIGURE 4.12. ^{13}C NMR spectrum of 2,2-difluoropentane

4.3. INFLUENCE OF SUBSTITUENTS/FUNCTIONAL GROUPS

Electronegative substituents, such as halogens, alcohol, and ether functions, deshield the fluorine nuclei of CF$_2$ groups when they are attached directly to the carbon bearing two fluorine substituents, whereas electronegative substituents at the β-position always have a shielding influence.

4.3.1. Halogen Substitution

As was the case for the monofluoro series, halogens attached directly to the CF$_2$ carbon deshield the fluorine nuclei (Tables 4.2 and 4.3). Iodine has the greatest deshielding effect on fluorine chemical shifts: I > Br > Cl > F. In contrast, iodine has its usual shielding effect upon carbon chemical shifts. When considering proton chemical shifts for the fluoromethanes, again one must keep in mind the significant solvent effects observed for all di- and trihalomethanes.

The chemical shifts for CF$_2$X groups attached to either aliphatic or aromatic systems are similar and are characteristic for X = Cl or Br, as seen in Scheme 4.10.

TABLE 4.2. ^{19}F, ^{1}H, and ^{13}C Chemical Shifts of X-CF$_2$H Compounds

			X-CF$_2$H			
X	H	CH$_3$	F	Cl	Br	I
δ	−144	−110	−78	−73	−69	−67
δ_H	5.42	5.75	6.47	7.79	8.05	7.62
$^2J_{HF}$	50	57	79	62	60	56
δ_C	109.4	116.9	118.4	118.0	a	83.4
$^1J_{FC}$	238	234	272	288	a	308

^{a}Unavailable.

TABLE 4.3. ^{19}F Chemical Shifts of X$_2$CF$_2$ Compounds, δ (ppm)

			X$_2$CF$_2$			
X	H	CH$_3$	F	Cl	Br	I
δ	−144	−84.5	−64.6	−6.8	+6.3	+16.3

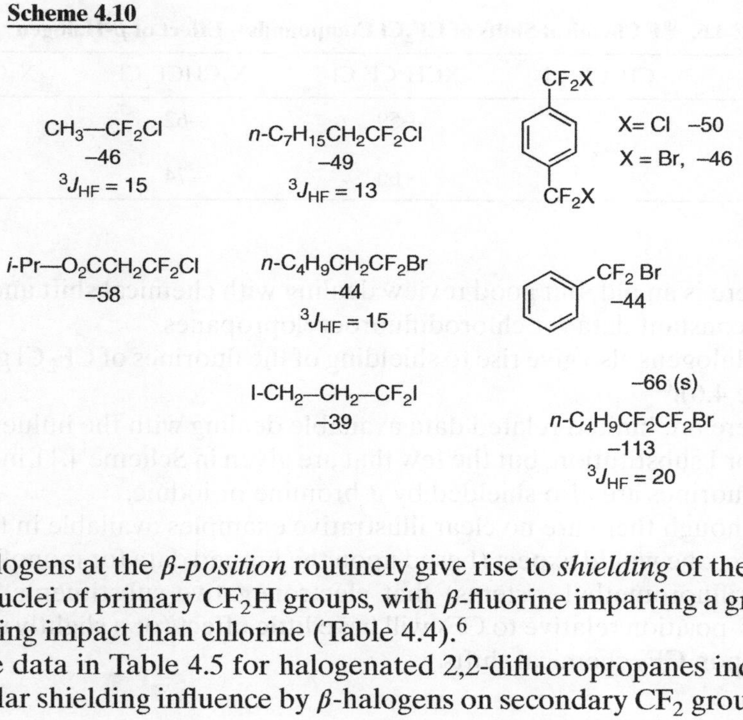

Scheme 4.10

CH$_3$—CF$_2$Cl
−46
$^3J_{HF}$ = 15

n-C$_7$H$_{15}$CH$_2$CF$_2$Cl
−49
$^3J_{HF}$ = 13

CF$_2$X / CF$_2$X (para-disubstituted benzene)
X= Cl −50
X = Br, −46

i-Pr—O$_2$CCH$_2$CF$_2$Cl
−58

n-C$_4$H$_9$CH$_2$CF$_2$Br
−44
$^3J_{HF}$ = 15

CF$_2$Br (benzene)
−44

I-CH$_2$—CH$_2$—CF$_2$I
−39

−66 (s)
n-C$_4$H$_9$CF$_2$CF$_2$Br
−113
$^3J_{HF}$ = 20

Halogens at the *β-position* routinely give rise to *shielding* of the fluorine nuclei of primary CF$_2$H groups, with β-fluorine imparting a greater shielding impact than chlorine (Table 4.4).[6]

The data in Table 4.5 for halogenated 2,2-difluoropropanes indicate a similar shielding influence by β-halogens on secondary CF$_2$ groups.[7, 8]

TABLE 4.4. ^{19}F Chemical Shifts of Primary CF$_2$H Compounds – Effect of β-Halogen

	CH$_3$CF$_2$H	XCH$_2$CF$_2$H	X$_2$CHCF$_2$H	X$_3$CCF$_2$H
X = Cl		−120	−124	−122
	−110			
X = F		−130	−138	−142
			Also: CHClFCF$_2$H	
			AB −131, −132	

TABLE 4.5. ^{19}F Chemical Shifts of Secondary CF$_2$ Groups – Effect of β-Halogen

	CH$_3$CF$_2$CH$_3$	XCH$_2$CF$_2$CH$_3$	X$_2$CHCF$_2$CH$_3$	X$_3$CCF$_2$CH$_3$
X = Cl		−95	−98	−100
	−85			
X = F		−103	−109	−111
				$^3J_{H,F} =$
				18.7 Hz

CF$_3$CF$_2$CH$_2$CH$_3$ −121 (appears in rightmost area of X = F row)

TABLE 4.6. ^{19}F Chemical Shifts of CF$_2$Cl Compounds – Effect of β-Halogen

	CH$_3$CF$_2$Cl	XCH$_2$CF$_2$Cl	X$_2$CHCF$_2$Cl	X$_3$CCF$_2$Cl
X = Cl		−59	−62	−65
	−47			
X = F		−66	−74	−75

There is an old, but good review dealing with chemical shift and coupling constant data for chlorodifluorocyclopropanes.[9]

β-Halogens also give rise to shielding of the fluorines of CF$_2$Cl groups (Table 4.6).[6]

There are limited related data available dealing with the influence of β-Br or I substitution, but the few that are given in Scheme 4.11 indicate that fluorines are also shielded by β-bromine or iodine.

Although there are no clear illustrative examples available in the literature, one would expect (based upon the limited data for monofluoro- and trifluoromethyl systems) that electronegative substituents at the γ- or δ-position relative to CF$_2$ will have little effect (or a slightly shielding) upon CF$_2$ chemical shifts.

4.3.1.1. ^{1}H and ^{13}C NMR Data.
Proton and carbon data are presented for some compounds bearing the CF$_2$Cl or CF$_2$Br groups (Scheme 4.12).

Scheme 4.11

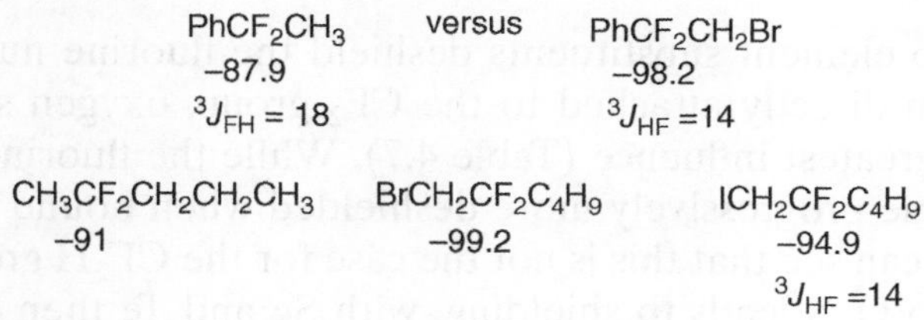

$$PhCF_2CH_3 \quad versus \quad PhCF_2CH_2Br$$
$$-87.9 \qquad\qquad\qquad -98.2$$
$$^3J_{FH} = 18 \qquad\qquad\qquad ^3J_{HF} = 14$$

$$CH_3CF_2CH_2CH_2CH_3 \qquad BrCH_2CF_2C_4H_9 \qquad ICH_2CF_2C_4H_9$$
$$-91 \qquad\qquad\qquad -99.2 \qquad\qquad\qquad -94.9$$
$$\qquad\qquad\qquad\qquad\qquad\qquad\qquad\qquad ^3J_{HF} = 14$$

Scheme 4.12

$$2.33 \qquad\qquad\qquad\qquad 2.10$$
$$n\text{-}C_4H_9CH_2CF_2Br \qquad CH_3{-}CF_2Cl \quad ClCH_2{-}CF_2Cl \quad ^2J_{FC} = 290$$
$$44.3 \quad 123.3 \qquad\qquad 47.3 \quad 126.5 \quad ^2J_{FC} = 30$$
$$^2J_{FC} = 21 \quad ^1J_{FC} = 304$$

$$125.3 \quad 125.7$$
$$\qquad\qquad CF_2Cl$$
$$\qquad\qquad 139.2$$

ClF_2C —⟨ring⟩

$$^1J_{FC} = 308$$
$$^2J_{FC} = 27$$
$$^3J_{FC} = 5$$

$$124.9 \quad 117.3$$
$$\qquad\qquad CF_2Br$$
$$\qquad\qquad 140.7$$

BrF_2C —⟨ring⟩

$$^1J_{FC} = 302$$
$$^2J_{FC} = 24$$
$$^3J_{FC} = 5$$

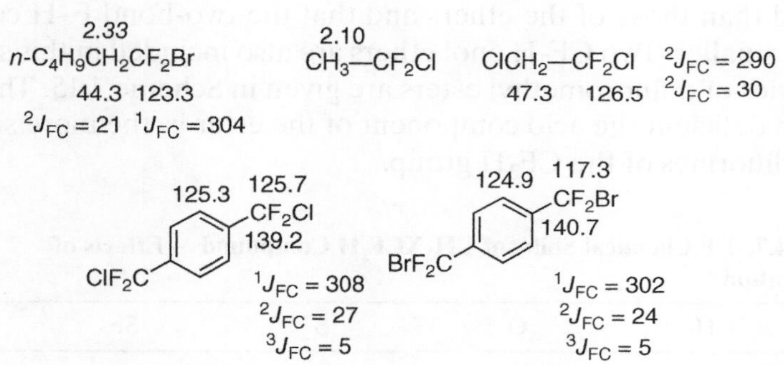

Notable are the much larger one-bond F–C coupling constants for CF_2X than CF_2H, which can probably be attributed to the greater degree of s character of the carbon orbitals bound to fluorine in CF_2Cl than to CF_2H fluorines.

Data related to the effect of β-halogen on the *proton* and carbon chemical shifts of CF_2H or CF_2 groups are scarce (Scheme 4.13), although there is a review of ^{13}C spectra of chlorofluorocyclopropanes.[5]

Scheme 4.13

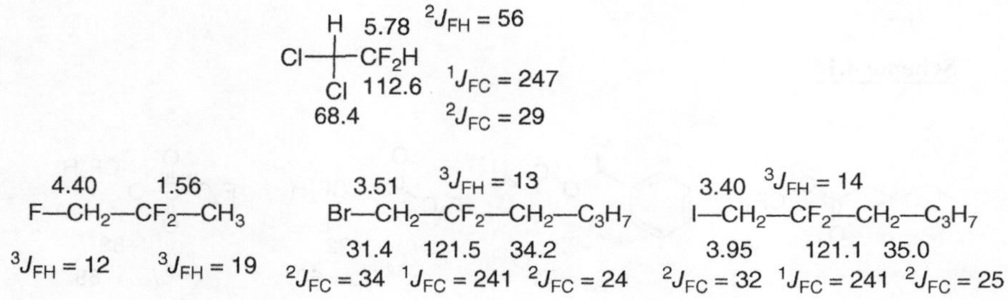

$$H \quad 5.78 \quad ^2J_{FH} = 56$$
$$Cl{-}\!\!\!|{-}CF_2H$$
$$Cl \quad 112.6 \quad ^1J_{FC} = 247$$
$$68.4 \qquad\quad ^2J_{FC} = 29$$

$$4.40 \qquad 1.56 \qquad\qquad 3.51 \quad ^3J_{FH} = 13 \qquad\qquad 3.40 \quad ^3J_{FH} = 14$$
$$F{-}CH_2{-}CF_2{-}CH_3 \qquad Br{-}CH_2{-}CF_2{-}CH_2{-}C_3H_7 \qquad I{-}CH_2{-}CF_2{-}CH_2{-}C_3H_7$$
$$^3J_{FH} = 12 \quad ^3J_{FH} = 19 \qquad 31.4 \quad 121.5 \quad 34.2 \qquad\qquad 3.95 \quad 121.1 \quad 35.0$$
$$\qquad\qquad\qquad\qquad ^2J_{FC} = 34 \; ^1J_{FC} = 241 \; ^2J_{FC} = 24 \quad ^2J_{FC} = 32 \; ^1J_{FC} = 241 \; ^2J_{FC} = 25$$

4.3.2. Alcohol, Ether, Esters, Thioether, and Related Substituents

All Group 6 element substituents deshield the fluorine nuclei of CF$_2$ groups when directly attached to the CF$_2$ group, oxygen substituents having the greatest influence (Table 4.7). While the fluorines of a CF$_3$ group became progressively more deshielded when bound to O, S, Se, and Te, one can see that this is not the case for the CF$_2$H group, where, compared to O, S leads to shielding, with Se and Te then deshielding relative to S, but with all deshielding less than O.

Some examples of alkyl difluoromethyl ethers and a sulfide are given in Scheme 4.14. Note that the fluorines of the sulfide are much more shielded than those of the ethers and that the two-bond F–H coupling is much smaller. Two CF$_2$H enol ethers are also included in this scheme.

A series of difluoromethyl esters are given in Scheme 4.15. The more electron deficient the acid component of the ester is, the more shielded are the fluorines of the CF$_2$H group.

TABLE 4.7. ^{19}F Chemical Shifts of CH$_3$XCF$_2$H Compounds – Effects of α-Substitution

X	CH$_2$	O	S	Se	Te
δ	−120	−86.9	−96.4	−94.4	−91.8

Scheme 4.14

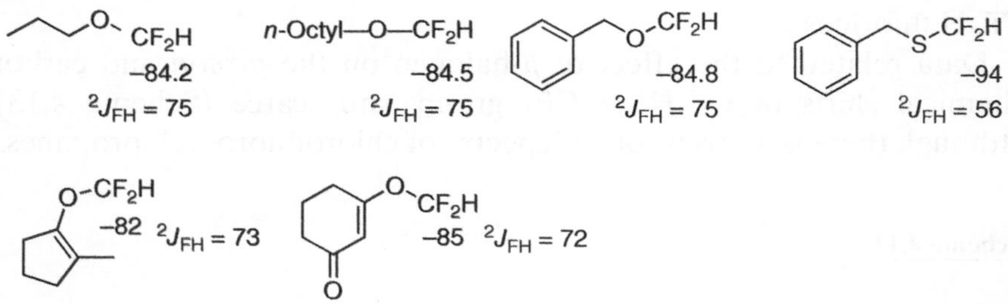

Scheme 4.15

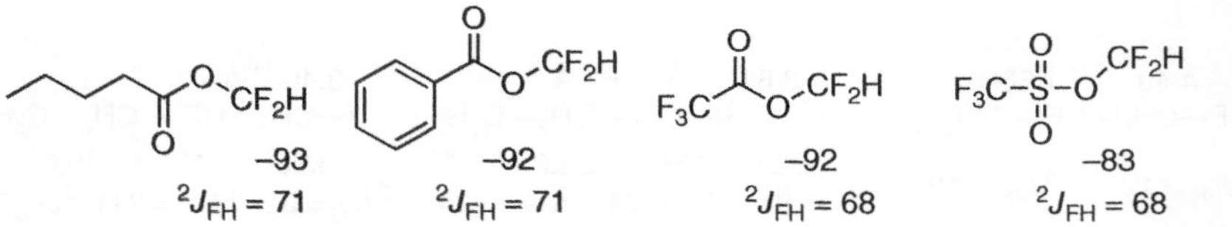

Aryl OCF_2H, SCF_2H, and $SeCF_2H$ compounds have been of much recent synthetic interest and the fluorines of arylOCF_2H are shielded about 10 ppm relative to alkyl analogs. The examples in Scheme 4.16 are representative of the fluorine chemical shifts and coupling constants that should be expected for such compounds. Examples of an alkyl-$S-CF_2H$ compound are also included. Note that the $^2J_{FH}$ coupling constants for OCF_2H compounds are not in the 56–58 Hz range that is characteristic of carbon-bound CF_2H groups but are significantly larger.

Scheme 4.16

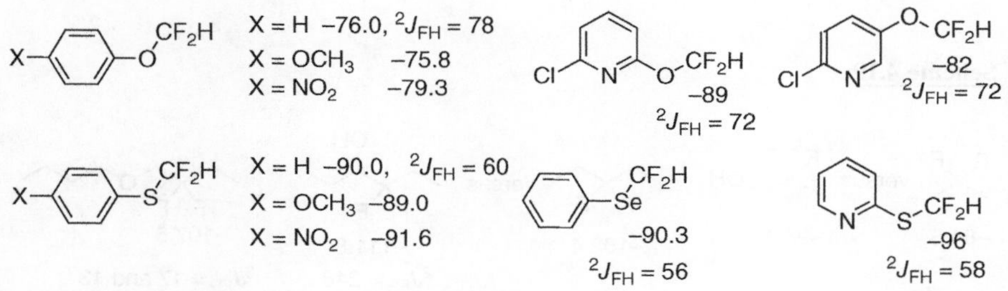

Secondary CF_2 groups are affected similarly by O, S, and Se substitution, as is exemplified by the examples in Scheme 4.17. Again the fluorines of the sulfide are shielded relative to those of the analogous ether.

Scheme 4.17

$$PhOCF_2Ph \qquad CH_3OCF_2Ph \qquad CH_3OCF_2C_7H_{15}$$
$$-66 \qquad\qquad -72 \qquad\qquad\qquad -79$$

$$PhSCF_2Ph$$
$$-72$$

$$PhSeCF_2Ph$$
$$-71$$

As was the case with β-halogens, β-hydroxy groups and ether functions shield both primary CF_2H and secondary CF_2 groups (Schemes 4.18 and 4.19).

Again, one would not expect hydroxy or ether substituents more distant (i.e., γ or δ) to the CF_2 group to have significant effect upon fluorine chemical shifts.

Scheme 4.18

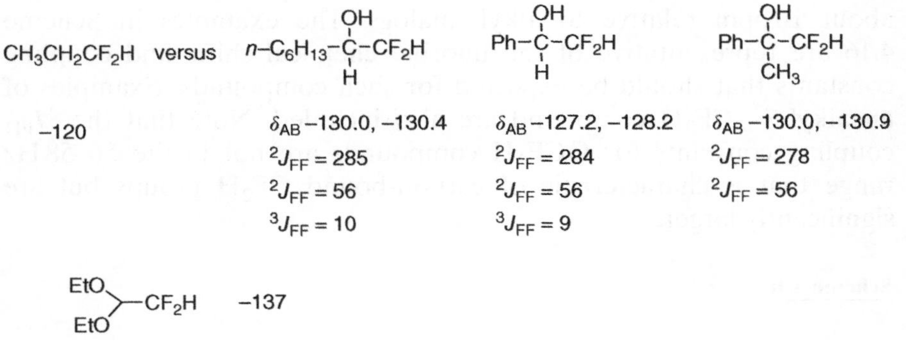

CH$_3$CH$_2$CF$_2$H versus n-C$_6$H$_{13}$-$\overset{\text{OH}}{\underset{\text{H}}{\text{C}}}$-CF$_2$H   Ph-$\overset{\text{OH}}{\underset{\text{H}}{\text{C}}}$-CF$_2$H   Ph-$\overset{\text{OH}}{\underset{\text{CH}_3}{\text{C}}}$-CF$_2$H

-120

δ_{AB} $-130.0, -130.4$ δ_{AB} $-127.2, -128.2$ δ_{AB} $-130.0, -130.9$
$^2J_{FF} = 285$ $^2J_{FF} = 284$ $^2J_{FF} = 278$
$^2J_{FF} = 56$ $^2J_{FF} = 56$ $^2J_{FF} = 56$
$^3J_{FF} = 10$ $^3J_{FF} = 9$

$\overset{\text{EtO}}{\underset{\text{EtO}}{\diagdown}}$-CF$_2$H -137

Scheme 4.19

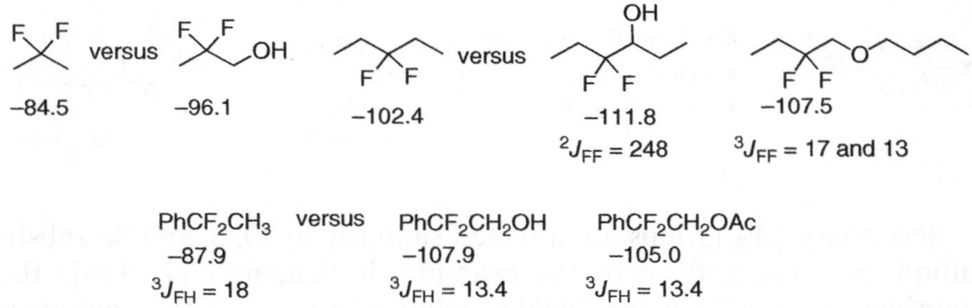

F$_2$C(CH$_3$)$_2$ versus F$_2$C(CH$_3$)CH$_2$OH   versus
-84.5 -96.1 -102.4

$\overset{\text{OH}}{|}$
-111.8 -107.5
$^2J_{FF} = 248$ $^3J_{FF} = 17$ and 13

PhCF$_2$CH$_3$ versus PhCF$_2$CH$_2$OH   PhCF$_2$CH$_2$OAc
-87.9 -107.9 -105.0
$^3J_{FH} = 18$ $^3J_{FH} = 13.4$ $^3J_{FH} = 13.4$

4.3.2.1. ^1H and ^{13}C NMR data.
The chemical shifts of protons of a CF$_2$H group bound to a carbinol carbon do not appear to be significantly affected by the presence of the OH group. Some proton chemical shifts of α-hydroxy CF$_2$H and CF$_2$R compounds are given in Scheme 4.20.

Scheme 4.20

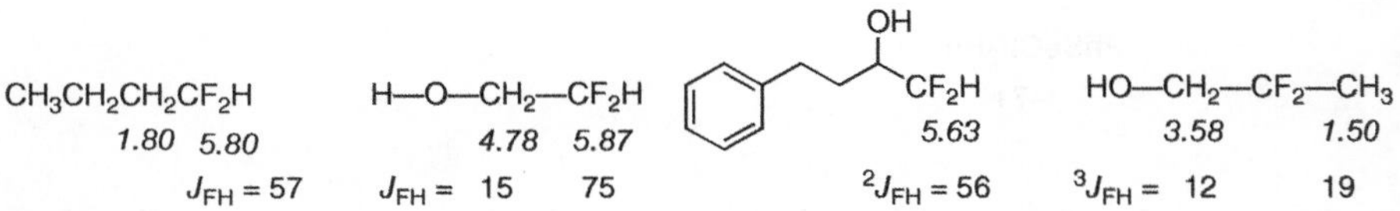

CH$_3$CH$_2$CH$_2$CF$_2$H   H—O—CH$_2$—CF$_2$H   [Ph]CH$_2$CH$_2$$\overset{\text{OH}}{\underset{}{\text{C}}}CF_2$H   HO—CH$_2$—CF$_2$—CH$_3$
1.80 *5.80*   *4.78* *5.87*   *5.63*   *3.58* *1.50*
$J_{FH} = 57$ $J_{FH} =$ *15* *75* $^2J_{FH} = 56$ $^3J_{FH} =$ *12* *19*

The ^1H chemical shifts of CF$_2$H protons of difluoromethyl ethers lie between 6.0 and 6.3 ppm, with a significantly enhanced F–H two-bond coupling constant of around 76 Hz (Scheme 4.21). The protons of

Scheme 4.21

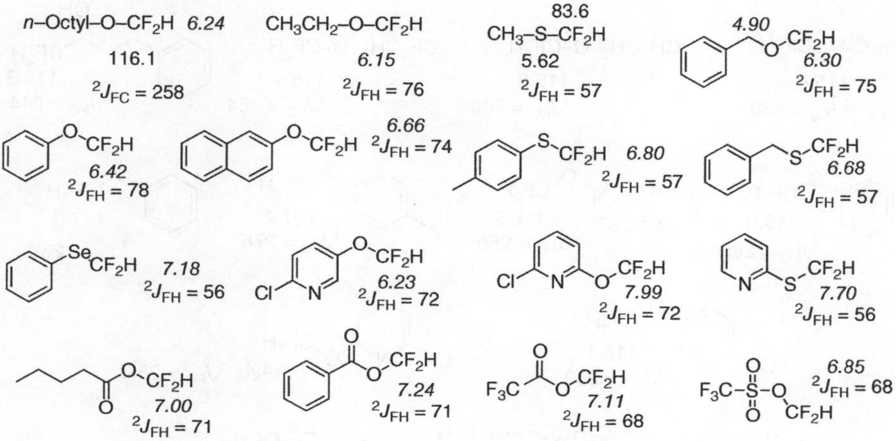

n–Octyl–O–CF$_2$H *6.24*

116.1

$^2J_{FC} = 258$

CH$_3$CH$_2$–O–CF$_2$H

6.15

$^2J_{FH} = 76$

83.6
CH$_3$–S–CF$_2$H
5.62

$^2J_{FH} = 57$

4.90
O–CF$_2$H
6.30

$^2J_{FH} = 75$

O–CF$_2$H
6.42
$^2J_{FH} = 78$

O–CF$_2$H *6.66* $^2J_{FH} = 74$

S–CF$_2$H *6.80*
$^2J_{FH} = 57$

S–CF$_2$H
6.68
$^2J_{FH} = 57$

Se–CF$_2$H *7.18*
$^2J_{FH} = 56$

O–CF$_2$H
6.23
$^2J_{FH} = 72$

O–CF$_2$H
7.99
$^2J_{FH} = 72$

S–CF$_2$H
7.70
$^2J_{FH} = 56$

O–CF$_2$H
7.00
$^2J_{FH} = 71$

O–CF$_2$H
7.24
$^2J_{FH} = 71$

F$_3$C–O–CF$_2$H
7.11
$^2J_{FH} = 68$

F$_3$C–S–O–CF$_2$H
6.85 $^2J_{FH} = 68$

difluoromethyl sulfides appear still farther downfield at about 6.8 ppm with a more "normal" F–H coupling constant of 57 Hz. Indeed, there is a noticeable decreasing trend of two-bond H–F coupling constants of CF$_2$H groups bound to O, S, and Se (78 > 60 > 56 Hz), similar to the trend observed for CF$_2$H groups bound to N and P (65 > 52 Hz) (see Section 4.3.5).

Surprisingly, the ^{13}C *chemical shifts* of CF$_2$H carbons of difluoromethyl ethers and of CF$_2$ carbons of 1,1-difluoroalkyl ethers are almost unchanged compared to those of the analogous nonethers. Compare this to the ~40 ppm downfield incremental shift that is generally observed for a hydrocarbon carbon bearing an ether substituent (Scheme 4.22). However, one can distinguish the ether-bound CF$_2$ groups from the nonether-bound CF$_2$ groups on the basis of the significantly (20–25 Hz) larger one-bond F–C coupling constants of the difluoromethyl ethers. Note also the trend in F–C coupling constants for ethers, thioethers, and selenoethers (260 < 276 < 289 Hz).

4.3.3. Epoxides

There is one example of a fluorine spectrum of an epoxide that incorporates a CF$_2$ group (Scheme 4.23). As was the case with *gem*-difluorocyclopropanes, the three-membered ring of the epoxide gives rise to significant shielding of the CF$_2$ fluorines relative to those of an RCF$_2$OR ether (~−79 ppm).

Scheme 4.22

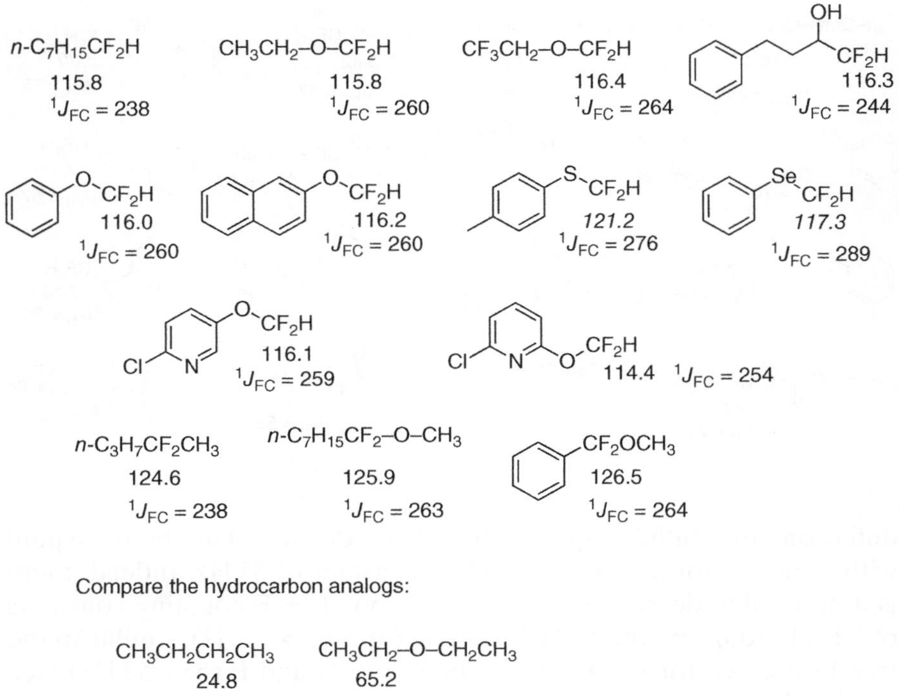

n-C$_7$H$_{15}$CF$_2$H
115.8
$^1J_{FC}$ = 238

CH$_3$CH$_2$–O–CF$_2$H
115.8
$^1J_{FC}$ = 260

CF$_3$CH$_2$–O–CF$_2$H
116.4
$^1J_{FC}$ = 264

OH
CF$_2$H
116.3
$^1J_{FC}$ = 244

O–CF$_2$H
116.0
$^1J_{FC}$ = 260

O–CF$_2$H
116.2
$^1J_{FC}$ = 260

S–CF$_2$H
121.2
$^1J_{FC}$ = 276

Se–CF$_2$H
117.3
$^1J_{FC}$ = 289

O–CF$_2$H
116.1
$^1J_{FC}$ = 259

O–CF$_2$H
114.4 $^1J_{FC}$ = 254

n-C$_3$H$_7$CF$_2$CH$_3$
124.6
$^1J_{FC}$ = 238

n-C$_7$H$_{15}$CF$_2$–O–CH$_3$
125.9
$^1J_{FC}$ = 263

CF$_2$OCH$_3$
126.5
$^1J_{FC}$ = 264

Compare the hydrocarbon analogs:

CH$_3$CH$_2$CH$_2$CH$_3$
24.8

CH$_3$CH$_2$–O–CH$_2$CH$_3$
65.2

Scheme 4.23

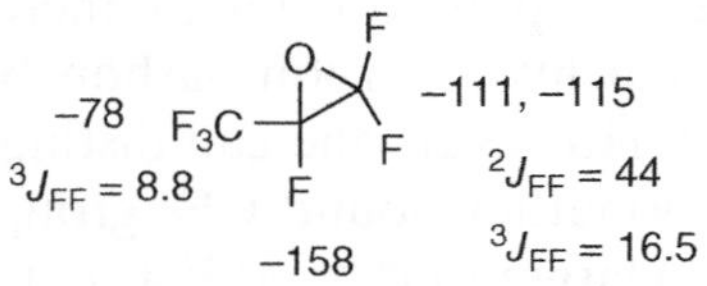

F
O   F
–78   F$_3$C   –111, –115
$^3J_{FF}$ = 8.8   F $^2J_{FF}$ = 44
–158 $^3J_{FF}$ = 16.5

4.3.4. Sulfoxides, Sulfones, Sulfoximines, and Sulfonic Acids

There has been much recent interest in difluoromethyl phenyl sulfones and sulfoximines regarding their use as difluoromethylating agents. Fluorine, proton, and carbon NMR data for these compounds, along with those of analogous sulfide, sulfoxide, and sulfonic acid are provided in Scheme 4.24. No obvious trends are observed for most of the data, the exception is as the oxidation state of the sulfur increases, the CF$_2$H carbons become shielded.

Scheme 4.24

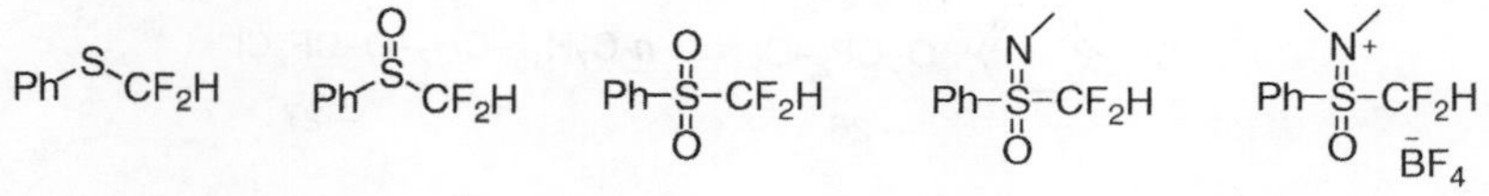

	$Ph\text{-}S\text{-}CF_2H$	$Ph\text{-}S(\!=\!O)\text{-}CF_2H$	$Ph\text{-}S(\!O_2)\text{-}CF_2H$	$Ph\text{-}S(\!=\!N)\text{-}CF_2H$	$Ph\text{-}S(\!=\!N^+)\text{-}CF_2H$ $\bar{B}F_4$
$\delta_F(^2J_{FF})$	-90	-119.6 and -120.3	-122	-118.0 and -120.5	-108.9 and -115.3
$\delta_C(^2J_{FC})$	121.2 (276)	120.9 (289)	114.6 (284)	115.6 (288)	117.2 (296)
$\delta_H(^2J_{FH})$	6.80 (57)	6.04 (55)	6.22 (53)	6.20 (54)	7.70 (52)

	HCF_2SO_2F	HCF_2SO_3H
δ_F	-119 $+37$	-121
$\delta_H(^2J_{FH})$	6.35 (52)	6.67 (52)

Scheme 4.25

$PhCF_2CH(OH)CO_2H$	$PhCF_2C(OH)_2CO_2H$
-104	-110
$^2J_{FF} = 253$	
$^3J_{FH} = 7.6$	

Scheme 4.26

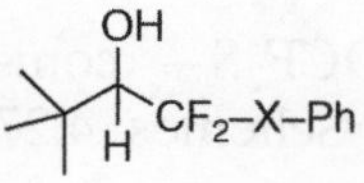

X = S $\delta F(AB) = -72.4$ and -83.5 $(^2J_{FF} = 209)$ $\delta_H = 3.52$, $^3J_{FH} = 18.6$

X = Se $\delta F(AB) = -68.7$ and -81.6 $(^2J_{FF} = 205)$ $\delta_H = 3.60$, $^3J_{FH} = 20.4$

X = Te $\delta F(AB) = -62.6$ and -76.7 $(^2J_{FF} = 218)$ $\delta_H = 3.63$, $^3J_{FH} = 23.6$

4.3.5. Multifunctional β,β-Difluoro Alcohols

Scheme 4.25 provides fluorine NMR data for some 3,3-difluoro-2-hydro-xycarboxylic acids, while Scheme 4.26 gives proton and fluorine data for examples of $RCH(OH)CF_2XPh$, where X = S, Se, and Te.

4.3.6. Compounds with Two Different Heteroatom Groups Attached to CF_2 Including Chloro- and Bromodifluoromethyl Ethers

An increasing number of compounds are prepared that contain two heteroatoms attached to a CF_2 group, in particular, halodifluoromethyl

Scheme 4.27

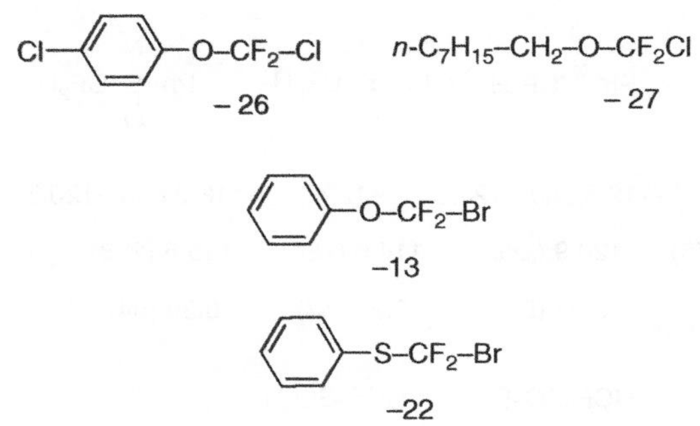

Cl—⟨benzene ring⟩—O–CF$_2$–Cl n-C$_7$H$_{15}$–CH$_2$–O–CF$_2$Cl
               – 26   – 27

⟨benzene ring⟩–O–CF$_2$–Br
               –13

⟨benzene ring⟩–S–CF$_2$–Br
               –22

Scheme 4.28

⟨benzene ring⟩–O-CF$_2$-O–⟨benzene ring⟩
               – 56

⟨benzene ring⟩–S-CF$_2$–S–⟨benzene ring⟩
               – 49

⟨benzene ring⟩–O-CF$_2$-S–⟨benzene ring⟩
               – 43

⟨fluorinated benzodioxole with PPh$_2$⟩ $_2$ – 50

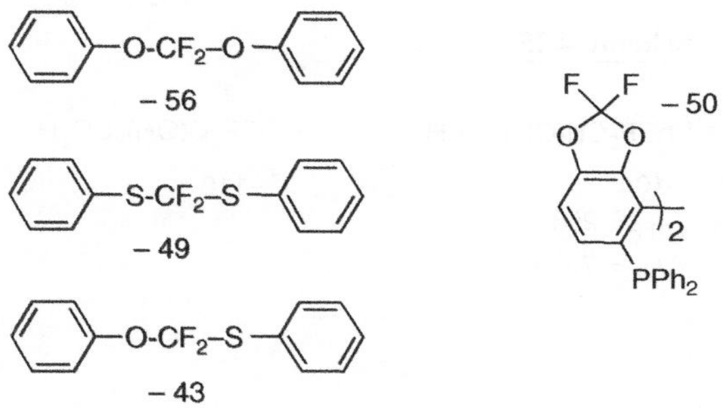

ethers and thioethers, but also $-OCF_2O-$ and $-OCF_2S-$ compounds. Some representative examples are given in Schemes 4.27
and 4.28.

4.3.6.1. ^{13}C NMR Data. Some carbon NMR data for OCF$_2$Cl groups
and others bearing two heteroatoms are provided in Scheme 4.29.

4.3.7. Amines, Azides, and Nitro Compounds

Unlike monofluoro systems, which could not tolerate an amino nitrogen
bound directly to the carbon bearing fluorine, the CF$_2$ group has greater
thermodynamic (and kinetic) stability, and although uncommon and
generally quite reactive, spectra of R$_2$NCF$_2$H compounds are known.
An amino nitrogen gives rise to less deshielding when α-substituted than
any of the Group 6 atoms. There are many more examples of CF$_2$H or
CF$_2$R groups being attached to less basic nitrogens, such as amides or
aromatic nitrogens (Scheme 4.30).

Scheme 4.29

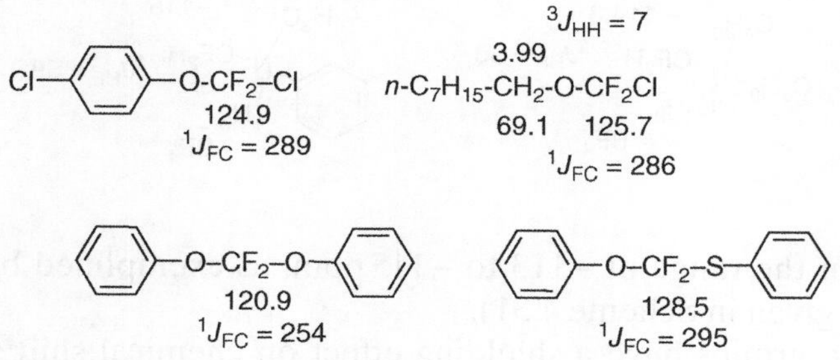

Cl—⟨ ⟩—O-CF$_2$–Cl
124.9
$^1J_{FC} = 289$

$^3J_{HH} = 7$
3.99
n-C$_7$H$_{15}$-CH$_2$-O-CF$_2$Cl
69.1   125.7
$^1J_{FC} = 286$

⟨ ⟩—O-CF$_2$-O—⟨ ⟩
120.9
$^1J_{FC} = 254$

⟨ ⟩—O-CF$_2$-S—⟨ ⟩
128.5
$^1J_{FC} = 295$

Scheme 4.30

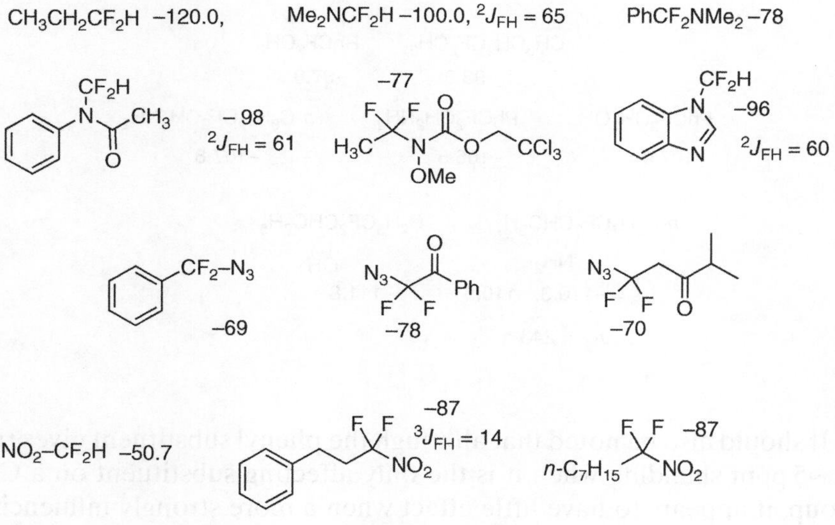

CH$_3$CH$_2$CF$_2$H –120.0,   Me$_2$NCF$_2$H –100.0, $^2J_{FH} = 65$ PhCF$_2$NMe$_2$ –78

CF$_2$H
⟨ ⟩N-CH$_3$ – 98
       ‖ $^2J_{FH} = 61$
       O

–77
F F O
H$_3$C   N   O   CCl$_3$
       |
      OMe

CF$_2$H
N
⟨ ⟩   –96
N $^2J_{FH} = 60$

⟨ ⟩CF$_2$–N$_3$

–69

N$_3$ O
   Ph
F F
–78

N$_3$
   |
   F F   O
–70

NO$_2$–CF$_2$H –50.7

–87
F F
⟨ ⟩ $^3J_{FH} = 14$
   NO$_2$

F F –87
n-C$_7$H$_{15}$ NO$_2$

Examples of a secondary CF$_2$ group bound to amino nitrogen are rare, with a chemical shift reported only for PhCF$_2$N(CH$_3$)$_2$.

Examples of azide bound to a CF$_2$ group also are unusual, but some examples are also provided in Scheme 4.30, as are examples of CF$_2$ bound to a nitro group. Interestingly, the highly electron-withdrawing NO$_2$ group *shields* the CF$_2$ fluorines considerably more than the azide group.

When a CF$_2$H group is attached to an *ammonium* nitrogen, its fluorines are considerably shielded relative to the respective amine. Thus, the fluorine chemical shifts of difluoromethyl trialkyl ammonium

Scheme 4.31

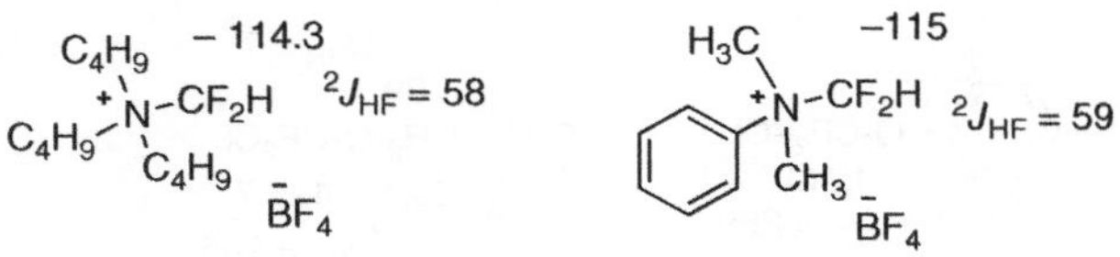

$$
\begin{array}{cc}
\underset{\substack{C_4H_9\\|\\C_4H_9{-}\overset{+}{N}{-}CF_2H\\|\\C_4H_9}}{} \quad -114.3 \;\; ^2J_{HF} = 58 &
\end{array}
$$

$$C_4H_9 \quad -114.3$$
$$\overset{+}{N}{-}CF_2H \quad ^2J_{HF} = 58$$
$$C_4H_9 \;\; C_4H_9$$
$$\overset{-}{BF_4}$$

$$H_3C \quad -115$$
$$\overset{+}{N}{-}CF_2H \quad ^2J_{HF} = 59$$
$$CH_3 \quad \overset{-}{BF_4}$$

salts are in the range of -113 to -115 ppm, as exemplified by the two examples given in Scheme 4.31).

β-amino groups have a shielding effect on chemical shift similar to that of an OH group (Scheme 4.32).

Scheme 4.32

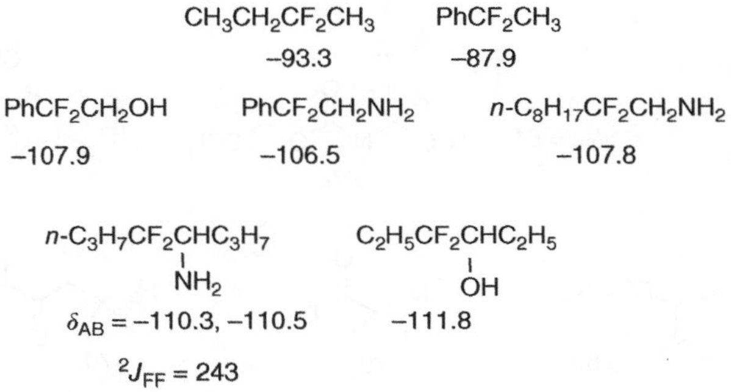

$$CH_3CH_2CF_2CH_3 \qquad PhCF_2CH_3$$
$$-93.3 \qquad\qquad -87.9$$

$$PhCF_2CH_2OH \qquad PhCF_2CH_2NH_2 \qquad n\text{-}C_8H_{17}CF_2CH_2NH_2$$
$$-107.9 \qquad\qquad -106.5 \qquad\qquad\qquad -107.8$$

$$n\text{-}C_3H_7CF_2CHC_3H_7 \qquad C_2H_5CF_2CHC_2H_5$$
$$\qquad\qquad |\qquad\qquad\qquad\qquad\qquad |$$
$$\qquad\quad NH_2 \qquad\qquad\qquad\qquad OH$$
$$\delta_{AB} = -110.3, -110.5 \qquad -111.8$$
$$^2J_{FF} = 243$$

It should also be noted that although the phenyl substituent gives rise to $\sim$5 ppm shielding when it is the only affecting substituent on a CF$_2$ group, it appears to have little effect when a more strongly influencing group such as OH, NH$_2$, or carbonyl is also proximate.

4.3.7.1. ^{1}H, ^{13}C and ^{15}N NMR Data. The protons of CF$_2$H groups bound to nitrogen are only slightly deshielded as compared to those at the end of an alkyl chain (Scheme 4.33). Again, note the somewhat large $^2J_{FH}$ coupling constants for such compounds.

4.3.7.2. Difluoromethyl-Substituted Imines, Imidoyl Chlorides, and Nitrones. These CF$_2$H nitrogen-bound building blocks have received much recent interest. A few examples are given in Scheme 4.34, along with their fluorine, proton, and carbon NMR spectral data.

Scheme 4.33

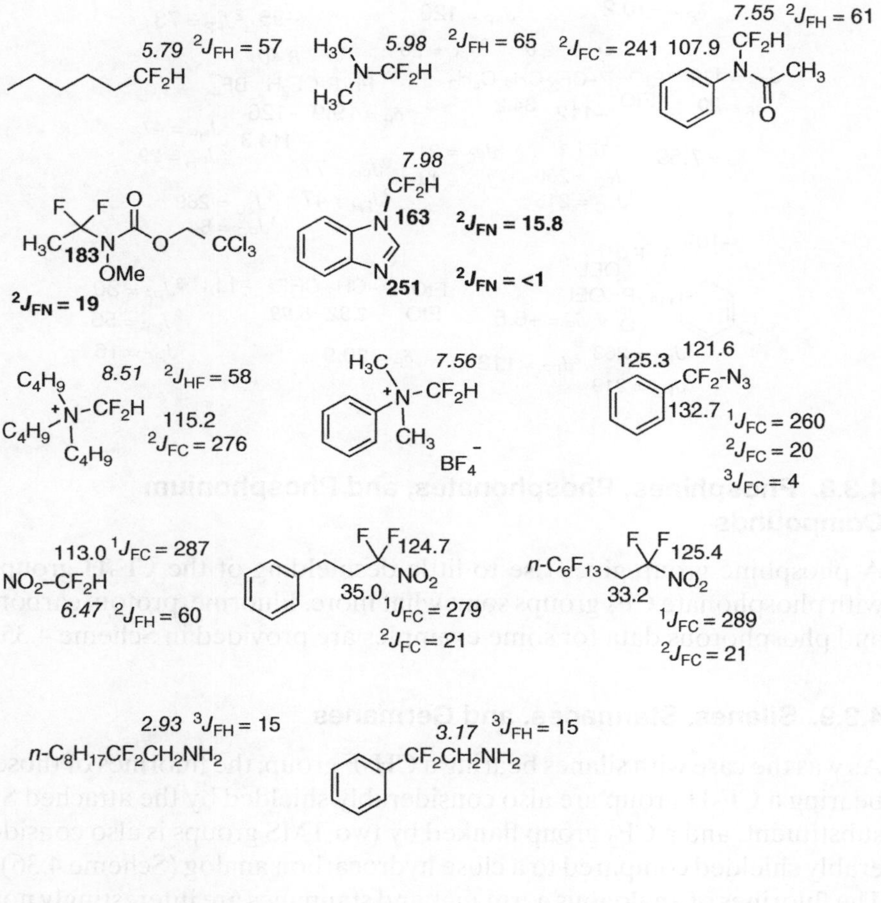

5.79 $^2J_{FH} = 57$
CF₂H (on hexyl chain)

H₃C 5.98 $^2J_{FH} = 65$ $^2J_{FC} = 241$ 107.9
H₃C—N—CF₂H

7.55 $^2J_{FH} = 61$
CF₂H
Ph—N—C(=O)CH₃

H₃C 183 C(F)(F)—N(OMe)—C(=O)—O—CH₂CCl₃
$^2J_{FN} = 19$

7.98
CF₂H
benzimidazole N 163 $^2J_{FN} = 15.8$
N 251 $^2J_{FN} = <1$

C₄H₉ 8.51 $^2J_{HF} = 58$
C₄H₉—N⁺—CF₂H 115.2
C₄H₉ $^2J_{FC} = 276$

H₃C 7.56
Ph—N⁺—CF₂H
CH₃
BF₄⁻

125.3 121.6
CF₂–N₃
132.7 $^1J_{FC} = 260$
$^2J_{FC} = 20$
$^3J_{FC} = 4$

113.0 $^1J_{FC} = 287$
NO₂–CF₂H
6.47 $^2J_{FH} = 60$

F F 124.7
Ph–CH₂CH₂–C–NO₂
35.0 $^1J_{FC} = 279$
$^2J_{FC} = 21$

F F 125.4
n-C₆F₁₃–C–NO₂
33.2
$^1J_{FC} = 289$
$^2J_{FC} = 21$

2.93 $^3J_{FH} = 15$
n-C₈H₁₇CF₂CH₂NH₂

3.17 $^3J_{FH} = 15$
CF₂CH₂NH₂

Scheme 4.34

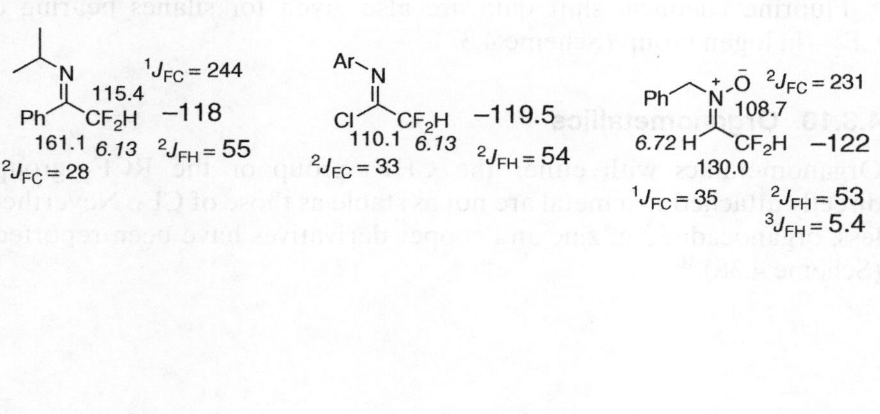

$^1J_{FC} = 244$
iPr–N=C(Ph)(CF₂H)
115.4
Ph— —CF₂H -118
161.1 6.13 $^2J_{FH} = 55$
$^2J_{FC} = 28$

Ar–N=C(Cl)(CF₂H)
-119.5
110.1 6.13 $^2J_{FH} = 54$
$^2J_{FC} = 33$

Ph–CH₂–N⁺(–O⁻)=CH–CF₂H
$^2J_{FC} = 231$ 108.7
6.72 H -122
130.0
$^1J_{FC} = 35$ $^2J_{FH} = 53$
$^3J_{FH} = 5.4$

Scheme 4.35

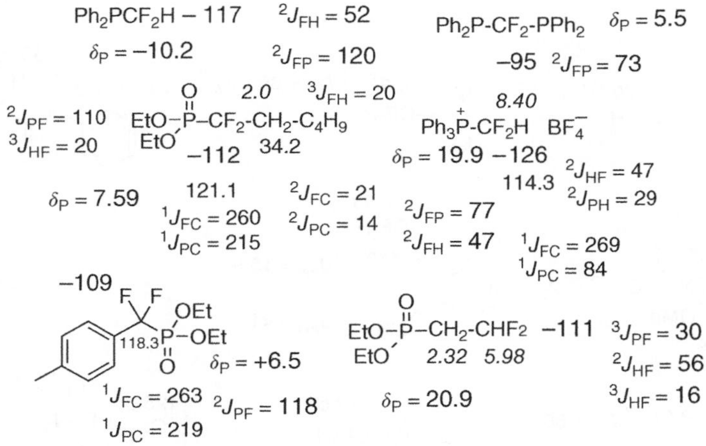

$Ph_2PCF_2H - 117$ $^2J_{FH} = 52$ $Ph_2P\text{-}CF_2\text{-}PPh_2$ $\delta_P = 5.5$
  $\delta_P = -10.2$ $^2J_{FP} = 120$ -95 $^2J_{FP} = 73$

   $^2J_{PF} = 110$ $\underset{EtO'}{EtO\text{-}\overset{\overset{O}{\parallel}}{P}}\text{-}\overset{2.0}{CF_2}\text{-}CH_2\text{-}C_4H_9$ $^3J_{FH} = 20$ 8.40
   $^3J_{HF} = 20$ -112 34.2 $Ph_3\overset{+}{P}\text{-}CF_2H$ BF_4^-
                $\delta_P = 19.9 -126$ $^2J_{HF} = 47$
   $\delta_P = 7.59$ 121.1 $^2J_{FC} = 21$ 114.3 $^2J_{PH} = 29$
           $^1J_{FC} = 260$ $^2J_{PC} = 14$ $^2J_{FP} = 77$
           $^1J_{PC} = 215$ $^2J_{FH} = 47$ $^1J_{FC} = 269$
                  $^1J_{PC} = 84$

   -109

   (aromatic ring structure with CF$_2$) 118.3 $\overset{\overset{F\,\,F}{\diagup}}{\underset{\overset{\parallel}{O}}{P}}\overset{\text{OEt}}{\underset{\text{OEt}}{}}$ $\delta_P = +6.5$ $EtO\text{-}\overset{\overset{O}{\parallel}}{P}\text{-}CH_2\text{-}CHF_2$ -111 $^3J_{PF} = 30$
                EtO' 2.32 5.98 $^2J_{HF} = 56$
   $^1J_{FC} = 263$ $^2J_{PF} = 118$ $\delta_P = 20.9$ $^3J_{HF} = 16$
   $^1J_{PC} = 219$

4.3.8. Phosphines, Phosphonates, and Phosphonium Compounds

A phosphine group gives rise to little deshielding of the CF$_2$H group, with phosphonate CF$_2$ groups somewhat more. Fluorine, proton, carbon and phosphorous data for some examples are provided in Scheme 4.35.

4.3.9. Silanes, Stannanes, and Germanes

As was the case with silanes bearing a CH$_2$F group, the fluorines of those bearing a CF$_2$H group are also considerably shielded by the attached Si substituent, and a CF$_2$ group flanked by two TMS groups is also considerably shielded compared to a close hydrocarbon analog (Scheme 4.36). The fluorines of analogous germanes and stannanes are interestingly not similarly shielded.

Fluorine chemical shift data are also given for silanes bearing a CF$_2$—halogen group (Scheme 4.37).

4.3.10. Organometallics

Organometallics with either the CHF$_2$ group or the RCF$_2$ group directly attached to a metal are not as stable as those of CF$_3$. Nevertheless, organocadmium, zinc and copper derivatives have been reported (Scheme 4.38).[10]

Scheme 4.36

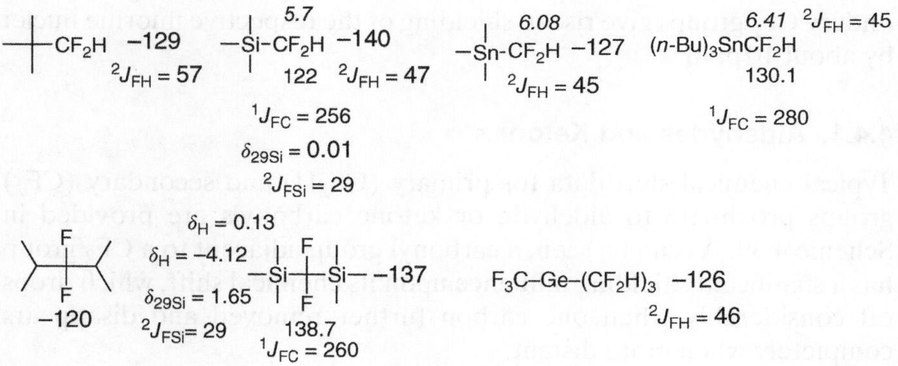

$$\begin{array}{llll}
\overset{5.7}{-}\!\!\!\!-CF_2H & -129 & -Si\!-\!CF_2H & -140 \\
& {}^2J_{FH} = 57 & \quad 122 \quad {}^2J_{FH} = 47
\end{array}$$

$$-Sn\!-\!CF_2H \;\; -127 \qquad (n\text{-Bu})_3SnCF_2H$$
$$^2J_{FH} = 45 \qquad\qquad 130.1$$

$$6.08 \qquad\qquad 6.41 \;\; {}^2J_{FH} = 45$$

$${}^1J_{FC} = 256 \qquad\qquad {}^1J_{FC} = 280$$
$$\delta_{29Si} = 0.01$$
$${}^2J_{FSi} = 29$$

$$\delta_H = 0.13$$
$$\delta_H = -4.12 \qquad -Si\!\!-\!\!\overset{F}{\underset{F}{|}}\!\!-\!\!Si\!\!-\!\!-137 \qquad F_3C\!-\!Ge\!-\!(CF_2H)_3 \;\; -126$$
$$\delta_{29Si} = 1.65 \qquad\qquad\qquad\qquad\qquad\qquad {}^2J_{FH} = 46$$
$$-120 \qquad {}^2J_{FSi} = 29 \qquad 138.7$$
$$\qquad\qquad\qquad\qquad {}^1J_{FC} = 260$$

Scheme 4.37

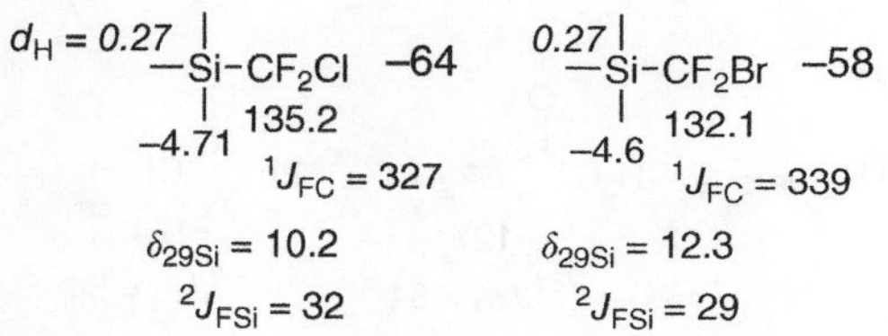

$$d_H = 0.27 \;\; -Si\!-\!CF_2Cl \;\; -64 \qquad\qquad 0.27 \;\; -Si\!-\!CF_2Br \;\; -58$$
$$-4.71 \qquad 135.2 \qquad\qquad\qquad\qquad -4.6 \qquad 132.1$$
$$\qquad\qquad {}^1J_{FC} = 327 \qquad\qquad\qquad\qquad\qquad {}^1J_{FC} = 339$$

$$\delta_{29Si} = 10.2 \qquad\qquad\qquad\qquad\qquad \delta_{29Si} = 12.3$$
$${}^2J_{FSi} = 32 \qquad\qquad\qquad\qquad\qquad {}^2J_{FSi} = 29$$

Scheme 4.38

$$\begin{array}{ll}
HCF_2CdI & -118 \\
146 & {}^2J_{FH} = 43 \\
{}^1J_{FC} = 283 & {}^2J_{113CdF} = 342 \\
& {}^2J_{113CdF} = 327 \\
6.2 \\
(HCF_2)_2Cd & -119 \\
& {}^2J_{FH} = 43 \\
& {}^2J_{113CdF} = 292 \\
& {}^2J_{113CdF} = 278 \\
1.5 \\
(CH_3CF_2)_2Cd & -77 \quad {}^3J_{FH} = 29 \\
30.8 \quad 151 \\
{}^2J_{FC} = 17 \quad {}^1J_{FC} = 287
\end{array}$$

$$\begin{array}{ll}
HCF_2ZnI & -126 \\
& {}^2J_{FH} = 44 \\
\\
(HCF_2)_2Zn & -126 \\
& {}^2J_{FH} = 44 \\
\\
6.1 \\
CHF_2Cu & -115 \\
148.5 \quad {}^2J_{FH} = 44 \\
{}^1J_{FC} = 265
\end{array}$$

(all in DMF)

4.4. CARBONYL FUNCTIONAL GROUPS

Carbonyl functional groups bound directly to primary CF$_2$H or secondary CF$_2$ groups give rise to shielding of the respective fluorine nuclei by about 10 ppm.

4.4.1. Aldehydes and Ketones

Typical chemical shift data for primary (CF$_2$H) and secondary (CF$_2$) groups proximate to aldehyde or ketone carbonyls are provided in Scheme 4.39. As can be seen, a carbonyl group adjacent to a CF$_2$ group has a significant shielding influence upon its chemical shift, which drops off considerably when one carbon further removed and disappears completely when more distant.

Examples of two 2,2-difluoro-1,3-diketones are given in Scheme 4.40. A few examples of X-CF$_2$ ketones are given in Scheme 4.41.

Scheme 4.39

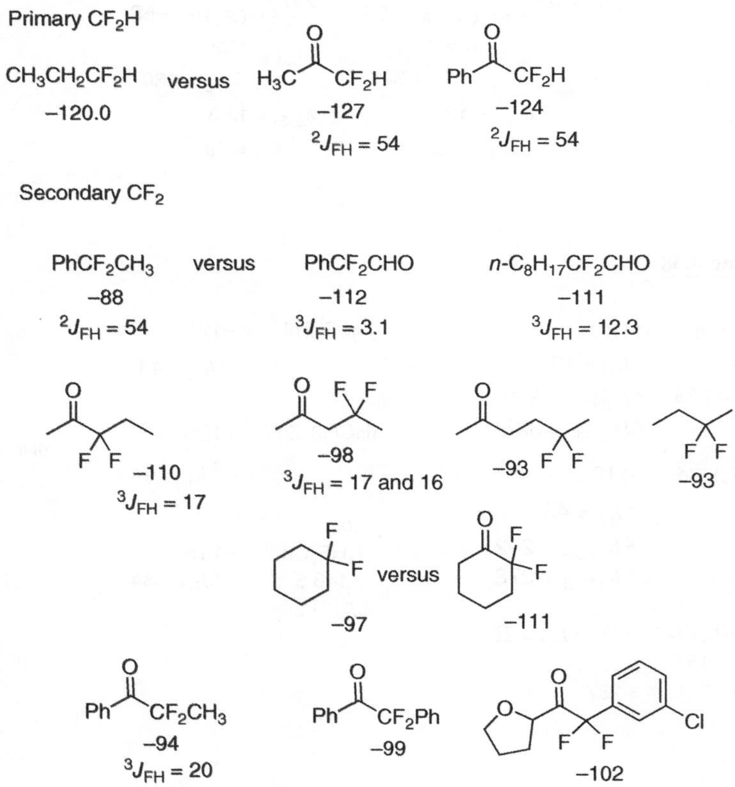

Primary CF$_2$H

CH$_3$CH$_2$CF$_2$H   versus   H$_3$C—CO—CF$_2$H   Ph—CO—CF$_2$H
−120.0   −127   −124
                               $^2J_{FH} = 54$ $^2J_{FH} = 54$

Secondary CF$_2$

PhCF$_2$CH$_3$ versus   PhCF$_2$CHO   n-C$_8$H$_{17}$CF$_2$CHO
−88   −112   −111
$^2J_{FH} = 54$ $^3J_{FH} = 3.1$ $^3J_{FH} = 12.3$

−110   −98   −93   −93
$^3J_{FH} = 17$ $^3J_{FH} = 17$ and 16

−97   versus   −111

Ph—CO—CF$_2$CH$_3$ Ph—CO—CF$_2$Ph
−94   −99
$^3J_{FH} = 20$

−102

Scheme 4.40

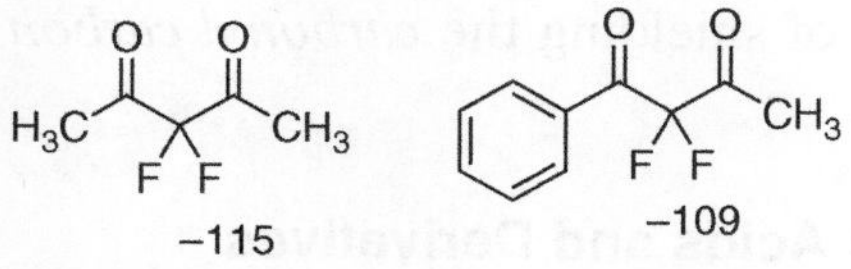

H_3C ... CH_3 ... F F
−115

Ph ... CH_3 ... F F
−109

Scheme 4.41

F F
$\diagdown$ O-Et
O −83

Ph$-CF_2Cl$
−93

Ph$-CF_2Br$
−58

Ph$-CF_2I$
−54

n-C_6H_{13}—CF_2Cl
−85

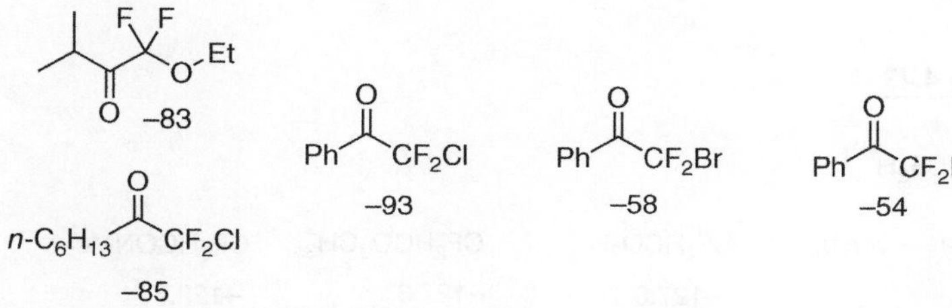

4.4.1.1. ^1H and ^{13}C NMR Data. A ketone or aldehyde carbonyl group bound to a CF_2H group shields its proton slightly (0.1 ppm), and even more surprisingly it also has a shielding effect upon its carbon chemical shift of about 8 ppm (Scheme 4.42). By comparison, a hydrocarbon

Scheme 4.42

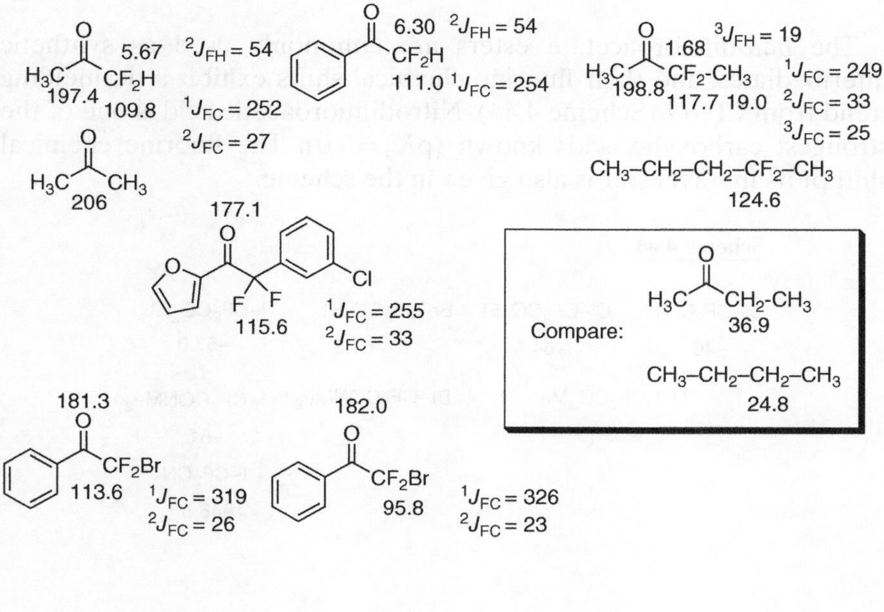

H_3C—CF_2H 5.67 $^2J_{FH} = 54$
197.4   109.8 $^1J_{FC} = 252$
$^2J_{FC} = 27$

H_3C—CH_3
206

Ph—CF_2H 6.30 $^2J_{FH} = 54$
111.0 $^1J_{FC} = 254$

H_3C—CF_2-CH_3 1.68 $^3J_{FH} = 19$
198.8   117.7   19.0 $^1J_{FC} = 249$
$^2J_{FC} = 33$
$^3J_{FC} = 25$

CH_3–CH_2–CH_2–CF_2–CH_3
124.6

177.1
CF_2 Cl
115.6 $^1J_{FC} = 255$
$^2J_{FC} = 33$

Compare:

H_3C ... CH_2-CH_3
36.9

CH_3–CH_2–CH_2–CH_3
24.8

181.3
CF_2Br
113.6 $^1J_{FC} = 319$
$^2J_{FC} = 26$

182.0
CF_2Br
95.8 $^1J_{FC} = 326$
$^2J_{FC} = 23$

ketone, as in 2-butanone, has the effect of *deshielding* the C-3 CH$_2$ carbon by about 12 ppm. A CF$_2$H group when bound to a ketone carbonyl also has the effect of shielding the *carbonyl carbon* relative to that in acetone.

4.4.2. Carboxylic Acids and Derivatives

As was the case for ketones and aldehydes, carboxylic acid functions act to shield the fluorine nuclei of both the primary CF$_2$H group and the secondary CF$_2$ group (Scheme 4.43).

<u>**Scheme 4.43**</u>

Primary CF$_2$H

CF$_2$HCH$_3$	versus	CF$_2$HCO$_2$H	CF$_2$HCO$_2$CH$_3$	CF$_2$HCONMe$_2$
−110		−127.0	−127.3	−122.7

Secondary CF$_2$

CH$_3$CH$_2$CF$_2$CH$_3$	versus	CH$_3$CF$_2$CO$_2$Et	CH$_3$CH$_2$CF$_2$CO$_2$CH$_3$	n-C$_7$H$_{15}$CF$_2$CO$_2$Et
−93		−100	−108	−105
		$^3J_{FH} = 19$	$^3J_{FH} = 17$	

PhCF$_2$CO$_2$CH$_3$
−104.3

The *halo*difluoroacetate esters are commonly used as synthetic intermediates, and their fluorine chemical shifts exhibit a deshielding trend from Cl to I (Scheme 4.44). Nitrodifluoroacetic acid is one of the strongest carboxylic acids known (pK_a = 0.0). The fluorine chemical shift of its methyl ester is also given in the scheme.

<u>**Scheme 4.44**</u>

Cl–CF$_2$CH$_3$	Cl–CF$_2$CO$_2$Et	Br–CF$_2$CO$_2$Et	I–CF$_2$CO$_2$Et
−46	−64.5	−61.3	−57.9

NO$_2$CF$_2$CO$_2$Me	Br–CF$_2$CONMe$_2$	I–CF$_2$CONMe$_2$
−53	−54	−51

I–CF$_2$CN
−46.5

Interestingly, although a sulfide linkage to an α,α-difluoro ester also gives rise to deshielding of the fluorines, a sulfoxide link actually *shields* the fluorines compared to RCF_2CO_2Et (Scheme 4.45)

Scheme 4.45

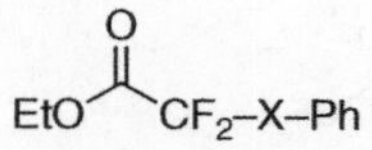
EtO
CF2–X–Ph
O

X = S $\delta_F = -82.8$ $\delta_C = 120.0$, $^1J_{FC} = 287$

X = SO δ_F (AB) = −110.0 and −111.8 ($^2J_{FF} = 227$)

Attaching a double bond to the α,α-difluoro ester leads to significant deshielding of the CF_2 group (Scheme 4.46).

Scheme 4.46

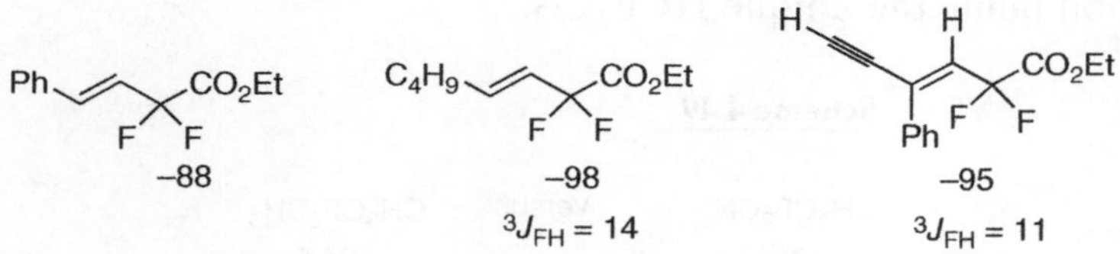
Ph
CO2Et
F F
–88
C4H9
CO2Et
F F
–98
3JFH = 14
H
H
CO2Et
Ph
F F
–95
3JFH = 11

An ester function one carbon removed from a CF_2 group slightly deshields the fluorines (Scheme 4.47).

Scheme 4.47

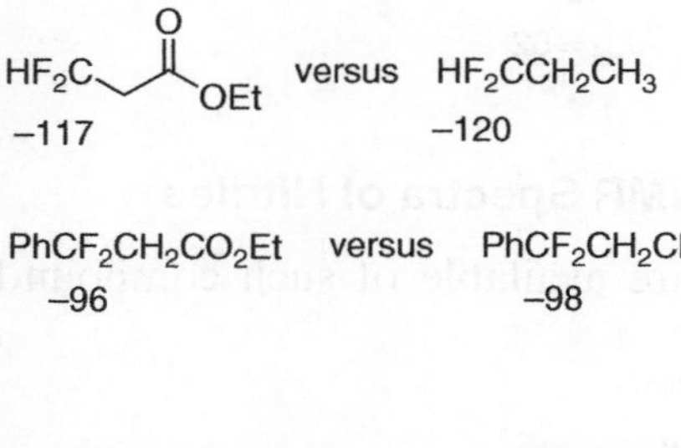
HF2C
O
OEt
versus HF2CCH2CH3
–117
–120
PhCF2CH2CO2Et versus PhCF2CH2CH3
–96
–98

4.4.2.1. ^1H and ^{13}C NMR Data. The ester function of ethyl difluoroacetate deshields the CF_2H proton slightly (about 0.1 ppm), whereas as was the case for ketones and aldehydes, it *shields* the carbon of either a CF_2H or a CF_2-alkyl group significantly (by about 10 ppm) (Scheme 4.48).

Scheme 4.48

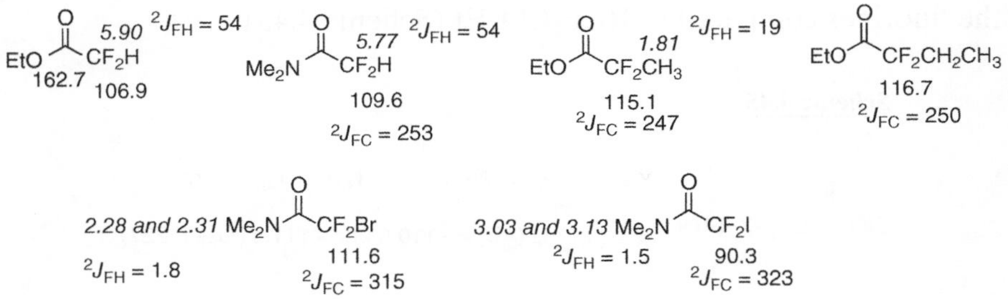

4.5. NITRILES

Unlike carbonyl functions, a nitrile function bound to a secondary CF$_2$ does not generally lead to shielding of the CF$_2$ group (Scheme 4.49), the exception being the unique HCF$_2$CN.

Scheme 4.49

$$CH_3CF_2CN \qquad \text{versus} \qquad CH_3CF_2CH_3$$
$$-85 \qquad\qquad\qquad\qquad -84.5$$
$$^3J_{FH} = 18.1$$

$$PhCF_2CN \qquad \text{versus} \qquad PhCF_2CH_3$$
$$-83.5 \qquad\qquad\qquad\qquad -87.9$$

$$HCF_2CN \qquad \text{versus} \qquad HCF_2CH_3$$
$$-120 \qquad\qquad\qquad\qquad -110$$
$$^2J_{HF} = 52$$

4.5.1. ^{1}H and ^{13}C NMR Spectra of Nitriles

The few data that are available of such compounds are provided in Scheme 4.50.

Scheme 4.50

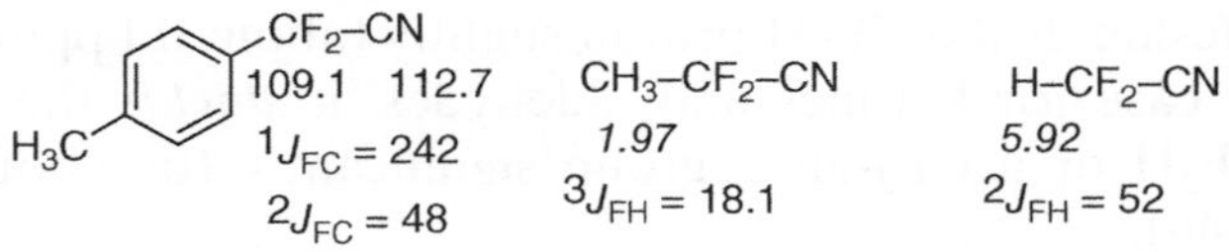

4.6. AMINO-, HYDROXYL-, AND KETO-DIFLUOROCARBOXYLIC ACID DERIVATIVES

The impact of combinations of functional groups on CF_2 chemical shifts depends on how they are arranged. If they are consecutive, then the closest one largely determines the chemical shift (Scheme 4.51).

Scheme 4.51

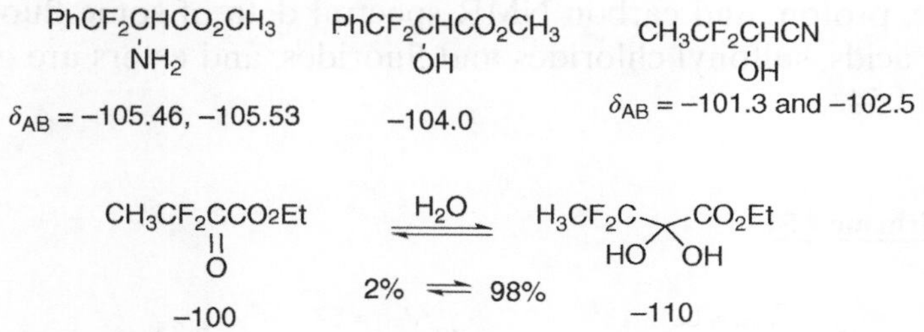

$$PhCF_2CHCO_2CH_3 \qquad PhCF_2CHCO_2CH_3 \qquad CH_3CF_2CHCN$$
$$\qquad | \qquad\qquad\qquad\quad | \qquad\qquad\qquad\qquad |$$
$$\qquad NH_2 \qquad\qquad\qquad\quad OH \qquad\qquad\qquad\qquad OH$$

$\delta_{AB} = -105.46, -105.53 \qquad -104.0 \qquad \delta_{AB} = -101.3$ and -102.5

$$CH_3CF_2CCO_2Et \qquad \xrightarrow{\;H_2O\;} \qquad H_3CF_2C\diagdown\;\diagup CO_2Et$$
$$\qquad\quad || \qquad\qquad\qquad\qquad\qquad\qquad HO \;\; OH$$
$$\qquad\quad O$$
$$\qquad\qquad\qquad\qquad 2\% \rightleftharpoons 98\%$$
$$\qquad -100 \qquad\qquad\qquad\qquad\qquad\qquad\qquad -110$$

Note that the latter compound, the 3,3-difluoro-α-keto ester, exists in aqueous solution at 98% *in the hydrate form,* with the CF_2 of the hydrate being more shielded than that of the keto form, with a chemical shift of -110 ppm.

On the other hand, when two functionalities are each directly bound to the CF_2 group, as in the following examples, the effect of each is felt (Scheme 4.52).

Scheme 4.52

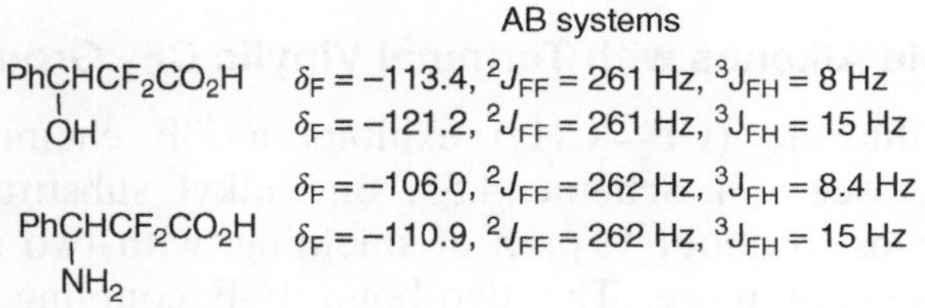

AB systems

$$PhCHCF_2CO_2H \qquad \delta_F = -113.4, \; ^2J_{FF} = 261 \text{ Hz}, \; ^3J_{FH} = 8 \text{ Hz}$$
$$\quad | \qquad\qquad\qquad \delta_F = -121.2, \; ^2J_{FF} = 261 \text{ Hz}, \; ^3J_{FH} = 15 \text{ Hz}$$
$$\quad OH$$

$$\qquad\qquad\qquad\qquad \delta_F = -106.0, \; ^2J_{FF} = 262 \text{ Hz}, \; ^3J_{FH} = 8.4 \text{ Hz}$$
$$PhCHCF_2CO_2H \qquad \delta_F = -110.9, \; ^2J_{FF} = 262 \text{ Hz}, \; ^3J_{FH} = 15 \text{ Hz}$$
$$\quad | $$
$$\quad NH_2$$

A couple of examples of β-ketoester systems are given in Scheme 4.53. In each case, the second carbonyl gives rise to additional shielding.

Scheme 4.53

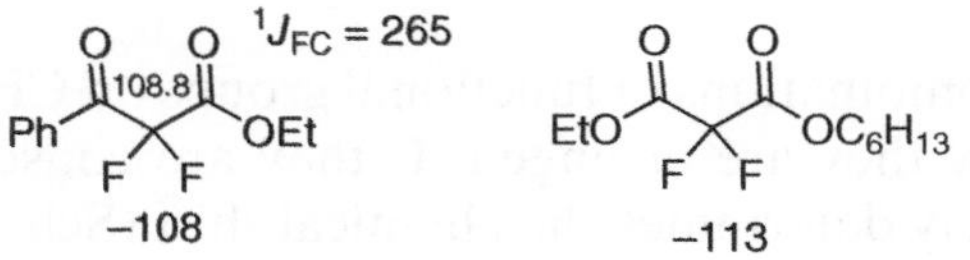

Ph—C(=O)—CF$_2$—C(=O)OEt $^1J_{FC} = 265$
108.8
F F
−108

EtO—C(=O)—CF$_2$—C(=O)OC$_6$H$_{13}$
F F
−113

4.7. SULFONIC ACID DERIVATIVES

Fluorine, proton, and carbon NMR spectral data of some fluorinated
sulfonic acids, sulfonyl chlorides and fluorides, and esters are given in
Scheme 4.54.

Scheme 4.54

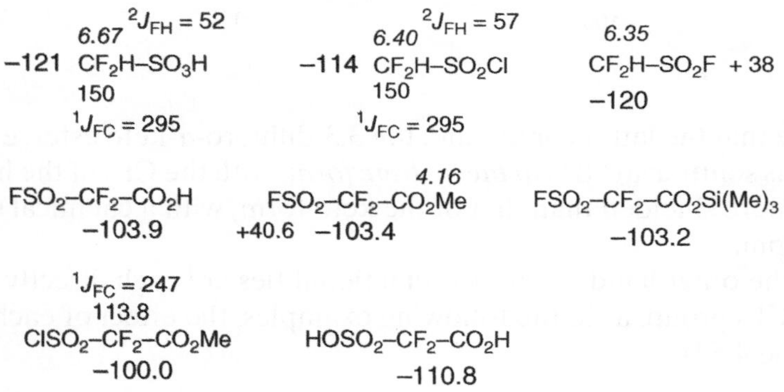

6.67 $^2J_{FH} = 52$
−121 CF$_2$H–SO$_3$H
150
$^1J_{FC} = 295$

6.40 $^2J_{FH} = 57$
−114 CF$_2$H–SO$_2$Cl
150
$^1J_{FC} = 295$

6.35
CF$_2$H–SO$_2$F + 38
−120

FSO$_2$–CF$_2$–CO$_2$H
−103.9

4.16
FSO$_2$–CF$_2$–CO$_2$Me
+40.6 −103.4

FSO$_2$–CF$_2$–CO$_2$Si(Me)$_3$
−103.2

$^1J_{FC} = 247$
113.8
ClSO$_2$–CF$_2$–CO$_2$Me
−100.0

HOSO$_2$–CF$_2$–CO$_2$H
−110.8

4.8. ALKENES AND ALKYNES

4.8.1. Simple Alkenes with Terminal Vinylic CF$_2$ Groups

Vinylidene fluoride (CF$_2$=CH$_2$) exhibits a ^{19}F chemical shift of
−82 ppm. As seen in Scheme 4.55, one alkyl substitution at the
2-position leads to about 10 ppm of shielding, with two alkyl groups
providing 6–7 ppm more. The two-bond F–F coupling constant in
such AB systems is typically around 50 Hz. Modest shielding of
the Z-fluorine is generally observed relative to the E-fluorine of
1,1-difluoroalkenes.

Figure 4.13 provides the ^{19}F NMR spectrum of 1,1-difluorobutene.
The chemical shifts for its Z- and E- fluorines are −92.8 and −90.8 ppm,

Scheme 4.55

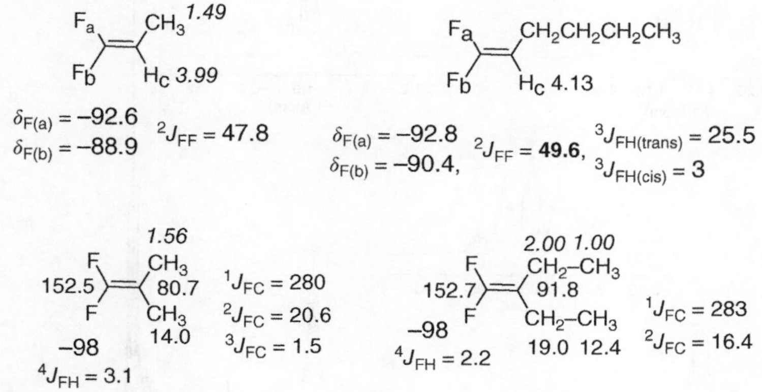

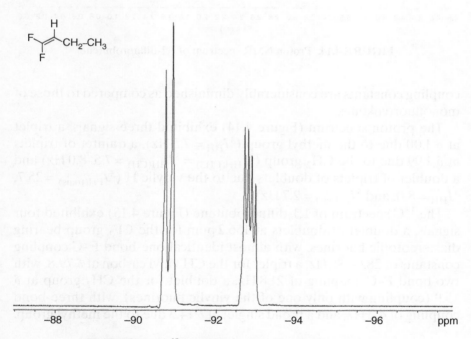

Top left structure:

F_a ⎯ CH_3 1.49

F_b ⎯ H_c 3.99

$\delta_{F(a)} = -92.6$
$\delta_{F(b)} = -88.9$ $^2J_{FF} = 47.8$

Top right structure:

F_a ⎯ $CH_2CH_2CH_2CH_3$

F_b ⎯ H_c 4.13

$\delta_{F(a)} = -92.8$ $^2J_{FF} = 49.6,$ $^3J_{FH(trans)} = 25.5$
$\delta_{F(b)} = -90.4,$ $^3J_{FH(cis)} = 3$

Bottom left structure:

1.56

F ⎯ CH_3 80.7 $^1J_{FC} = 280$

152.5

F ⎯ CH_3 14.0 $^2J_{FC} = 20.6$

−98 $^3J_{FC} = 1.5$

$^4J_{FH} = 3.1$

Bottom right structure:

2.00 1.00

F ⎯ CH_2–CH_3 91.8 $^1J_{FC} = 283$

152.7

F ⎯ CH_2–CH_3 $^2J_{FC} = 16.4$

−98   19.0 12.4

$^4J_{FH} = 2.2$

Structure near spectrum:

H

F₂C= (CH with) CH_2–CH_3

FIGURE 4.13. ^{19}F NMR spectrum of 1,1-difluorobutene

−88   −90   −92   −94   −96   ppm

respectively, with the geminal $^2J_{FF}$ coupling constant of 50 Hz, and the *trans* $^3J_{HF}$ coupling constant of 25.5 Hz. The *cis* coupling was too small to be seen in the fluorine spectrum but was determined to be 2.7 Hz from the proton spectrum shown in Figure 4.14. The magnitudes of these F–H

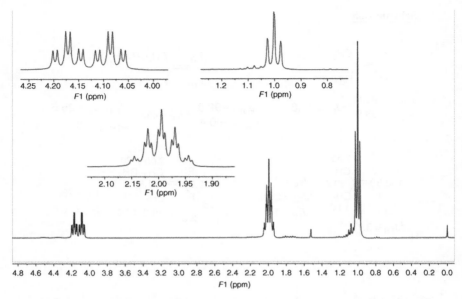

FIGURE 4.14. Proton NMR spectrum of 1,1-difluorobutene

coupling constants are considerably diminished as compared to those of monofluoroalkenes.

The proton spectrum (Figure 4.14) exhibited three signals, a triplet at δ 1.00 due to the methyl group ($^3J_{HH} = 7.5$ Hz), a quintet of triplets at δ 1.99 due to the CH$_2$ group ($^3J_{HH(CH3)} = {}^3J_{HH(CH)} = 7.5$–8.0 Hz) and a doublet of triplets of doublets due to the vinylic H ($^3J_{FH(trans)} = 25.7$, $^3J_{HH} = 8.0$, and $^3J_{FH(cis)} = 2.7$ Hz).

The ^{13}C spectrum of 1,1-difluorobutene (Figure 4.15) exhibited four signals, a doublet of doublets at 156.2 ppm for the CF$_2$ group bearing diastereotopic fluorines, with almost identical one-bond F–C coupling constants of 282–285 Hz, a triplet for the CH vinyl carbon at δ 79.8, with two-bond F–C coupling of 21.8 Hz, a doublet for the CH$_2$ group at δ 15.9 (coupling with only one of the vinylic fluorines), with three-bond coupling of 4.2 Hz, and a broad singlet at δ 14.3 due to the methyl group.

4.8.2. Conjugated Alkenes with Terminal Vinylic CF$_2$ Group

Conjugation in the form of a phenyl substituent at the 2-position leads to 8–10 ppm deshielding of the fluorines of a terminal vinylic CF$_2$ group, whereas for a conjugating vinyl group at the 2-position (as in 1,1-difluoro-1,3-butadiene) such deshielding is somewhat less

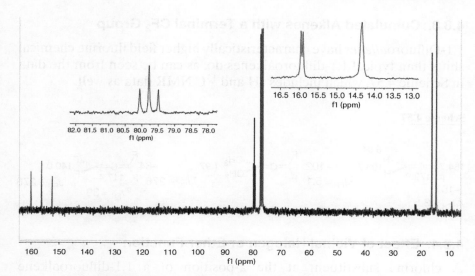

FIGURE 4.15. ^{13}C NMR spectrum of 1,1-difluorobutene

Scheme 4.56

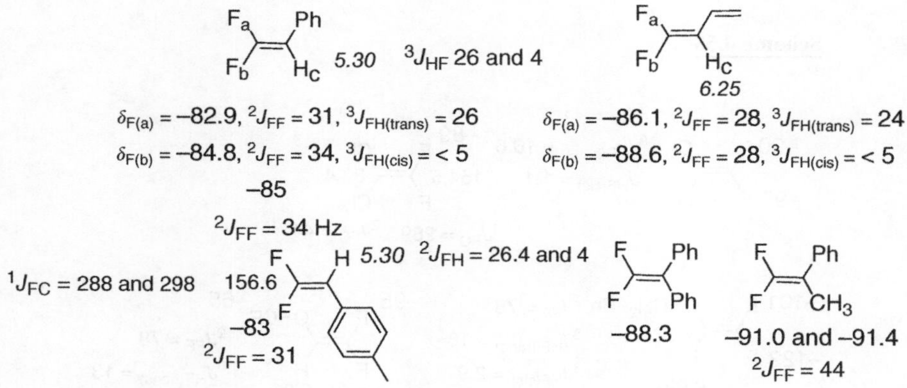

5.30 $^3J_{HF}$ 26 and 4

6.25

$\delta_{F(a)} = -82.9,\ ^2J_{FF} = 31,\ ^3J_{FH(trans)} = 26$

$\delta_{F(b)} = -84.8,\ ^2J_{FF} = 34,\ ^3J_{FH(cis)} = <5$

-85

$\delta_{F(a)} = -86.1,\ ^2J_{FF} = 28,\ ^3J_{FH(trans)} = 24$

$\delta_{F(b)} = -88.6,\ ^2J_{FF} = 28,\ ^3J_{FH(cis)} = <5$

$^2J_{FF} = 34\ Hz$

$^1J_{FC} = 288$ and 298 156.6 5.30 $^2J_{FH} = 26.4$ and 4

-83

$^2J_{FF} = 31$

-88.3

-91.0 and -91.4

$^2J_{FF} = 44$

(Scheme 4.56). The two-bond, F–F coupling constant observed in such conjugated systems is much smaller than that of the nonconjugated alkene systems.

Pertinent ^{1}H and ^{13}C NMR data are included, where available, in Schemes 4.55 and 4.56. As expected, *trans* F–H coupling is characteristically much larger than the analogous *cis* coupling. Conjugation does not noticeably affect the carbon chemical shift of the CF$_2$ carbon.

4.8.3. Cumulated Alkenes with a Terminal CF$_2$ Group

1,1-Difluoro*allenes* have characteristically higher field fluorine chemical shifts than typical 1,1-difluoroalkenes do, as can be seen from the data in Scheme 4.57, which includes ^{1}H and ^{13}C NMR data as well.

Scheme 4.57

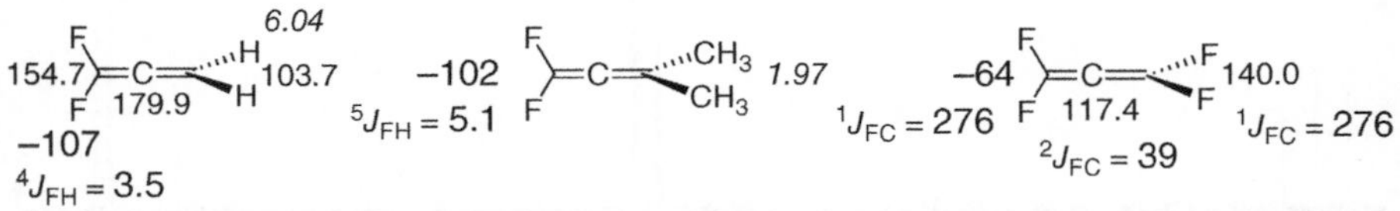

4.8.4. Effect of Vicinal Halogen or Ether Function

A chlorine substituent at the 2-position of a 1,1-difluoroalkene deshields the fluorines modestly, but as seen in Scheme 4.58, a vicinal alkoxy or siloxy group shields both fluorines, the *trans* fluorine more greatly. With the additional electronegative substituent, three-bond F–H coupling constants for such compounds become even smaller.

Scheme 4.58

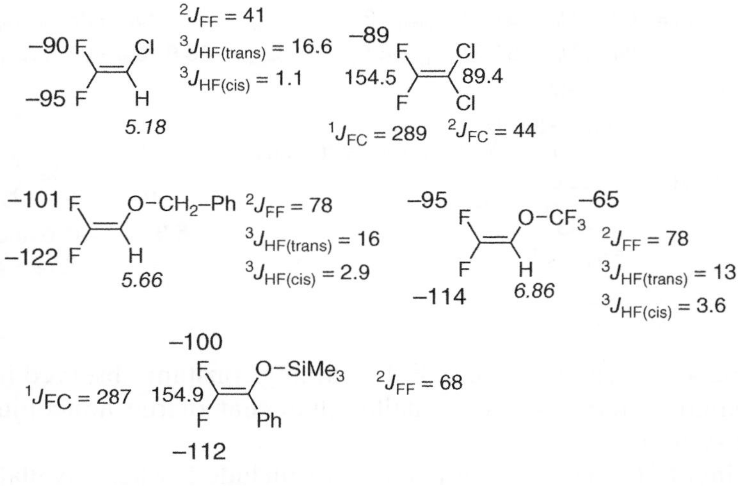

4.8.5. Effect of Allylic Substituents

There are few examples of 1,1-difluoroalkenes that bear allylic substituents. Scheme 4.59 provides those examples for which spectra exist.

Scheme 4.59

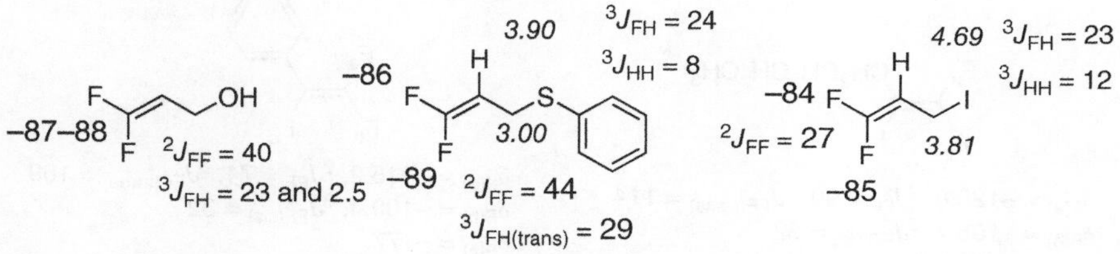

4.8.6. Polyfluoroethylenes

Fluorine chemical shift and coupling constant data are provided in Scheme 4.60 for all of the hydrofluoroethylenes.

Scheme 4.60

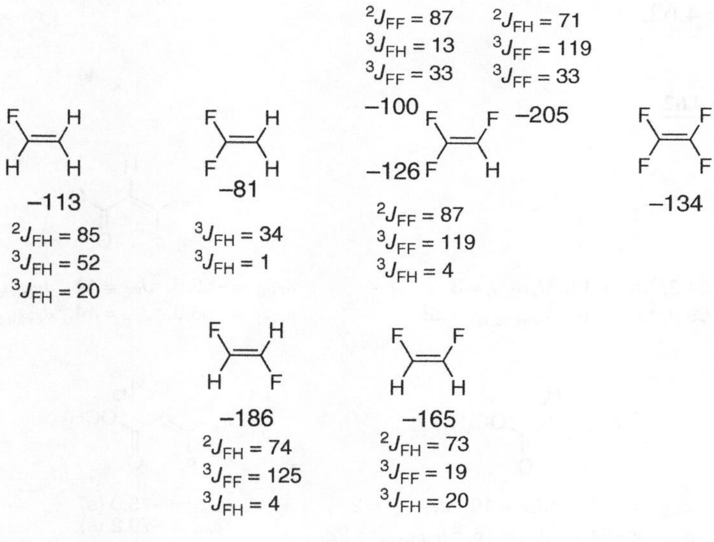

4.8.7. Trifluorovinyl Group

Although the presence of a chlorine at the 2-position of a 1,1-difluoro-alkene has almost no influence upon the chemical shifts of the fluorine nuclei, a fluorine substituent at the 2-position gives rise to very significant shielding, and it causes a much greater "split" of the diastereotopic fluorines at the 1-position and much greater coupling constants, both geminal and vicinal (Scheme 4.61). Further data for trifluorovinyl compounds can be found in Chapter 6.

Scheme 4.61

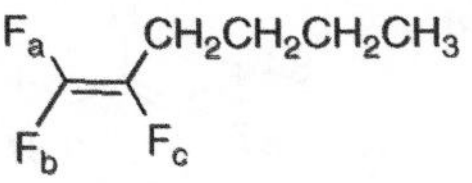

$\delta_{F(a)} = -125.8$, $^2J_{FF} = 90$, $^3J_{FF(trans)} = 114$
$\delta_{F(b)} = -106.7$, $^3J_{FF(cis)} = 32$
$\delta_{F(c)} = -174.8$

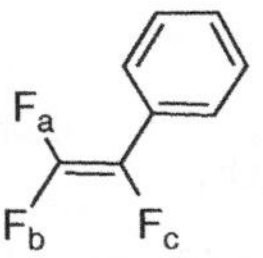

$\delta_{F(a)} = -115.2$, $^2J_{FF} = 71$, $^3J_{FF(trans)} = 109$
$\delta_{F(b)} = -100.4$, $^3J_{FF(cis)} = 32$
$\delta_{F(c)} = -177$

4.8.8. α,β-Unsaturated Carbonyl Systems with a Terminal Vinylic CF$_2$ Group

The fluorines of a CF$_2$= group of an α,β-unsaturated carbonyl system are considerably deshielded, and the geminal and vicinal coupling constants dramatically diminished, as seen from the examples given in Scheme 4.62.

Scheme 4.62

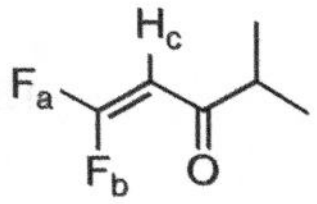

$\delta_{F(a)} = -64.2$, $^2J_{FF} = 14$, $^3J_{FH(cis)} = 3$
$\delta_{F(b)} = -59.0$, $^2J_{FF} = 14$, $^3J_{FH(trans)} = 22$

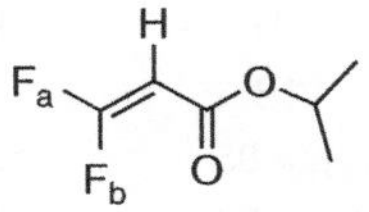

$\delta_{F(a)} = -63.5$, $^2J_{FF} = 14$, $^3J_{FH(cis)} = 3$
$\delta_{F(b)} = -58.0$, $^2J_{FF} = 14$, $^3J_{FH(trans)} = 22$

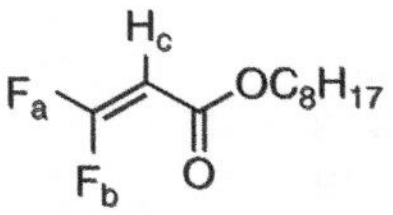

$\delta_{F(a)} = -70.7$, $^2J_{FF} = 16$, $^3J_{FH(cis)} = 2$
$\delta_{F(b)} = -64.7$, $^2J_{FF} = 16$, $^3J_{FH(trans)} = 22$

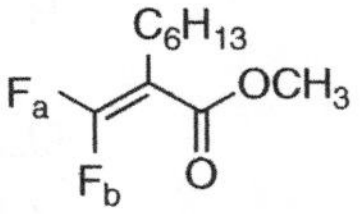

$\delta_{F(a)} = -75.0$ (s)
$\delta_{F(b)} = -70.2$ (s)

4.8.8.1. ^1H and ^{13}C NMR Data.
Typical proton and carbon NMR data for α,β-unsaturated carbonyl compounds with a terminal vinylic CF$_2$ group are given in Scheme 4.63. The pertinent F–H coupling constants have been given in Scheme 4.62. Conjugation with a carbonyl group deshields the β-CF$_2$ carbon by 4–5 ppm.

Scheme 4.63

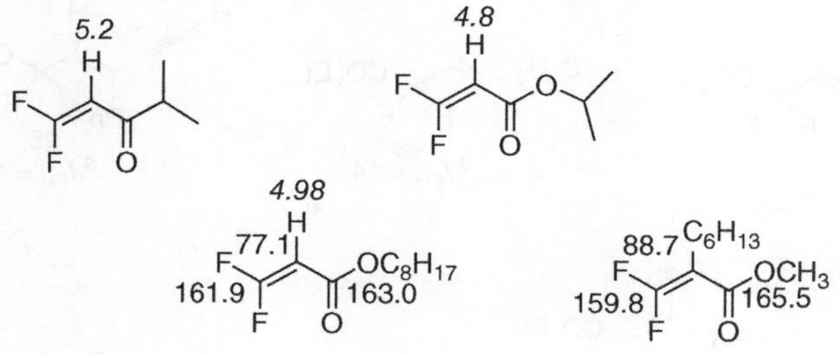

5.2
H
F
F
O

4.8
H
F
F
O
O

4.98
H
77.1
F
161.9
F
OC$_8$H$_{17}$
163.0
O

88.7 C$_6$H$_{13}$
F
159.8
F
OCH$_3$
165.5
O

4.8.9. Allylic and Propargylic CF$_2$ Groups

A *vinyl* substituent deshields both primary (CF$_2$H) and secondary (CF$_2$) groups by between 6 and 10 ppm. There is also one example of an allylic CF$_2$I compound. Few data on the impact of an *acetylenic* group are available, but it seems to have slightly greater deshielding influence than either a vinyl or a phenyl substituent on a CF$_2$H group (Scheme 4.64).

Scheme 4.64

CH$_3$CH$_2$CF$_2$H –120

CH$_2$=CHCF$_2$H –113 $^2J_{FH} = 57$, $^3J_{FH} = 8.4$

PhCH=CHCF$_2$H –108 $^2J_{FH} = 56$

n-C$_3$H$_7$CH=CHCF$_2$H –110, $^2J_{FH} = 56$

CF$_2$H
–116, $^2J_{FH} = 56$

CH$_3$CH$_2$CF$_2$CH$_3$ –93.3

CH$_3$CH=CHCF$_2$CH$_3$ –83.8 (*Z*-isomer)
–87.3 (*E*-isomer)

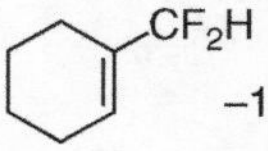

CF$_2$I   –42 $^3J_{FH} = 12$

Ph—≡—CF$_2$H   –106, $^2J_{FH} = 55$

Notice that the CF$_2$ fluorines of the Z-isomer of 4,4-difluoro-2-butene are significantly deshielded relative to those of the *E*-isomer. This is another probable example of "steric deshielding" by a proximate, in this case, *cis*-alkyl group (see Section 2.2.1).

Placing a carbonyl function next to an allylic CF$_2$ group leads to the usual shielding, but its impact appears to be dampened by the influence of the double bond. The fluorines of a terminal allylic CF$_2$H group to an α,β-unsaturated ester appear to be slightly shielded (Scheme 4.65).

Scheme 4.65

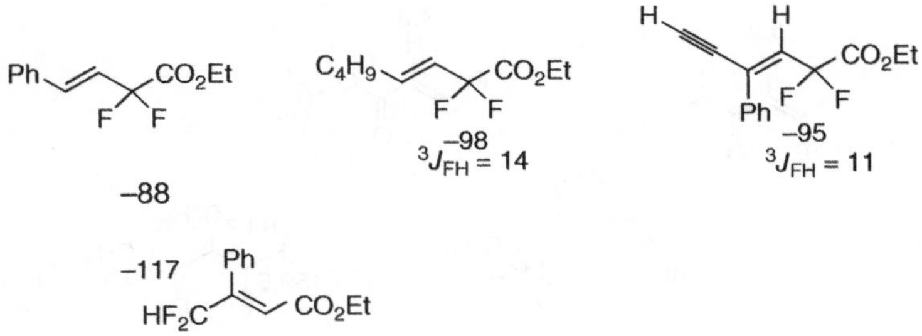
Ph
CO_2Et
F F
−88
C_4H_9
CO_2Et
F F
−98
$^3J_{FH}$ = 14
H
H
CO_2Et
Ph F F
−95
$^3J_{FH}$ = 11
−117
Ph
HF_2C
CO_2Et

4.8.9.1. ¹H and ¹³C NMR Data. The protons of allylic CF_2H groups are deshielded by the vinylic group to the extent of ~0.2 ppm, and its carbon is *shielded* by about 2 ppm. The carbons of allylic secondary CF_2 groups are shielded to the extent of about 4 ppm (Scheme 4.66).

Scheme 4.66

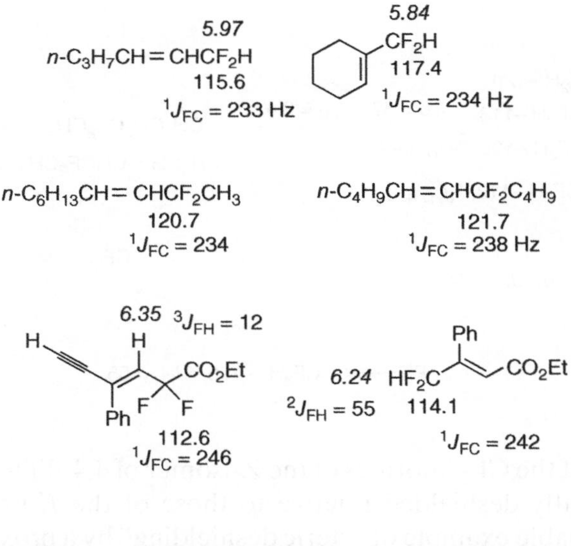
5.97
n-$C_3H_7CH{=}CHCF_2H$
115.6
$^1J_{FC}$ = 233 Hz
5.84
CF_2H
117.4
$^1J_{FC}$ = 234 Hz
n-$C_6H_{13}CH{=}CHCF_2CH_3$
120.7
$^1J_{FC}$ = 234
n-$C_4H_9CH{=}CHCF_2C_4H_9$
121.7
$^1J_{FC}$ = 238 Hz
6.35 $^3J_{FH}$ = 12
H
H
CO_2Et
Ph F F
112.6
$^1J_{FC}$ = 246
Ph
6.24 HF_2C
CO_2Et
$^2J_{FH}$ = 55
114.1
$^1J_{FC}$ = 242

4.9. BENZENOID AROMATICS BEARING A CF_2H OR CF_2R GROUP

Although direct phenyl substitution on a CF_2H group leads to more than 10 ppm deshielding relative to $CH_3CH_2CF_2H$, such "benzylic" systems are also subject to smaller but similar hyperconjugative π–σ_{CF}^* effects to those that were observed for the phenyl-substituted CH$_2$F group (see Section 3.10), with electron-donating and -withdrawing groups giving rise to deshielding and shielding effects, respectively. Direct phenyl substitution on a secondary CF$_2$ group has a similar deshielding influence, and when the phenyl is one carbon farther away, its effect is less than 5 ppm. Examples of each of these types of compounds are given in Scheme 4.67.

Scheme 4.67

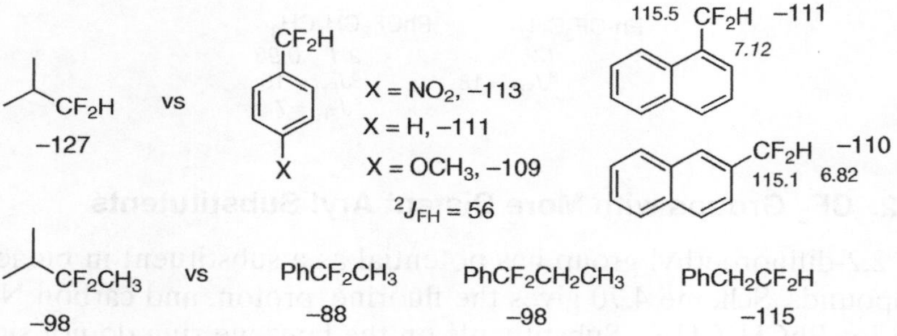

The fluorines of a phenyl-CF$_2$ group appended to an ester function exhibit the usual shielding that is imparted by the proximate ester function (Scheme 4.68).

Scheme 4.68

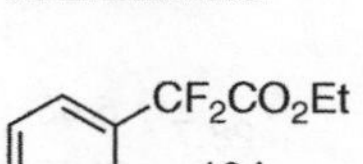

4.9.1. ^{1}H and ^{13}C NMR Data

The characteristic ^{1}H chemical shifts for ArCF$_2$*H* protons lie between 6.6 and 7.0 ppm, and the characteristic ^{13}C chemical shifts for the C̲F$_2$H carbons of such compounds are in the range of −113 to −115 ppm (Table 4.8).

Few data that are available for PhCF$_2$R compounds are given in Scheme 4.69. There do not appear to be any carbon data for such compounds.

TABLE 4.8. Proton and Carbon Data for Aryl CF$_2$H Groups

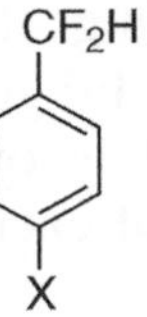

		δ_H	δ_C
CF$_2$H	X = OCH$_3$	6.65	114.9
	X = H	6.55	114.8
	X = NO$_2$	6.80	113.2

Scheme 4.69

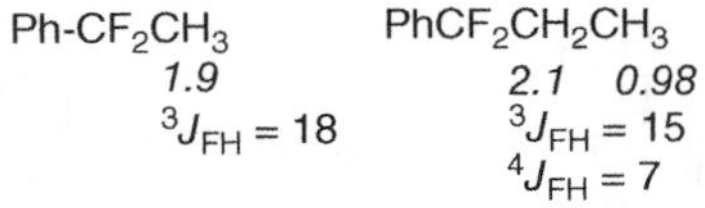

Ph-CF$_2$CH$_3$ PhCF$_2$CH$_2$CH$_3$
    1.9   2.1   0.98
$^3J_{FH}$ = 18 $^3J_{FH}$ = 15
                    $^4J_{FH}$ = 7

4.9.2. CF$_2$ Groups with More Distant Aryl Substitutents

The 2,2-difluoroethyl group has potential as a substituent in bioactive compounds. Scheme 4.70 gives the fluorine, proton, and carbon NMR data for PhCH$_2$CHF$_2$. Substituents on the benzene ring do not significantly affect the fluorine chemical shifts. The fluorine chemical shifts are not significantly affected by the presence of the phenyl substituent in these compounds, with the CF$_2$H proton in 2,2-difluoroethylbenzene being only slightly deshielded.

Scheme 4.70

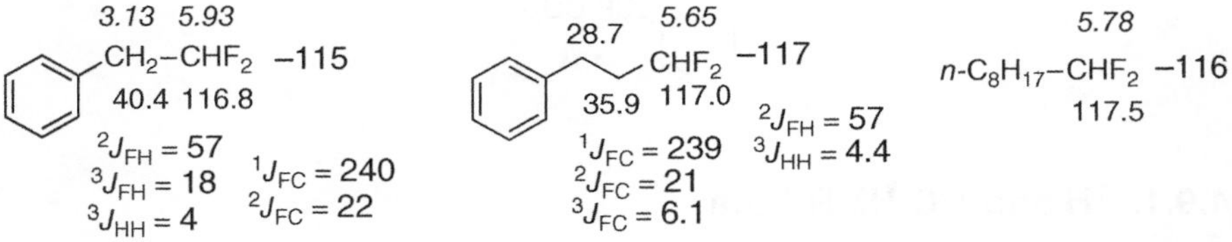

      3.13 5.93   28.7 5.65   5.78
    CH$_2$–CHF$_2$ –115   CHF$_2$ –117 n-C$_8$H$_{17}$–CHF$_2$ –116
    40.4 116.8   35.9 117.0   117.5
                                              $^2J_{FH}$ = 57
$^2J_{FH}$ = 57 $^1J_{FC}$ = 239 $^3J_{HH}$ = 4.4
$^3J_{FH}$ = 18 $^1J_{FC}$ = 240 $^2J_{FC}$ = 21
$^3J_{HH}$ = 4 $^2J_{FC}$ = 22 $^3J_{FC}$ = 6.1

4.10. HETEROAROMATIC CF$_2$ GROUPS

There can be more variation in the ^{19}F chemical shifts of CF$_2$H groups on heteroaromatic than on benzenoid systems, depending on the position of substitution.

4.10.1. Pyridines, Quinolones, Phenanthridines, and Acridines

The fluorines of CF$_2$H groups attached at the 2- or 4-position of a pyridine ring appear at approximately −116 ppm, whereas a CF$_2$H substituent at the 3-position appears at −113 ppm. A secondary CF$_2$ substituent exhibits a similar trend in chemical shift (Scheme 4.71).

Scheme 4.71

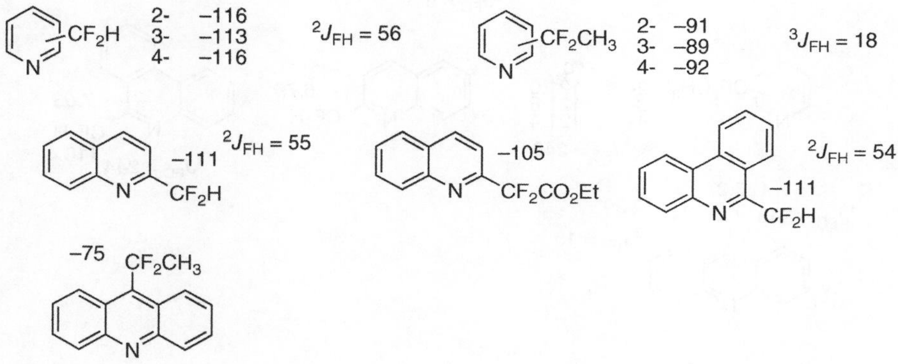

4.10.1.1. Proton and Carbon NMR Data.
Proton and carbon NMR data for pyridines, quinolones, phenanthridines, and acridines are presented in Scheme 4.72.

4.10.2. Furans, Benzofurans, Thiophenes, Pyrroles, and Indoles

On the basis of the few data available, thiophene CF$_2$H groups appear at a lower field than furan CF$_2$H groups and indoles at a lower field than benzofurans (Scheme 4.73). This is consistent with the trend for CF$_3$ groups. On the basis of trends observed for CF$_3$ chemical shifts, one would also expect that CF$_2$H groups in the 2-position should appear at lower fields (less negative) than CF$_2$H groups at the 3-position of furans, thiophenes, and pyrroles, although this can only be confirmed for the thiophene case.

4.10.2.1. ^{1}H, ^{13}C, and ^{15}N NMR Data for Five-Membered Ring Systems with One Hetero Atom.
Proton and carbon NMR data, although limited, are available for CF$_2$H-substituted furans, thiophenes, benzofurans, and indoles (Scheme 4.74).

Scheme 4.72

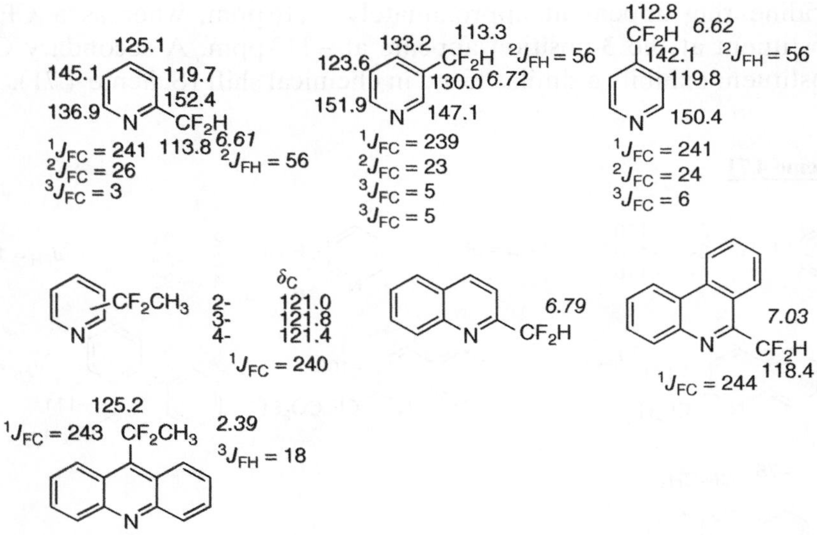

125.1
145.1 / 119.7
136.9 / 152.4
$^1J_{FC} = 241$ 113.8 6.61
$^2J_{FC} = 26$ CF$_2$H $^2J_{FH} = 56$
$^3J_{FC} = 3$

133.2 113.3
123.6 CF$_2$H $^2J_{FH} = 56$
130.0 6.72
151.9 147.1
$^1J_{FC} = 239$
$^2J_{FC} = 23$
$^3J_{FC} = 5$
$^3J_{FC} = 5$

112.8
CF$_2$H 6.62
142.1 $^2J_{FH} = 56$
119.8
150.4
$^1J_{FC} = 241$
$^2J_{FC} = 24$
$^3J_{FC} = 6$

CF$_2$CH$_3$
	δ_C
2-	121.0
3-	121.8
4-	121.4

$^1J_{FC} = 240$

6.79
CF$_2$H

7.03
CF$_2$H
118.4
$^1J_{FC} = 244$

125.2
$^1J_{FC} = 243$ CF$_2$CH$_3$ 2.39
$^3J_{FH} = 18$

Scheme 4.73

CF$_2$H
-115 $^2J_{FH} = 54$

CF$_2$CO$_2$Et
-103

CF$_2$SO$_2$Ph
-102

CF$_2$H -106

CF$_2$H
-115 $^2J_{FH} = 54$

CF$_2$CO$_2$Et
-105

CF$_2$H
-102 $^2J_{FH} = 56$

-108 $^2J_{FH} = 57$
CF$_2$H

CF$_2$SO$_2$Ph
-93

CH$_3$
CF$_2$SO$_2$Ph
-97

CH$_3$
CF$_2$H
-109 $^2J_{FH} = 54$

H
CF$_2$H
-110 $^2J_{FH} = 55$

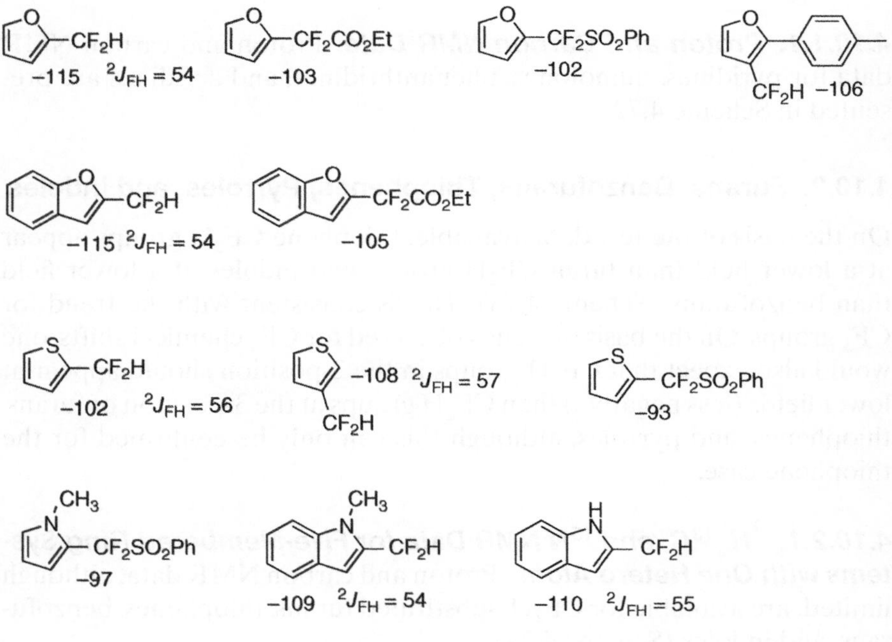

Scheme 4.74

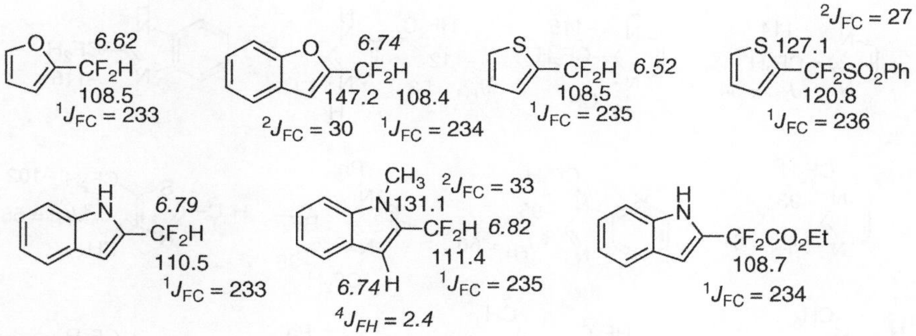

$^2J_{FC} = 27$

6.62
-CF$_2$H
108.5
$^1J_{FC} = 233$

6.74
-CF$_2$H
147.2 108.4
$^2J_{FC} = 30$ $^1J_{FC} = 234$

-CF$_2$H 6.52
108.5
$^1J_{FC} = 235$

S 127.1
-CF$_2$SO$_2$Ph
120.8
$^1J_{FC} = 236$

6.79
-CF$_2$H
110.5
$^1J_{FC} = 233$

CH$_3$ $^2J_{FC} = 33$
N 131.1
-CF$_2$H 6.82
111.4
6.74 H $^1J_{FC} = 235$
$^4J_{FH} = 2.4$

-CF$_2$CO$_2$Et
108.7
$^1J_{FC} = 234$

4.10.3. Pyrimidines

One example of a difluoromethylpyrimidine could be found (Scheme 4.75) but unfortunately no pyrazines or pyridazines.

Scheme 4.75

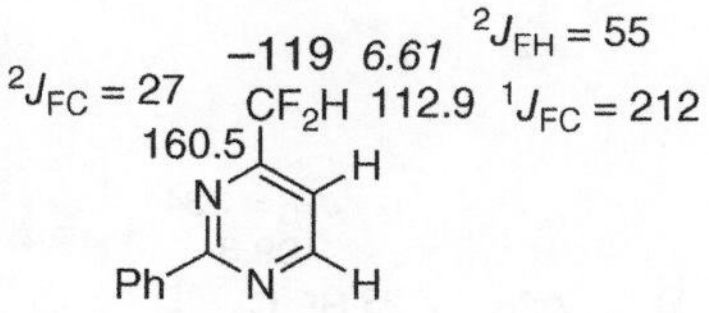

$^2J_{FC} = 27$
-119 6.61 $^2J_{FH} = 55$
CF$_2$H 112.9 $^1J_{FC} = 212$
160.5 H
Ph N H

4.10.4. Five-Membered Ring Heterocycles with Two Hetero Atoms: Imidazoles, Benzimidazoles, 1*H*-pyrazoles, Oxazoles, Isoxazoles, Thiazoles, and Indazoles

The fluorines of a CF$_2$H group at the 2-position of an imidazole or benzimidazole are more highly shielded than those at the 2-position of indole (Scheme 4.76). Included are compounds with the CF$_2$H group bound to nitrogen.

Proton, carbon, and nitrogen NMR data for such systems are provided in Scheme 4.77.

4.10.5. Five-Membered Ring Heterocycles with Three or More Heteroatoms: Sydnones, Triazoles, and Benzotriazoles

Fluorine, proton, and carbon NMR data for these types of compounds are presented in Scheme 4.78.

Scheme 4.76

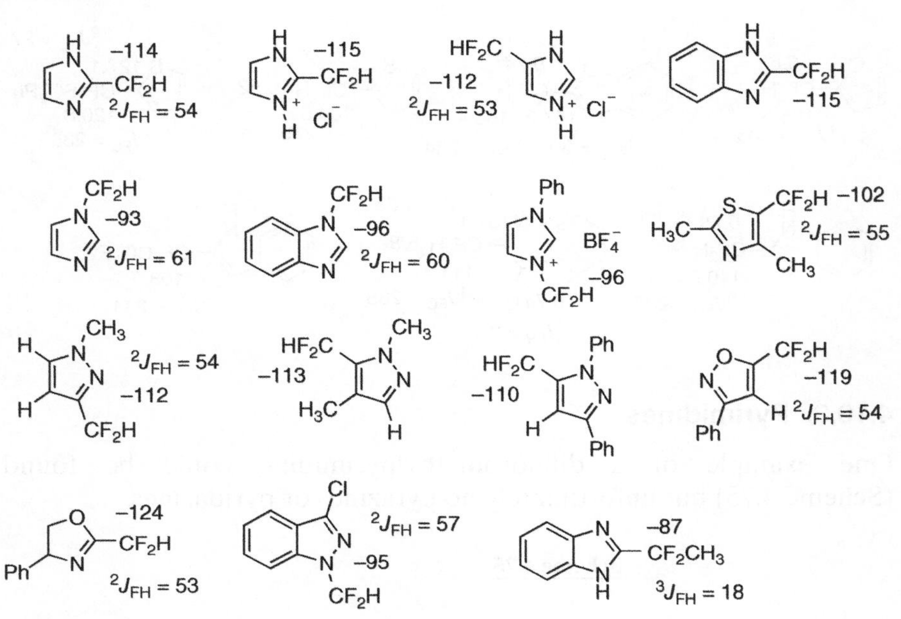

-114 CF_2H $^2J_{FH} = 54$

-115 CF_2H Cl$^-$

HF_2C -112 $^2J_{FH} = 53$ Cl$^-$

CF_2H -115

CF_2H -93 $^2J_{FH} = 61$

CF_2H -96 $^2J_{FH} = 60$

Ph BF$_4^-$ -96 CF_2H

H_3C CF_2H -102 $^2J_{FH} = 55$ CH_3

CH_3 $^2J_{FH} = 54$ -112 CF_2H

HF_2C CH_3 -113 H_3C

HF_2C Ph -110 Ph

CF_2H -119 $^2J_{FH} = 54$ Ph

-124 CF_2H Ph $^2J_{FH} = 53$

Cl $^2J_{FH} = 57$ -95 CF_2H

-87 CF_2CH_3 $^3J_{FH} = 18$

Scheme 4.77

$^2J_{FC} = 28$ 140.3 CF_2H 6.78 108.7 $^1J_{FC} = 236$

7.38 CF_2H Cl$^-$

$^1J_{FC} = 234$ 109.3 7.23 HF_2C 128.1 $^2J_{FC} = 29$ Cl$^-$

CF_2H 7.64 109.9 $^1J_{FC} = 288$

2.22 CF_2CH_3 118.0 $^1J_{FC} = 234$

CF_2H 7.10

102 $^1J_{FC} = 234$ CF_2H 6.83 H_3C 123.9 $^2J_{FC} = 25$ CH_3 Ph BF$_4^-$ CF_2H 7.61

3.90 7.37 H CH_3 $^1J_{FC} = 235$ 108.5 6.70 6.44 H CF_2H 6.68

HF_2C CH_3 137.6 $^2J_{FC} = 30$ H_3C

$^1J_{FC} = 240$ 107.2 HF_2C 159.7 Ph $^2J_{FC} = 28$ Ph

107.9 $^1J_{FC} = 239$ CF_2H 6.80 Ph 6.89

$^2J_{FC} = 28$ 159.7 CF_2H 6.34 Ph 107.2 $^1J_{FC} = 240$

7.98 CF_2H 163 $^2J_{FN} = 15.8$ 251 $^4J_{FN} = <1$

Cl 302 $^2J_{FN} = 1.2$ 184 $^2J_{FN} = 18.2$ CF_2H

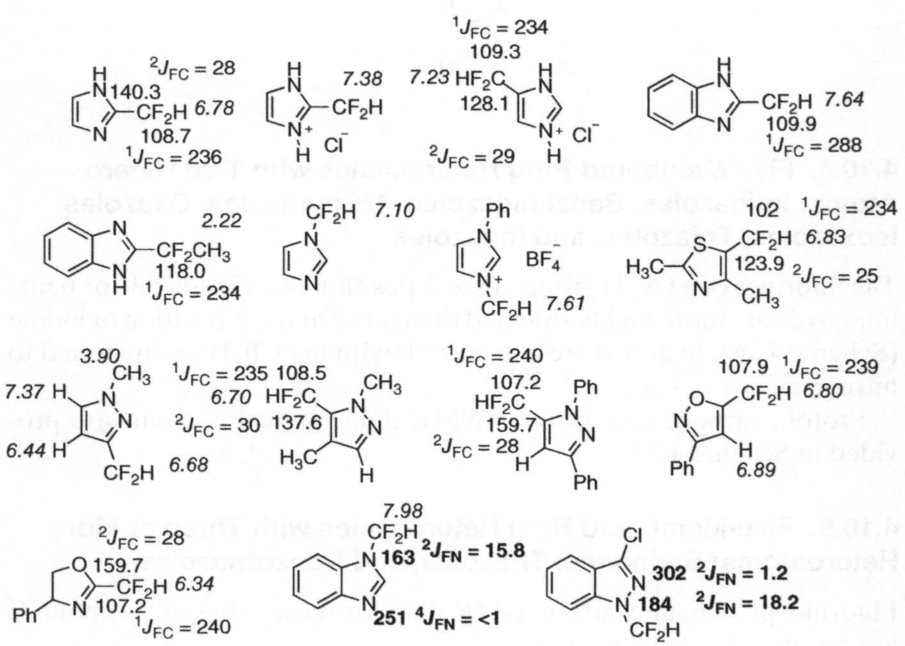

Scheme 4.78

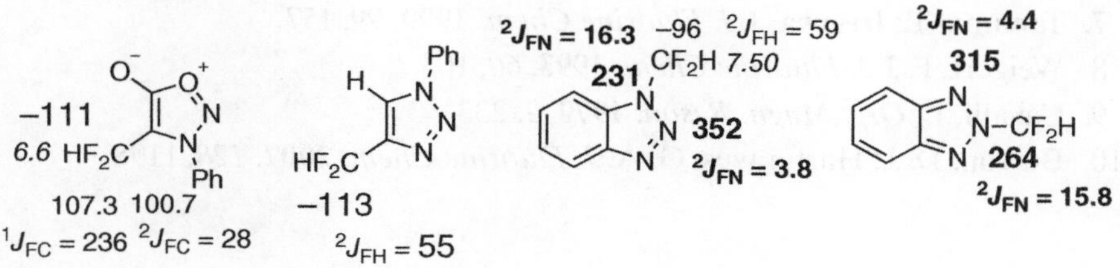

4.10.6. Various Other Difluoromethyl-Substituted Heterocyclic Systems

NMR data for a number of heterocycles with NCF_2H, OCF_2H, or SCF_2H groups are provided in Scheme 4.79.

Scheme 4.79

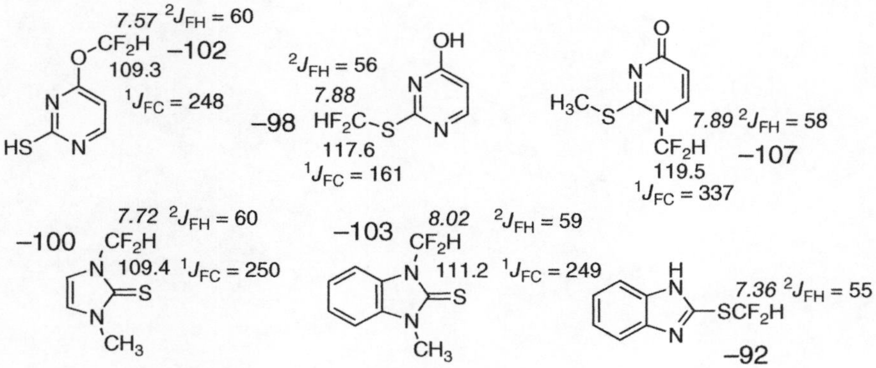

REFERENCES

1. Percy, J. M. *Chimica Oggi* **2004**, *22*, 18.
2. Wiberg, K. B.; Zilm, K. W. *J. Org. Chem.* **2001**, *66*, 2809.
3. Smith, W. H.; Irig, A. M. *J. Phys. Chem.* **1969**, *75*, 497.
4. Cox, R. H.; Smith, S. L. *J. Magn. Res.* **1969**, *1*, 432.

5. Brey, W. S. *Magn. Res. Chem.* **2008**, *46*, 480.
6. Weigert, F. J. *J. Fluorine Chem.* **1990**, *46*, 375.
7. Tanuma, T.; Irisawa, J. *J. Fluorine Chem.* **1999**, *99*, 157.
8. Weigert, F. J. *J. Fluorine Chem.* **1993**, *60*, 103.
9. Cavalli, L. *Org. Magn. Reson.* **1970**, *2*, 233.
10. Burton, D. J.; Hartgraves, G. A. *J. Fluorine Chem.* **2007**, *128*, 1198.

CHAPTER 5

THE TRIFLUOROMETHYL GROUP

5.1. INTRODUCTION

The trifluoromethyl group has become an important structural component of many bioactive compounds, mainly because of its polar influence and its effect on the lipophilicity of compounds.

A few illustrative examples of important agrochemicals and pharmaceuticals that contain a CF_3 group are given as follows. They include the insecticide, triflumuron (**5-1**), the neuroleptic, Fluphenazine (**5-2**), which is used in the treatment of schizophrenia, and the well-known antidepressant drug, Prozac® (**5-3**) (Figure 5.1).

5.1.1. NMR Spectra of Compounds Containing the CF_3 Group – General Considerations

The *fluorine NMR spectra* of carbon-bound trifluoromethyl-containing compounds are generally unmistakably distinctive, with the chemical shifts of most CF_3 groups lying in the range of −60 to −80 ppm. Exceptions are alkynyl CF_3 groups that absorb at the lowest field, typified by 3,3,3-trifluoropropyne ($\delta_F = -52.1$), and CF_3 groups contained within perfluorocarbons, which can absorb at a higher field than −80 ppm, or CF_3s attached to a cyclopropane ring, which also absorb above −80 ppm.

Guide to Fluorine NMR for Organic Chemists, Second Edition. William R. Dolbier, Jr.
© 2016 John Wiley & Sons, Inc. Published 2016 by John Wiley & Sons, Inc.

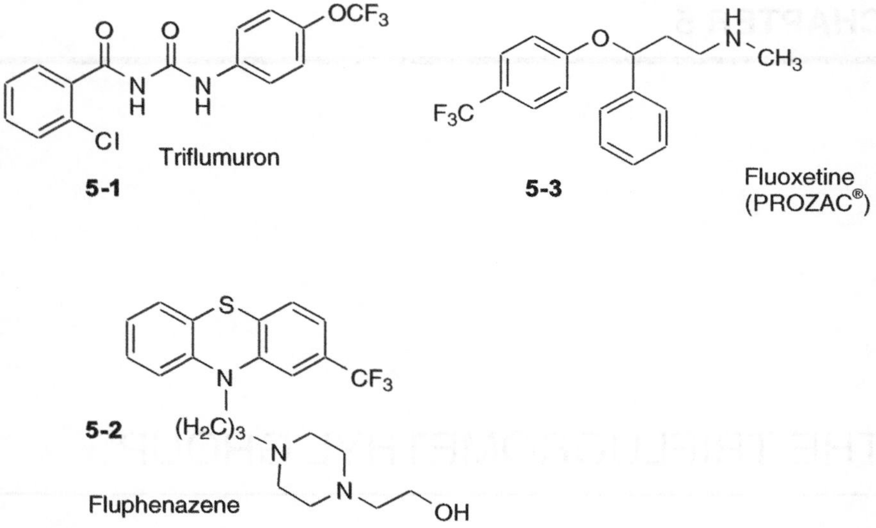

FIGURE 5.1. Examples of bioactive compounds containing a CF_3 group

The ^{13}C *NMR spectra* of compounds bearing a carbon-bound trifluoromethyl group are also characteristic, although because of the combination of weak signal and multiple couplings, the carbon signals deriving from CF_3 groups are often difficult to discern and thus sometimes are unfortunately not even reported. It is not uncommon for ^{13}C NMR spectra of compounds containing trifluoromethyl groups to require overnight accumulation, even when the compounds are relatively soluble. With a few exceptions, the ^{13}C chemical shifts of CF_3 groups lie generally in the relatively narrow range of 107–130 ppm. The trifluoromethyl group splits its own carbon as well as all carbons in its vicinity into characteristic quartets, with one-bond F–C couplings usually in the range of 275–285 Hz for carbon-bound CF_3 groups. Such one-bond couplings are much larger than those of $-CF_2-$ or CF_2H groups (234–250 Hz), which are themselves greater than the one-bond coupling constants of $-CHF-$ and CH_2F groups (162–170 Hz). Two-bond couplings are generally in the range of 25–35 Hz, and three-bond couplings can also be observed with values in the range of 2–3 Hz.

Regarding ^1H *NMR spectra*, trifluoromethyl groups have but a modest influence on the chemical shifts of nearby protons, certainly much less than do single fluorine substituents. Deshielding of protons on carbons next to a CF_3 group is usually less than 1 ppm, whereas those farther away feel almost no effect. The three-bond F–H *coupling* between trifluoromethyl groups and vicinal hydrogen in trifluoromethyl hydrocarbons is much smaller (between 7 and 11 Hz) than those observed for CF$_2$ (15–22 Hz) and single fluorine compounds (21–27 Hz).

5.2. SATURATED HYDROCARBONS BEARING A CF$_3$ GROUP

As was the case for single fluorines and CF$_2$ groups, branching near the CF$_3$ group leads to increased shielding of the CF$_3$ fluorines and thus more negative chemical shifts. The few examples available seem to bear that out.

5.2.1. Alkanes Bearing a CF$_3$ Group

The CF$_3$ groups of 1,1,1-trifluoro-n-alkanes generally absorb at about −68 ppm, as typified by 1,1,1-trifluorohexane and 1,1,1-trifluorooctane, which are reported to absorb at −67.8 and −67.7 ppm, respectively, each appearing as a triplet with a $^3J_{FH} = 11$ Hz. As can be seen in Scheme 5.1, branching provides additional shielding to the CF$_3$ fluorines.

Scheme 5.1

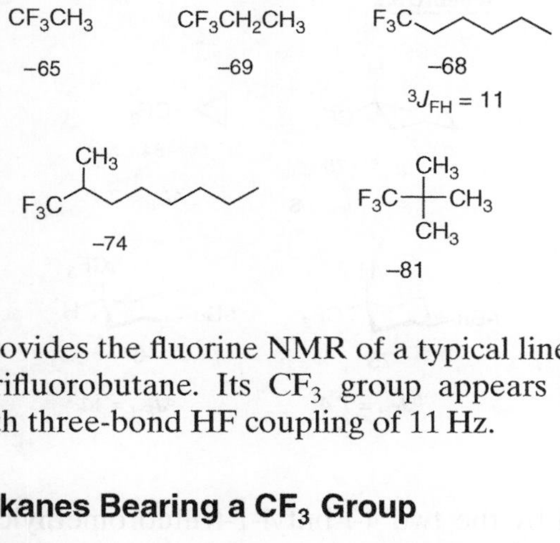

Figure 5.2 provides the fluorine NMR of a typical linear CF$_3$ alkane, that of 1,1,1-trifluorobutane. Its CF$_3$ group appears as a triplet at −66.94 ppm with three-bond HF coupling of 11 Hz.

5.2.2. Cycloalkanes Bearing a CF$_3$ Group

A trifluoromethyl group attached to a cyclohexane ring is unremarkable with respect to its chemical shift, absorbing at −75 ppm, with a $^3J_{FH} = 8$ Hz (Scheme 5.2). There are no data available for trifluoromethylcyclopentane or cyclobutane. The fluorine chemical shift for trifluoromethylcyclopropane reflects additional shielding, such CF$_3$ groups appearing the farthest upfield of any CF$_3$-substituted hydrocarbon.

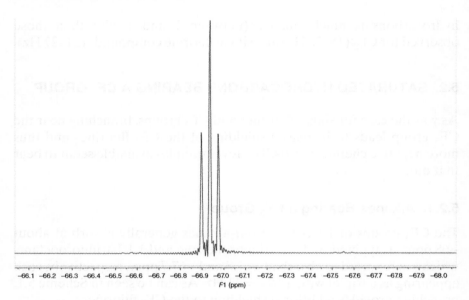

FIGURE 5.2. ^{19}F NMR spectrum of 1,1,1-trifluorobutane

Scheme 5.2

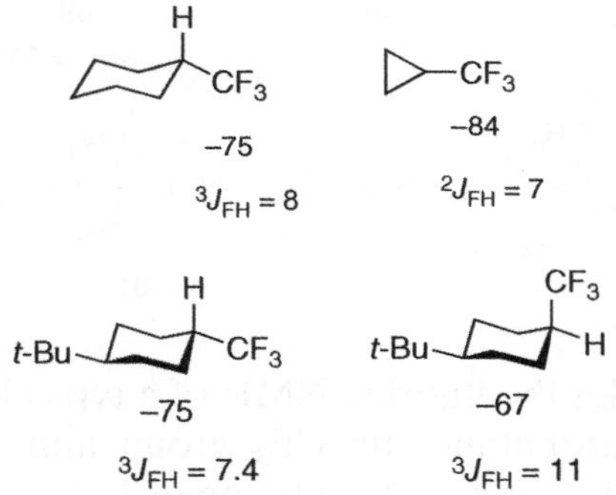

As indicated by the two 4-*t*-butyl-1-trifluoromethylcyclohexanes in Scheme 5.2, a CF_3 group in the equatorial position is more greatly shielded than that in an axial position. This may be another example of what is known as steric deshielding of fluorine atoms.

The trifluoromethylcycloalkane systems *do* provide, however, some examples of CF_3 bound to a *tertiary* center, with 1-methyl-1-trifluoromethylcyclohexane, cyclopentane, and cyclobutane all absorbing at higher fields (-81, -78, and -80 ppm, respectively) than the secondary systems mentioned earlier, as would be expected based on the branching principle.

5.2.3. ¹H and ¹³C NMR Data, General Information

As seen in Scheme 5.3, the trend of anomalous behavior of fluorine compared to other halogens with respect to its effect upon the chemical shift of hydrogens bound to the same carbon continues in the series of trihalomethanes. It is seen that a CCl_3 group has a greater inductive effect on the chemical shifts of β-hydrogens as well.

Scheme 5.3

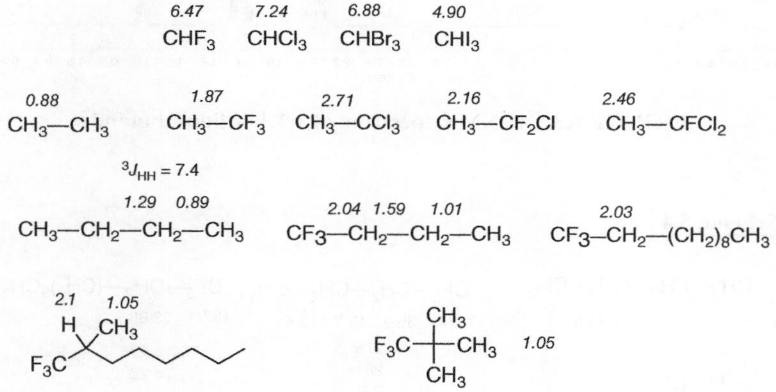

6.47	7.24	6.88	4.90
CHF_3	$CHCl_3$	$CHBr_3$	CHI_3

0.88	1.87	2.71	2.16	2.46
CH_3—CH_3	CH_3—CF_3	CH_3—CCl_3	CH_3—CF_2Cl	CH_3—$CFCl_2$

$^3J_{HH} = 7.4$

1.29 0.89

CH_3—CH_2—CH_2—CH_3

2.04 1.59 1.01

CF_3—CH_2—CH_2—CH_3

2.03

CF_3—CH_2—$(CH_2)_8CH_3$

2.1 1.05

F_3C H CH₃ (structure)

CH₃
F₃C—C—CH₃   1.05
CH₃

Generally, the presence of a CF_3 group induces relatively mild deshielding of vicinal *hydrogens* (<1 ppm) (Scheme 5.3). As mentioned earlier, the impact is less than that from a CCl_3 group.

The proton NMR spectrum of 1,1,1-trifluorobutane is provided as an example in Figure 5.3. The proton chemical shift data derived from this spectrum are as follows: δ 1.01 (t, $^3J_{HH} = 7$, 3H), 1.59 (sextet, $^3J_{HH} = 8$, 2H), 2.04 (m, 2H).

The *carbon* of a CF_3 group bound to a *hydrocarbon* chain generally appears at about 127–128 ppm, as seen in Scheme 5.4. It deshields the adjacent carbon by 10–11 ppm, whereas it appears to actually *shield* the next carbon by about 9 ppm.

As was the case for their fluorine spectra, the carbon NMR spectra of trifluoromethyl cyclohexanes show some slight differences depending on whether the CF_3 group is equatorial or axial, with the axial CF_3 appearing at slightly (~1 ppm) higher field.

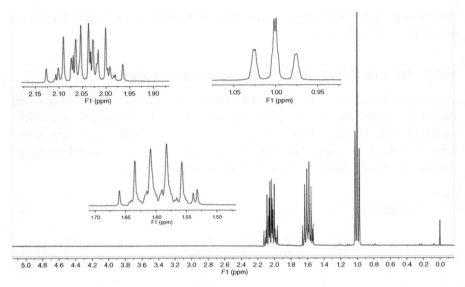

FIGURE 5.3. ^{1}H NMR spectrum of 1,1,1-trifluorobutane

Scheme 5.4

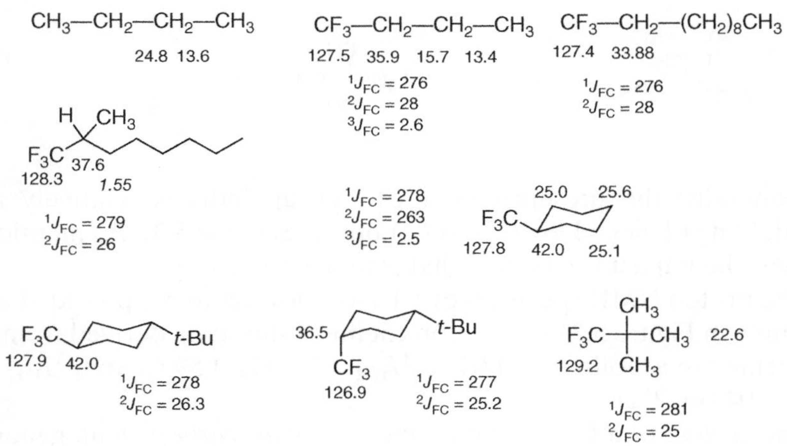

CH$_3$—CH$_2$—CH$_2$—CH$_3$
   24.8 13.6

CF$_3$—CH$_2$—CH$_2$—CH$_3$
127.5 35.9 15.7 13.4
   $^1J_{FC} = 276$
   $^2J_{FC} = 28$
   $^3J_{FC} = 2.6$

CF$_3$—CH$_2$—(CH$_2$)$_8$CH$_3$
   127.4 33.88
   $^1J_{FC} = 276$
   $^2J_{FC} = 28$

F$_3$C $\overset{H}{\underset{37.6}{\diagup}}$ $\overset{CH_3}{}$
128.3   1.55
   $^1J_{FC} = 279$
   $^2J_{FC} = 26$

   $^1J_{FC} = 278$
   $^2J_{FC} = 263$
   $^3J_{FC} = 2.5$

F$_3$C   25.0 25.6
127.8 42.0 25.1

F$_3$C —⬡— t-Bu
127.9 42.0
   $^1J_{FC} = 278$
   $^2J_{FC} = 26.3$

36.5 —⬡— t-Bu
   CF$_3$
   126.9
   $^1J_{FC} = 277$
   $^2J_{FC} = 25.2$

$$F_3C\!-\!\!\overset{CH_3}{\underset{CH_3}{\overset{|}{C}}}\!\!-\!CH_3 \quad 22.6$$
129.2 CH$_3$
   $^1J_{FC} = 281$
   $^2J_{FC} = 25$

The spectrum of 1,1,1-trifluorobutane, in Figure 5.4, provides a typical example of carbon NMR spectra of trifluoromethylalkanes.

The ^{13}C chemical shift data for the above spectrum of 1,1,1-trifluorobutane is as follows: δ 127.5 (q, $^1J_{FC} = 276$), 35.9 (q, $^2J_{FC} = 28$), 15.7 (q, $^3J_{FC} = 2.6$), 13.4 (s).

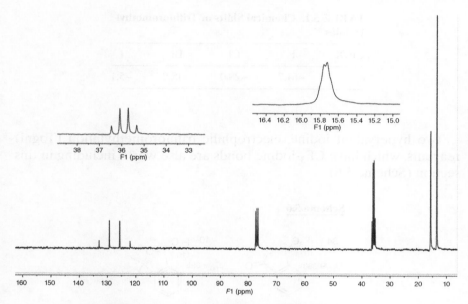

FIGURE 5.4. ^{13}C NMR spectrum of 1,1,1-trifluorobutane

5.3. INFLUENCE OF SUBSTITUENTS AND FUNCTIONAL GROUPS

Virtually any group headed by an atom in Groups 5–7 (i.e., halogen, O, S, N, or P) that is attached *directly* to the CF_3 group gives rise to *deshielding* of the CF_3 fluorines compared to those of CF_3CH_3. In contrast, the *electropositive* $SiMe_3$ group shields the CF_3 group slightly relative to CF_3CH_3. Note the strong relative shielding effect of the single hydrogen atom on the fluorines of CF_3H (Scheme 5.5).

Scheme 5.5

n-alkyl—O—CF$_3$	CH$_3$—CF$_3$	Me$_3$Si—CF$_3$	H—CF$_3$
–60	–65.0	–67	–78

Electronegative substituents, as well as carbonyl groups or other functional groups on the carbon β to a CF_3 group will be seen to shield its fluorines.

5.3.1. Impact of Halogens

As was the case for single fluorines and CF_2 groups, the deshielding of CF_3 by directly bound halogen decreases in the order: $I > Br > Cl > F$ (Table 5.1).

TABLE 5.1. Chemical Shifts of Trifluoromethyl Halides

CF_3X	F	Cl	Br	I
δ_F	−61.7	−28.0	−18.0	−5.1

Two hypervalent iodine, electrophilic trifluoromethylation (Togni) reagents, which have CF_3-iodine bonds are also worth including in this section (Scheme 5.6).

Scheme 5.6

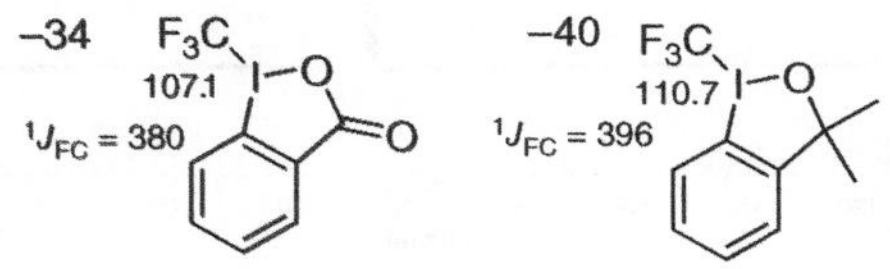

Also, β-substituted halogens shield a CF_3 group, much as they did single fluorines and CF_2 groups (Table 5.2).

In contrast to their effect at the β-position, halogens at the γ-*position* deshield the CF_3 group (Scheme 5.7).

5.3.1.1. ^1H and ^{13}C NMR Data. Halogens bound directly to a CF_3 group have a considerable effect upon the respective ^{13}C chemical shifts, ranging from 78 to 125 ppm (Table 5.3).[1] The one-bond F–C coupling constants increase progressively going down the group, ranging from a low of 260 Hz for CF_4 to a high of 344 Hz for CF_3I.

Chlorine substitution at the β-position leads to CF_3 *carbon* chemical shifts at progressively higher field (Scheme 5.8), whereas the *proton* and carbon chemical shifts for the CH_2 group of CF_3CH_2Cl are affected more by the chlorine substituent than by the CF_3 group.

TABLE 5.2. Fluorine Chemical Shifts of 2,2,2-Trifluoroethyl Halides

X	F	Cl	Br	I	H
CF_3CH_2X	−77.6	−72.1	−69.3	−65.6	−68.0
CF_3CHX_2	−86.2	−78.5			
CF_3CX_3	−88.2	−82.2			

Scheme 5.7

g-halogen substitution

$CF_3CH_2CH_3$	-69
$CF_3CH_2CH_2Cl$	-66
$CF_3CH_2CF_2CH_3$	-63
$CF_3CH_2CCl_3$	-62
$CF_3CH_2CF_3$	-61

TABLE 5.3. ^{13}C NMR Data for CF_3 Halides

	CHF_3	CF_4	CF_3Cl	CF_3Br	CF_3I
$^{13}C\ \delta$	119.1	119.8	125.4	112.6	78.1
$^1J_{FC}$ (Hz)	274	260	299	324	344

Scheme 5.8

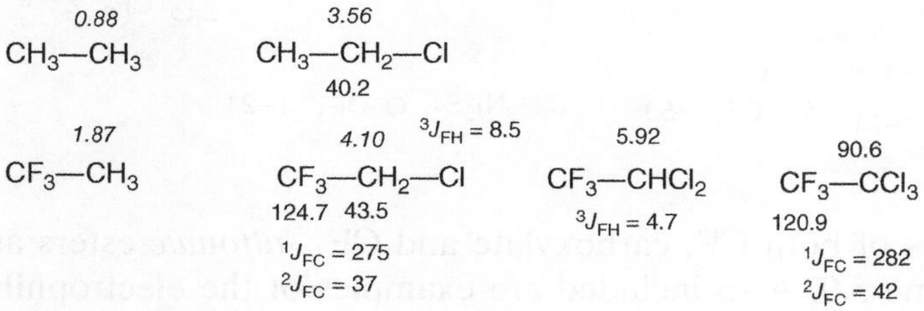

0.88
$CH_3—CH_3$

3.56
$CH_3—CH_2—Cl$
40.2

1.87
$CF_3—CH_3$

4.10 $^3J_{FH} = 8.5$
$CF_3—CH_2—Cl$
124.7 43.5
$^1J_{FC} = 275$
$^2J_{FC} = 37$

5.92
$CF_3—CHCl_2$
$^3J_{FH} = 4.7$
120.9

90.6
$CF_3—CCl_3$
$^1J_{FC} = 282$
$^2J_{FC} = 42$

5.3.2. Ethers, Alcohols, Esters, Sulfides, and Selenides

As indicated in Chapter 3, compounds with fluorines bound directly to a carbon bearing a hydroxyl group are usually very unstable relative to the carbonyl compound plus HF. Nevertheless, trifluoromethanol has been prepared, but it spontaneously loses HF at temperatures above $-30\,°C$. Its fluorine and carbon NMR data are given in Schemes 5.8 and 5.11, and they resemble the respective data of trifluoromethyl ethers.

An oxygen bound to a trifluoromethyl group has much less effect upon its chemical shift than a chlorine substituent. Thus, the fluorines of trifluoromethyl ethers ($\sim-58\,ppm$) are not as deshielded as those of CF_3Cl ($\sim-28\,ppm$). Those of CF_3 sulfides and selenides are deshielded still further (~-42 and $-37\,ppm$, respectively). Also, aryl and alkyl trifluoromethyl ethers have similar fluorine chemical shifts, as do aryl and alkyl trifluoromethyl sulfides.

Because of the high reactivity of such compounds toward acyl substitution, CF_3 carboxylate esters are rarely encountered. Nevertheless,

Scheme 5.9

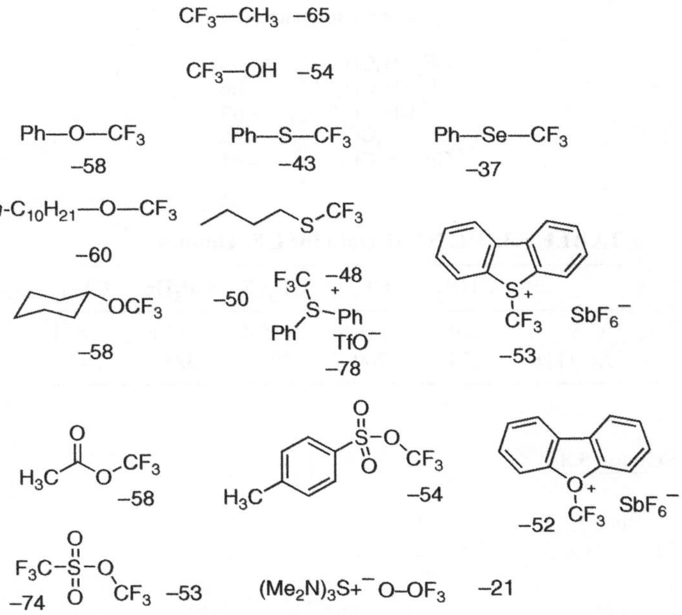

CF_3—CH_3 −65

CF_3—OH −54

Ph—O—CF_3 Ph—S—CF_3 Ph—Se—CF_3
    −58   −43   −37

n-$C_{10}H_{21}$—O—CF_3 ⋀⋀⋀$_S$-CF_3
      −60   −50

F_3C −48
$\overset{+}{S}$—Ph
Ph TfO$^-$
    −78

(dibenzothiophene S$^+$—CF_3 SbF$_6^-$)
    −53

⬡—OCF_3
    −58

H$_3$C$\overset{O}{\underset{}{C}}$—O-$CF_3$
        −58

H_3C—⬡—$\overset{O}{\underset{O}{S}}$-O$CF_3$
              −54

(dibenzofuran O$^+$—CF_3 SbF$_6^-$)
    −52

F_3C-$\overset{O}{\underset{O}{S}}$-O$CF_3$ −53
−74

$(Me_2N)_3S^+$ $^-$O–OF_3 −21

examples of both CF_3 carboxylate and CF_3 *sulfonate* esters are given in Scheme 5.9. Also included are examples of the electrophilic trifluoromethylating agents, in which the CF_3 groups are bound to either a positively charged sulfur or oxygen atom.

Although the effects are only relatively small, there appears to be a consistent trend in the chemical shifts of oxygen-bound CF_3 groups, with increasing donor ability of the oxygen giving rise to greater shielding of the CF_3 fluorines. Thus, the chemical shift for the trifluoromethyl oxonium ion reflects the greatest deshielding, whereas those of alkyl-O CF_3 groups reflect the greatest shielding.

Scheme 5.10

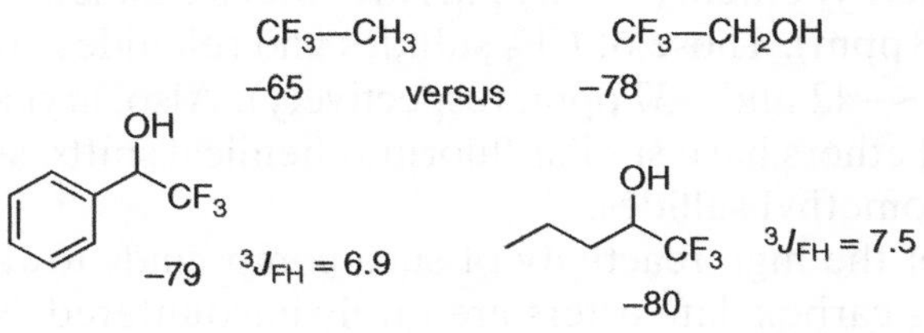

CF_3—CH_3 CF_3—CH_2OH
    −65   versus   −78

(phenyl)$\overset{OH}{\underset{}{CH}}$$CF_3$
        −79 $^3J_{FH} = 6.9$

⋀⋀$\overset{OH}{\underset{}{CH}}$$CF_3$ $^3J_{FH} = 7.5$
        −80

Scheme 5.11

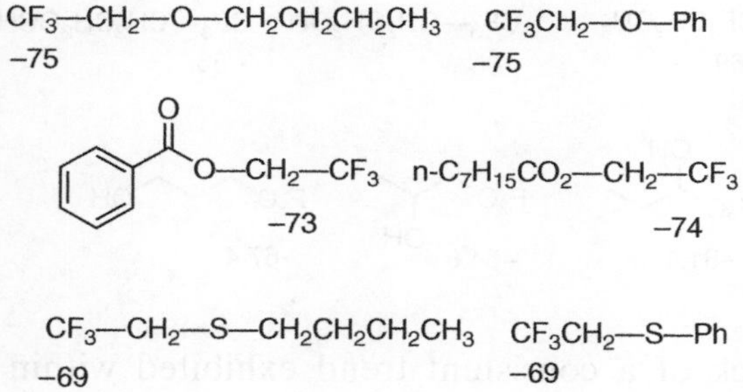

CF3—CH2—O—CH2CH2CH2CH3
–75
CF3CH2—O—Ph
–75
O
O—CH2—CF3
–73
n-C7H15CO2—CH2—CF3
–74
CF3—CH2—S—CH2CH2CH2CH3
–69
CF3CH2—S—Ph
–69

Interestingly, the trifluoromethyl oxonium compound is deshielded relative to the analogous ether, whereas those of the sulfonium CF_3 groups are actually shielded compared to the simple sulfides.

The fluorines of a CF_3 group vicinal to an alcohol (Scheme 5.10), ether or sulfide function (Scheme 5.11) are significantly *shielded*, relative to those of a trifluoromethylalkane.

The fluorine NMR spectrum of 1,1,1-trifluoro-2-propanol (Figure 5.5) provides an example of a compound with a hydroxyl group vicinal to a CF_3 group.

Move the alcohol function one carbon farther from the CF_3 group and one loses most of the shielding effect (Scheme 5.12). Note the

1.40
–81.45
–81.50
–81.55
–81.60
–81.65
–81.70
–81.75
–81.80
–81.85
–81.90
F1 (ppm)

FIGURE 5.5. ^{19}F NMR spectrum of 1,1,1-trifluoro-2-propanol

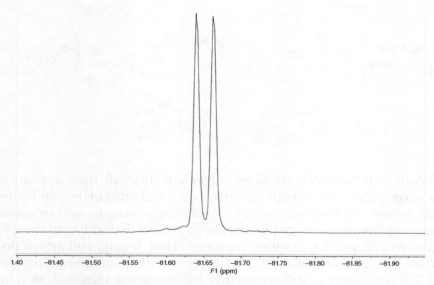

Scheme 5.12

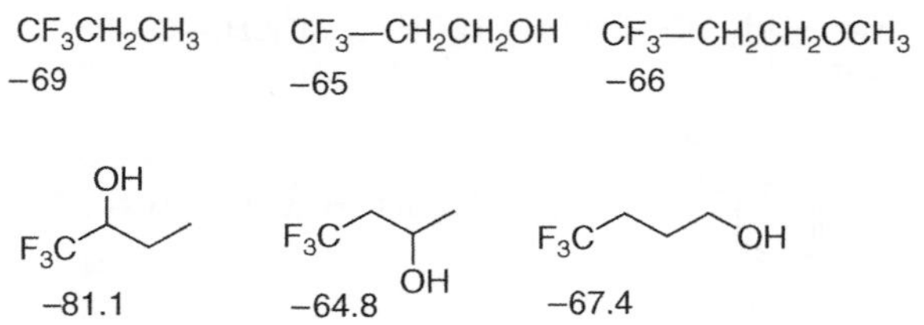

$CF_3CH_2CH_3$ $CF_3—CH_2CH_2OH$ $CF_3—CH_2CH_2OCH_3$
−69   −65   −66

F_3C ⟨OH⟩ F_3C ⟨OH⟩ F_3C ⟨OH⟩
−81.1   −64.8   −67.4

interesting lack of a consistent trend exhibited within the series of 1,1,1-trifluorobutanols given in the scheme.

5.3.2.1. ^1H and ^{13}C NMR Data. Some typical proton and carbon NMR data for trifluoromethyl ethers, sulfides, and esters are given in Scheme 5.13.

Scheme 5.13

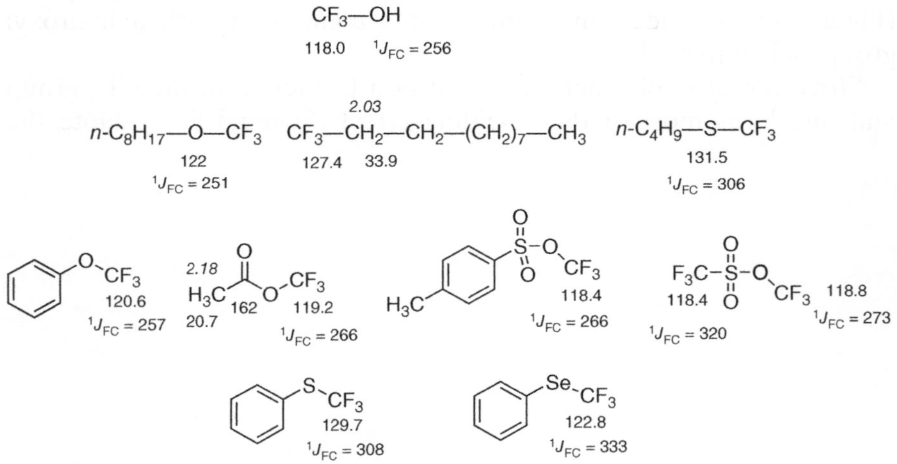

$CF_3—OH$
118.0 $^1J_{FC} = 256$

2.03
$n-C_8H_{17}—O—CF_3$ $CF_3—CH_2—CH_2—(CH_2)_7—CH_3$ $n-C_4H_9—S—CF_3$
122   127.4   33.9   131.5
$^1J_{FC} = 251$ $^1J_{FC} = 306$

$O—CF_3$ (phenyl)   2.18 H_3C—C(=O)—O—CF_3 (tolyl)—S(=O)(=O)—O—CF_3 F_3C—S(=O)(=O)—O—CF_3
120.6   162   119.2   118.4   118.4   118.8
$^1J_{FC} = 257$ 20.7 $^1J_{FC} = 266$ $^1J_{FC} = 266$ $^1J_{FC} = 320$ $^1J_{FC} = 273$

(phenyl)—S—CF_3 (phenyl)—Se—CF_3
129.7   122.8
$^1J_{FC} = 308$ $^1J_{FC} = 333$

Although the effects are again only relatively small, there appears to be a consistent trend in both the carbon chemical shifts of oxygen-bound CF_3 groups and their one-bond C–F coupling constants, with *increased donor ability* of the oxygen giving rise to greater *deshielding* of the CF_3 carbons and smaller coupling constants. Thus, n-octyl trifluoromethyl ether has a chemical shift of 122 ppm, with a coupling constant of 251 Hz, whereas trifluoromethyl triflate has a chemical shift of 118.8 ppm, with a coupling constant of 273 Hz.

Continuing the trend observed going from CH_2F to CF_2H to CF_3 carbons, the ^{13}C chemical shift of a trifluoromethyl ether is actually *more shielded* (by about 5 ppm) than that of a trifluoromethyl *hydrocarbon*. Scheme 5.14 summarizes the relative impact of an ether substituent upon the chemical shifts of various fluorinated carbons.

Scheme 5.14

Hydrocarbon	Ether	Δδppm
$CH_3CH_2CH_2CH_2CH_3$	$CH_3CH_2CH_2—O—CH_3$	
13.9	57.6	+43.7
$CH_3CH_2CH_2CH_2CH_2F$	$CH_3—O—CH_2F$	
83.6	104.8	+21.2
$n\text{-}C_7H_{15}CF_2H$	$CH_3CH_2—O—CF_2H$	
117.5	115.8	−1.7
$CH_3CH_2CH_2CF_3$	$n\text{-}C_8H_{17}—O—CF_3$	
127.5	122.0	−5.5

Regarding proton spectra, as was the case with 2,2,2-trifluoroethyl chloride (Scheme 5.8), the chemical shifts of the CH_2 protons of 2,2,2-trifluoroethanol and of 2,2,2-trifluoroethyl ethers are more affected by the OH or ether substituents than they are by the CF_3 group.

Trends in proton chemical shifts as the alcohol function is moved progressively farther from the CF_3 function can be seen in the 1,1,1-trifluoropentan-2- and -3-ols, and 4,4,4-trifluorobutan-1-ol series of compounds in Scheme 5.15. The proton NMR spectrum of 1,1,1-trifluoro-2-propanol provides an example of an alcohol with a vicinal OH group (Figure 5.6).

The proton at carbon 2 appears as a septet at 4.14 ppm, which indicates that all of the three-bond couplings must be virtually the same (about 6 Hz). This includes the observed coupling of the OH proton, which appears as a doublet at 2.13 ppm. The CH_3 protons also appear as a doublet at 1.40 ppm.

Regarding carbon spectra, an alcohol, ether, or ester function on the adjacent carbon shields the carbons of a CF_3 group (Scheme 5.15). Again, 1,1,1-trifluoro-2-propanol is used to provide an example of a carbon NMR spectrum (Figure 5.7). Three signals are seen, with the CF_3 carbon appearing as a quartet at 125 ppm with a one-bond C–F

Scheme 5.15

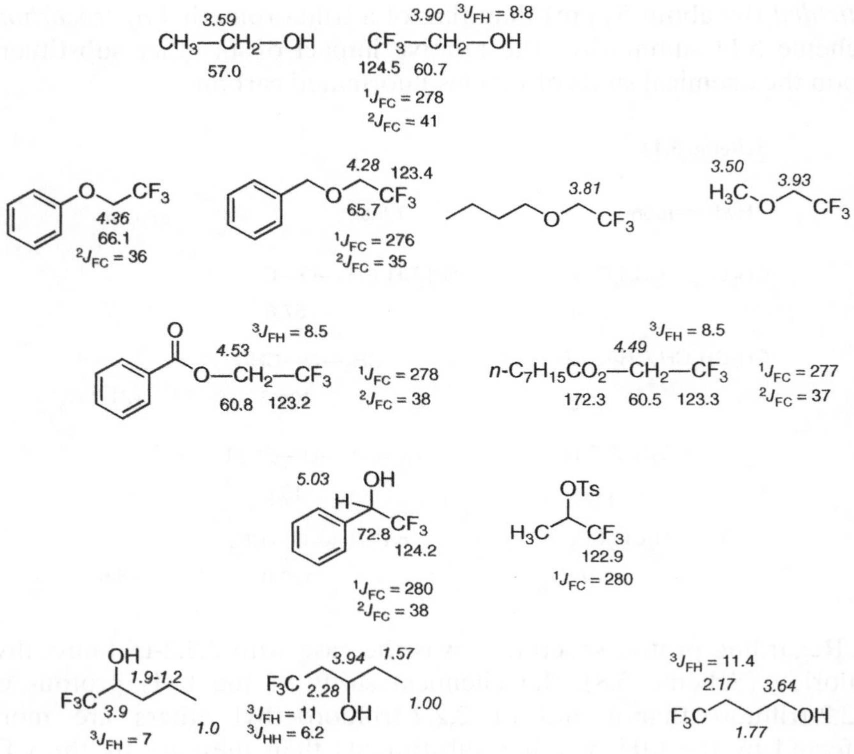

coupling constant of 280 Hz. The other two carbon signals appear at 67.0 (q, $^2J_{FC} = 32$ Hz) and 15.5 (br s) ppm.

A couple of examples of 3,3,3-trifluoropropyl alcohols and an ether are provided in Scheme 5.16, which have a slight shielding influence on the CF$_3$ fluorines.

5.3.3. Sulfones, Sulfoxides, and Sulfoximines

Other classes of sulfur-bound CF$_3$ groups include sulfones, sulfoxides, and sulfoximines, examples of which are given in Scheme 5.17.

5.3.4. Amines and Nitro Compounds

Unlike fluoromethyl- and difluoromethyl amines, trifluoromethy-lamines have reasonable kinetic stability when the nitrogen does not bear a hydrogen. As is the case with other relatively electronegative

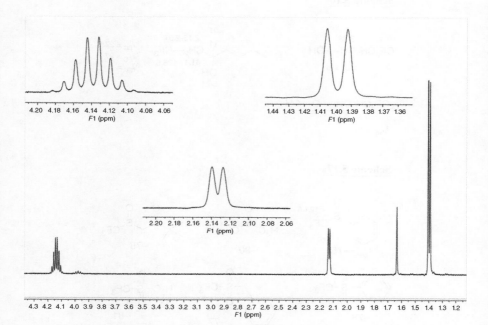

FIGURE 5.6. ^{1}H NMR spectrum of 1,1,1-trifluoro-2-propanol

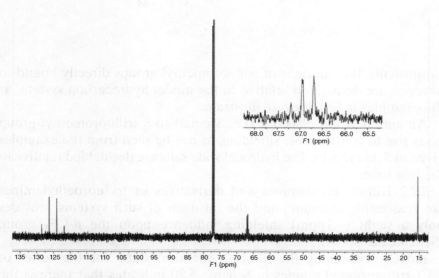

FIGURE 5.7. ^{13}C NMR spectrum of 1,1,1-trifluoro-2-propanol

Scheme 5.16

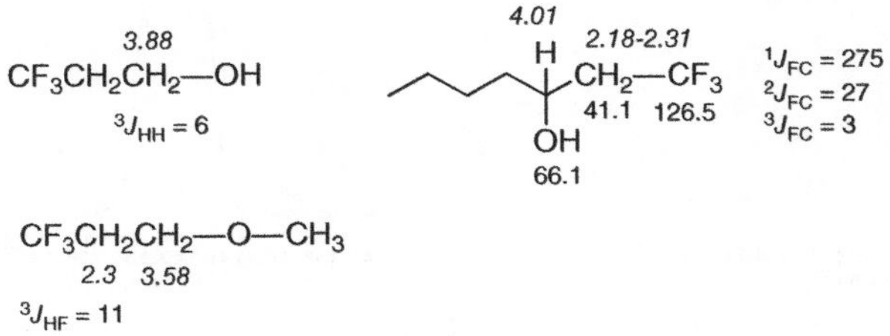
3.88
CF3CH2CH2—OH
3JHH = 6
4.01
H
2.18-2.31
CH2—CF3
41.1 126.5
OH
66.1
1JFC = 275
2JFC = 27
3JFC = 3
CF3CH2CH2—O—CH3
2.3 3.58
3JHF = 11

Scheme 5.17

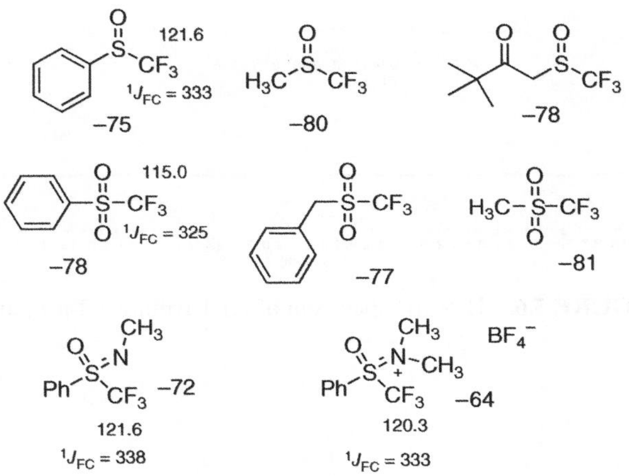
121.6
1JFC = 333
−75
H3C S CF3
−80
−78
115.0
1JFC = 325
−78
−77
H3C-S-CF3
−81
Ph
CF3 −72
121.6
1JFC = 338
BF4−
CF3 −64
120.3
1JFC = 333

substituents, the fluorines of trifluoromethyl groups directly bound to nitrogen are deshielded relative to the model hydrocarbon system, as the examples in Scheme 5.18 illustrate.

An amino group placed β (i.e., vicinal) to a trifluoromethyl group gives rise to considerable shielding, as can be seen from the examples given in Scheme 5.19. The hydrochloride salts are deshielded relative to the free base.

2,2,2-Trifluoroethylamines and derivatives of trifluoroethylamines are reasonably common, and the nitrogen of such systems provides only a slight (~3 ppm) shielding influence upon the β-CF_3 group (Scheme 5.20).

Indeed, examining the fluorine chemical shifts for the series of 1,1,1-trifluorobutyl amines in Scheme 5.20 indicates that there is the same lack of consistent trend that was observed for the analogous alcohols.

Scheme 5.18

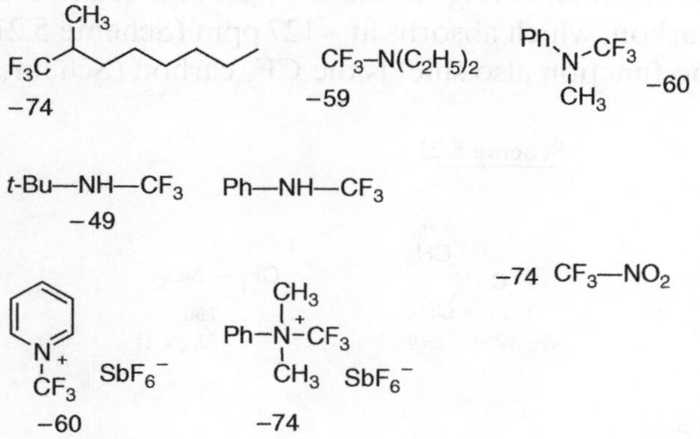

CH$_3$
F$_3$C—CH—(CH$_2$)$_n$CH$_3$
−74

CF$_3$—N(C$_2$H$_5$)$_2$
−59

Ph—N(CH$_3$)—CF$_3$
−60

t-Bu—NH—CF$_3$ Ph—NH—CF$_3$
−49

−74 CF$_3$—NO$_2$

Pyridinium-N—CF$_3$ SbF$_6^-$
−60

Ph—N$^+$(CH$_3$)$_2$—CF$_3$ SbF$_6^-$
−74

Scheme 5.19

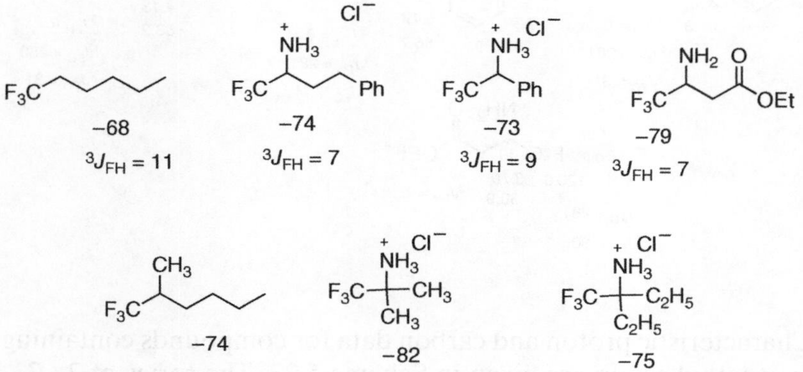

F$_3$C—(CH$_2$)$_4$CH$_3$
−68
$^3J_{FH} = 11$

F$_3$C—CH(NH$_3^+$)—CH$_2$CH$_2$Ph Cl$^-$
−74
$^3J_{FH} = 7$

F$_3$C—CH(NH$_3^+$)—Ph Cl$^-$
−73
$^3J_{FH} = 9$

F$_3$C—CH(NH$_2$)—CH$_2$—C(=O)OEt
−79
$^3J_{FH} = 7$

CH$_3$
F$_3$C—CH—(CH$_2$)$_2$CH$_3$
−74

F$_3$C—C(NH$_3^+$)(CH$_3$)—CH$_3$ Cl$^-$
−82

F$_3$C—C(NH$_3^+$)(C$_2$H$_5$)—C$_2$H$_5$ Cl$^-$
−75

Scheme 5.20

CF$_3$CH$_2$CH$_3$
−69

CF$_3$CH$_2$NH$_2$
−72

Ph—C(=O)—NH—CH$_2$—CF$_3$
−71

CF$_3$CH$_2$CH$_2$NH$_2$
−66

F$_3$C—CH(NH$_3^+$)—CH$_2$CH$_3$ Cl$^-$
−75.5

F$_3$C—CH$_2$—CH(NH$_3^+$)— Cl$^-$
−61.8

F$_3$C—CH$_2$—CH$_2$—NH$_3^+$
−67.1

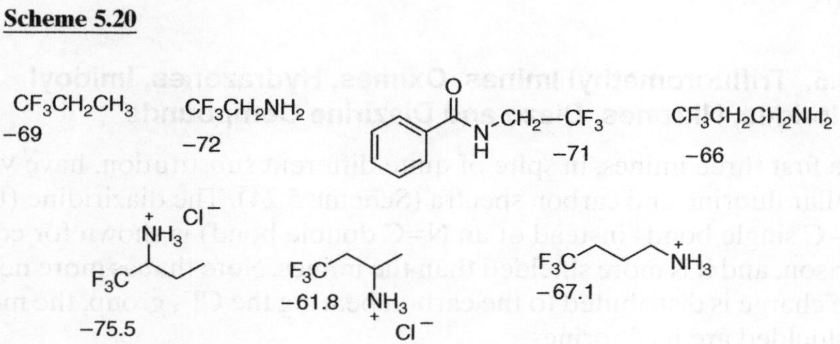

5.3.4.1. *¹H, ¹⁵N, and ¹³C NMR Data.* Similarly to oxygen, but to a lesser degree, an amino nitrogen bound directly to a CF_3 group gives rise to slight *shielding* of the CF_3 carbon relative to that of a CF_3-hydrocarbon, which absorbs at ~127 ppm (Scheme 5.21).

A β-amino function also shields the CF_3 carbon (Scheme 5.22).

Scheme 5.21

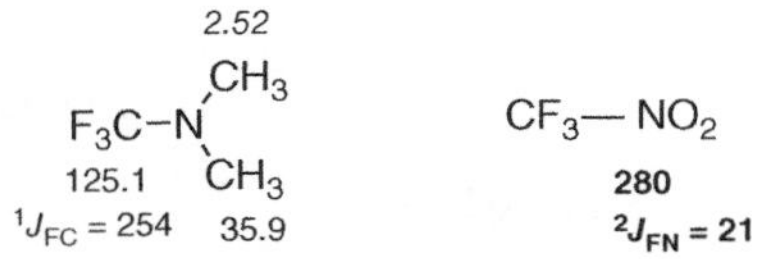

Scheme 5.22

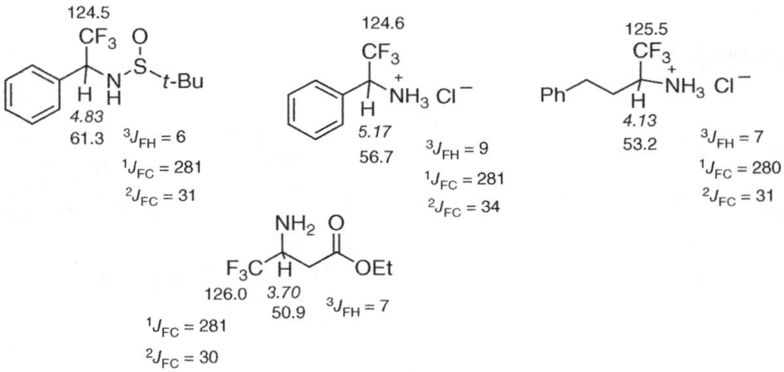

Characteristic proton and carbon data for compounds containing the trifluoroethyl group are given in Scheme 5.23. The series of 2-, 3-, and 4-butanamines provides insight regarding trends in proton chemical shifts.

5.3.5. Trifluoromethyl Imines, Oximes, Hydrazones, Imidoyl Chlorides, Nitrones, Diazo and Diazirine Compounds

The first three imines, in spite of quite different substitution, have very similar fluorine and carbon spectra (Scheme 5.24). The diaziridine (two N—C single bonds instead of an N=C double bond) is shown for comparison, and it is more shielded than the imines. Note that as more negative charge is distributed to the carbon bearing the CF_3 group, the more deshielded are its fluorines.

Scheme 5.23

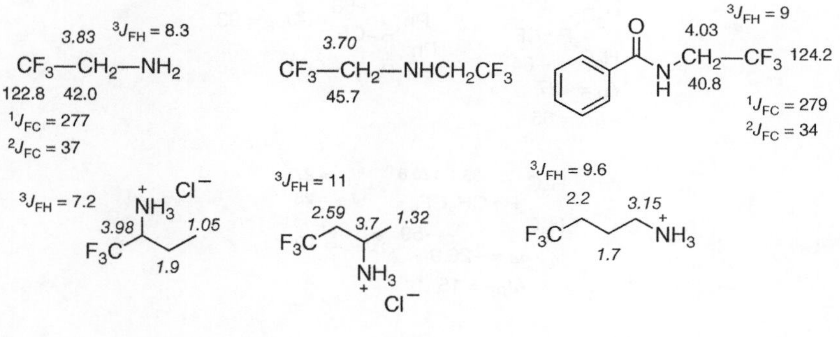

$$3.83 \quad {}^3J_{FH} = 8.3$$
$$CF_3{-}CH_2{-}NH_2$$
$$122.8 \quad 42.0$$
$${}^1J_{FC} = 277$$
$${}^2J_{FC} = 37$$

$$3.70$$
$$CF_3{-}CH_2{-}NHCH_2CF_3$$
$$45.7$$

$${}^3J_{FH} = 9$$
Ph—C(=O)—N(H)—CH$_2$—CF$_3$ 124.2
4.03 40.8
$${}^1J_{FC} = 279$$
$${}^2J_{FC} = 34$$

$${}^3J_{FH} = 7.2$$
$$\overset{+}{NH_3} \quad Cl^-$$
$$3.98 \quad 1.05$$
$$F_3C \quad 1.9$$

$${}^3J_{FH} = 11$$
$$2.59 \quad 3.7 \quad 1.32$$
$$F_3C$$
$$\overset{NH_3}{\underset{+}{}} \quad Cl^-$$

$${}^3J_{FH} = 9.6$$
$$2.2 \quad 3.15$$
$$F_3C \quad \overset{+}{NH_3}$$
$$1.7$$

Scheme 5.24

−71 F$_3$C—CH=N—Ph 7.81
118.9 147.2 ${}^3J_{FH} = 3.5$
$${}^1J_{FC} = 272 \quad {}^2J_{FC} = 39$$

−70 F$_3$C—C(Ph)=N—Ph
119.8
$${}^1J_{FC} = 278$$

−72 F$_3$C—C(Ph)=N—butyl
119.7
$${}^1J_{FC} = 279$$

−76 F$_3$C—C(Ph)=N—NH(HN)
HN–NH

$${}^3J_{FH} = 7.7$$
N–N–Ts
2.33 −70.5
CF$_3$
145.3 120.4
$${}^2J_{FC} = 34 \quad {}^1J_{FC} = 275$$

−65.5 F$_3$C—C(H)=N—OH
119.7 139.7 7.5
$${}^1J_{FC} = 270 \quad {}^2J_{FC} = 38$$

N=N
Ph—C(CF$_3$)=N=N −66
$${}^3J_{FH} = 4.1$$

−66 F$_3$C—C(H)=N(→O)—CH$_2$—Ph
119.5 123.5 6.87
$${}^1J_{FC} = 268$$
$${}^2J_{FC} = 37$$
$${}^3J_{FH} = 5.5$$

Ph—C(CF$_3$)=N$_2$ −57
126.3
$${}^1J_{FC} = 269$$

−72 F$_3$C—C(Cl)=N—Ph
117.1 132.1
$${}^1J_{FC} = 275 \quad {}^2J_{FC} = 43$$

−73 F$_3$C—C(Cl)=N—S(O$_2$)—Ph

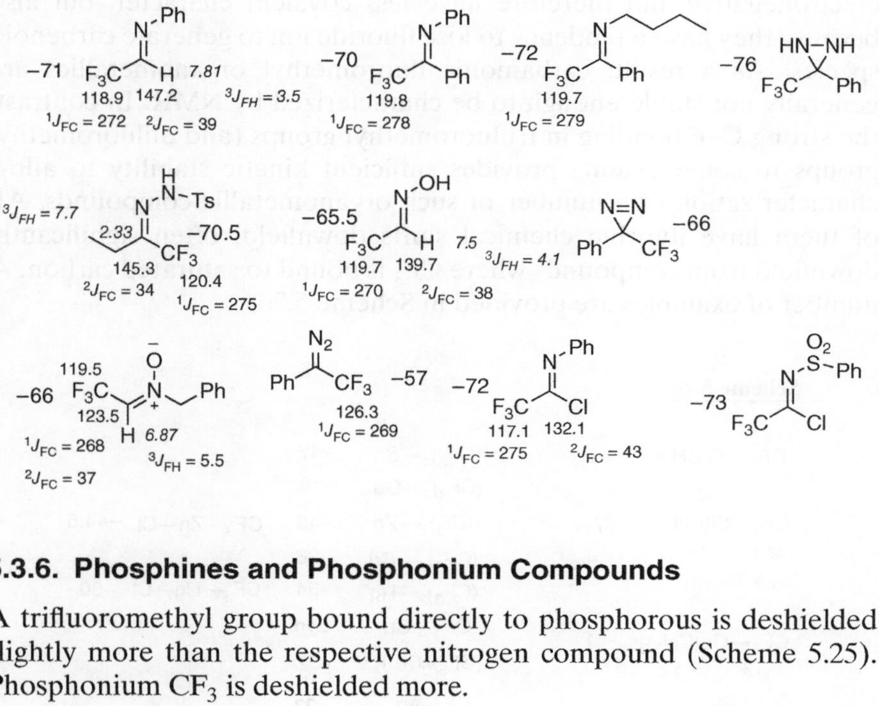

5.3.6. Phosphines and Phosphonium Compounds

A trifluoromethyl group bound directly to phosphorous is deshielded slightly more than the respective nitrogen compound (Scheme 5.25). Phosphonium CF_3 is deshielded more.

5.3.7. Organometallics[2]

Although one generally thinks of CH_3 organometallic compounds, such as methyl lithium or methyl Grignards, as being carbanionic in character, such compounds are generally covalent in character but

Scheme 5.25

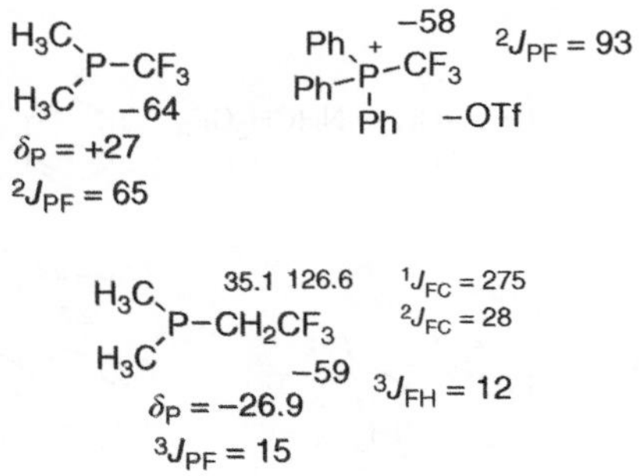

$$
\begin{array}{cc}
\underset{\overset{H_3C}{\displaystyle H_3C}}{\searrow}\!\!P\!-\!CF_3 & \underset{\overset{Ph}{\displaystyle Ph}}{\searrow}\!\!\overset{+}{P}\!-\!CF_3 \\
\end{array}
$$

$\overset{H_3C}{\underset{H_3C}{}}\!P\!-\!CF_3$ −64

$\delta_P = +27$
$^2J_{PF} = 65$

$\underset{Ph}{Ph}\overset{+}{-}\!\!P\!-\!CF_3$ −58 $^2J_{PF} = 93$
        |
       Ph   −OTf

$\overset{H_3C}{\underset{H_3C}{}}\!P\!-\!CH_2CF_3$
35.1 126.6 $^1J_{FC} = 275$
              $^2J_{FC} = 28$
         −59 $^3J_{FH} = 12$
$\delta_P = -26.9$
$^3J_{PF} = 15$

behave chemically like carbanions. Analogous fluorinated methyl organometallics are much less stable because such groups are more electronegative and therefore have less covalent character, but also because they have a tendency to lose fluoride ion to generate carbenoid species. As a result, carbanionic fluoromethyl organometallics are generally not stable enough to be characterized by NMR. In contrast, the strong C–F bonding in trifluoromethyl groups (and difluoromethyl groups to some extent) provides sufficient kinetic stability to allow characterization of a number of such organometallic compounds. All of them have fluorine chemical shifts downfield, often significantly downfield from compounds where CF_3 is bound to saturated carbon. A number of examples are provided in Scheme 5.26.

Scheme 5.26

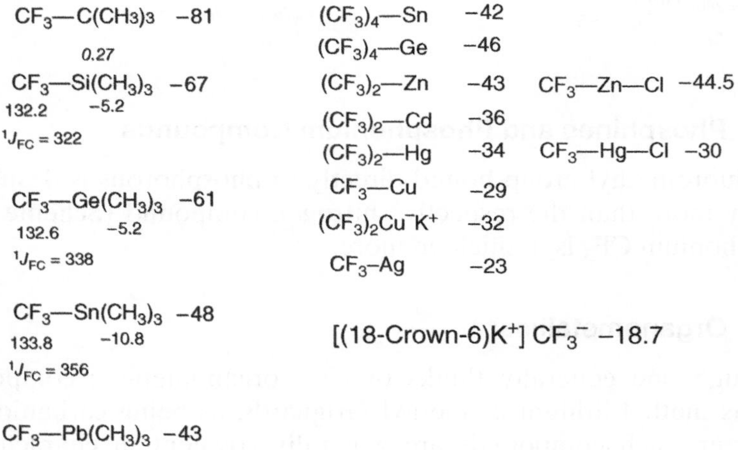

$CF_3\!-\!C(CH_3)_3$ −81

0.27

$CF_3\!-\!Si(CH_3)_3$ −67
132.2   −5.2
$^1J_{FC} = 322$

$CF_3\!-\!Ge(CH_3)_3$ −61
132.6   −5.2
$^1J_{FC} = 338$

$CF_3\!-\!Sn(CH_3)_3$ −48
133.8   −10.8
$^1J_{FC} = 356$

$CF_3\!-\!Pb(CH_3)_3$ −43

$(CF_3)_4\!-\!Sn$ −42
$(CF_3)_4\!-\!Ge$ −46
$(CF_3)_2\!-\!Zn$ −43 $CF_3\!-\!Zn\!-\!Cl$ −44.5
$(CF_3)_2\!-\!Cd$ −36
$(CF_3)_2\!-\!Hg$ −34 $CF_3\!-\!Hg\!-\!Cl$ −30
$CF_3\!-\!Cu$ −29
$(CF_3)_2Cu^-K^+$ −32
$CF_3\!-\!Ag$ −23

$[(18\text{-Crown-6})K^+]\, CF_3^-$ −18.7

Data for the group 4 CF_3-$X(CH_3)_3$ series, including $TMSCF_3$ (the Ruppert–Prakash reagent), show an increasing trend of deshielding of the CF_3 group as one proceeds down the group.

The trifluoromethyl anion itself, which has no covalent character, has been shown to have the most deshielded fluorines of any group in the scheme.

5.4. BORONIC ESTERS

An example of CF_3 bound to boron in a boronic ester is provided in Scheme 5.27, as is an example of a boronic ester with the CF_3 group one carbon away. The fluorines of this trifluoroethyl group are perhaps the most deshielded of any CF_3 group bound to a saturated carbon.

Scheme 5.27

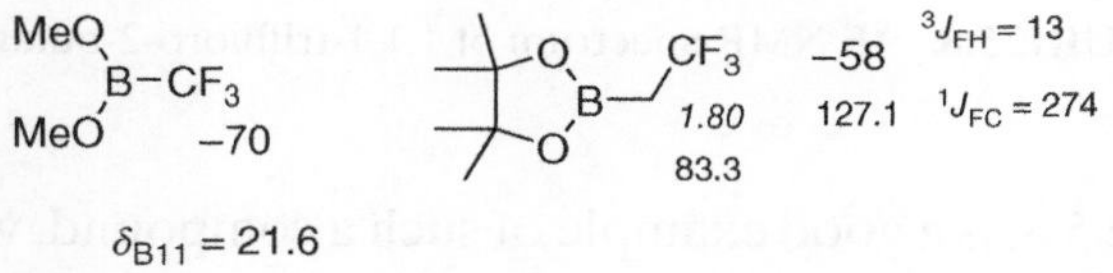

MeO
    B–CF_3
MeO   –70

$\delta_{B11} = 21.6$

$^3J_{FH} = 13$

CF_3 –58
1.80   127.1 $^1J_{FC} = 274$
83.3

5.5. CARBONYL COMPOUNDS

The impact of a carbonyl group on the chemical shift of a CF_3 group somewhat emulates the effect of other electronegative substituents. As indicated by the examples of aldehyde and ketones in Scheme 5.28, when the carbonyl carbon is *directly attached* to the CF_3 group, it causes shielding of the CF_3 fluorines. Note that aryl trifluoromethyl ketones are considerably less shielded than are alkyl trifluoromethyl ketones. The fluorine signal of trifluoromethyl ketones of course always appears as a sharp singlet. The spectrum of 1,1,1-trifluoromethyl-2-butanone, as

Scheme 5.28

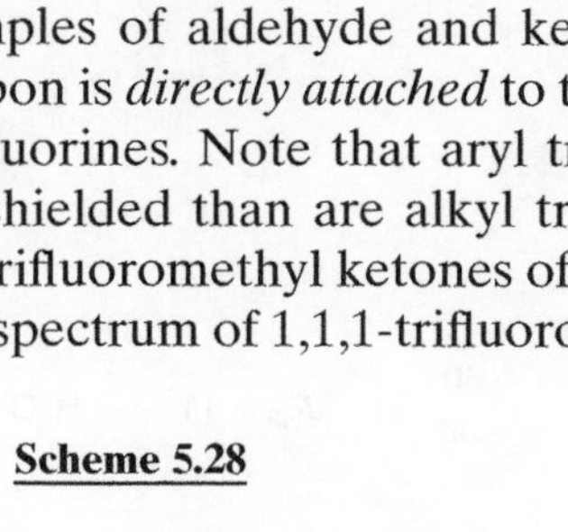

CF_3—CH_2CH_3 CF_3CHO
–69   –82

CH_3 CF_3 Ph CF_3
–81   –72

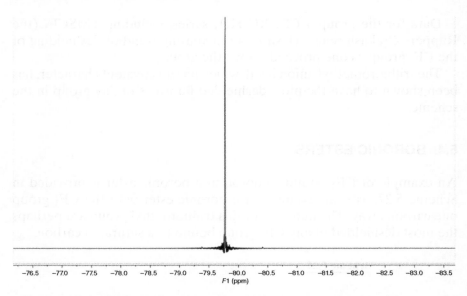

-76.5 -77.0 -77.5 -78.0 -78.5 -79.0 -79.5 -80.0 -80.5 -81.0 -81.5 -82.0 -82.5 -83.0 -83.5
F1 (ppm)

FIGURE 5.8. ^{19}F NMR spectrum of 1,1,1-trifluoro-2-butanone

seen in Figure 5.8, is a good example of such a compound, with its singlet fluorine signal at −80.8 ppm.

When the trifluoromethyl group is at the β-position, as in a tri-fluoroethyl group, the carbonyl group then leads to *de*shielding of the fluorines, and three-bond H–F coupling in the range of 10 Hz is observed (Scheme 5.29). When the trifluoromethyl group is at the γ position (or presumably further from the carbonyl group), little

Scheme 5.29

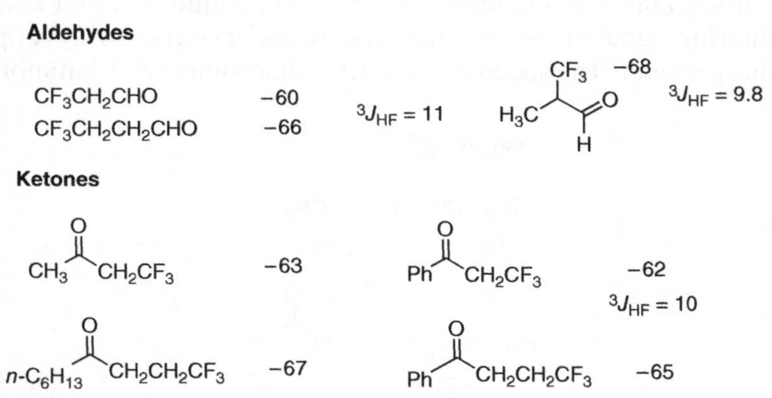

Aldehydes

CF$_3$CH$_2$CHO   −60 $^3J_{HF} = 11$

CF$_3$CH$_2$CH$_2$CHO   −66

CF$_3$ −68 $^3J_{HF} = 9.8$

Ketones

CH$_3$—C(O)—CH$_2$CF$_3$ −63

n-C$_6$H$_{13}$—C(O)—CH$_2$CH$_2$CF$_3$ −67

Ph—C(O)—CH$_2$CF$_3$ −62 $^3J_{HF} = 10$

Ph—C(O)—CH$_2$CH$_2$CF$_3$ −65

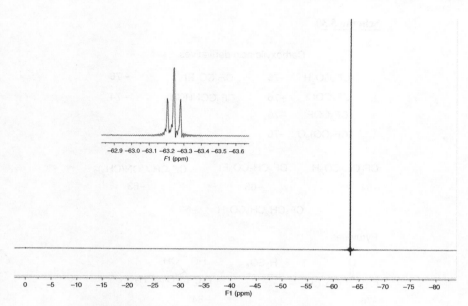

FIGURE 5.9. ^{19}F NMR spectrum of 4,4,4-trifluoro-2-butanone

influence relative to $CF_3CH_2CH_3$ is observed. The fluorine spectrum of 4,4,4-trifluoro-2-butanone (Figure 5.9) is a good example, exhibiting a triplet at −63.8 ppm with three-bond H–F coupling of 10.4 Hz.

With regard to carboxylic derivatives, there is generally little difference in chemical shift among the various trifluoroacetic acid derivatives, as exemplified by the examples in Scheme 5.30. Also, the effect of moving the CF_3 farther from the carboxylic acid function is similar to that seen with the aldehydes and ketones. Trifluoromethyl ketones will often be in equilibrium with their hydrated form, in which case signals from both the hydrate and water-free ketone will be observed, as is the case for the following pyruvate example.

5.5.1. ^{1}H and ^{13}C NMR Data

Some typical carbon NMR data for trifluoromethyl compounds are provided in Scheme 5.31.

There is no noticeable four-bond coupling between the CF_3 fluorines and the CH_2 protons of 1,1,1-trifluoro-2-butanone, as can be seen in the ^{1}H spectrum of this compound given in Figure 5.10.

The carbon spectrum of this compound (Figure 5.11) is typical of such spectra, where the CF_3 group and the carbonyl group have chemical shifts of δ 115.7 and 192.1, respectively, and F–C couplings of

Scheme 5.30

Carboxylic acid derivatives

CF_3CO_2H	−76	CF_3CO_2Et	−76
CF_3COCl	−76	$CF_3CONHCH_3$	−74
CF_3COF	−75		
$(CF_3CO)_2O$	−76		

$CF_3CH_2CO_2H$ $CF_3CH_2CO_2Et$ $CF_3CH_2CON(CH_3)_2$

−64   −65   −63

$CF_3CH_2CH_2CO_2H$ −68

Pyruvates

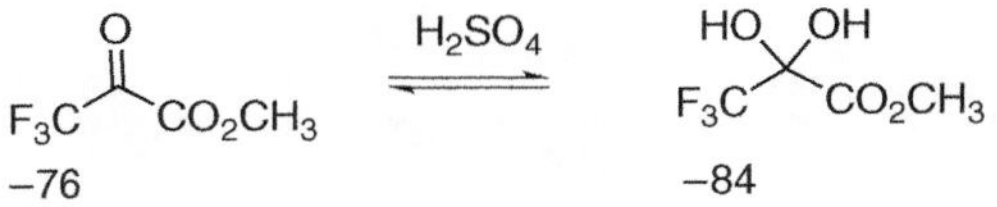

$$F_3C \overset{O}{\underset{}{\|}} CO_2CH_3 \quad \underset{\rightleftharpoons}{\overset{H_2SO_4}{}} \quad F_3C \overset{HO \quad OH}{\underset{}{}} CO_2CH_3$$

−76   −84

Scheme 5.31

$$\underset{115.0}{F_3C} \overset{O}{\underset{}{\|}} \overset{154.4}{\underset{2.72}{CH_2}} \overset{}{\underset{1.13}{-CH_3}} \quad {}^1J_{FC} = 285 \quad {}^2J_{FC} = 46$$
30.7   5.9

$$\overset{O}{\underset{180.3}{\|}} \overset{116.7}{CF_3}$$
$${}^1J_{FC} = 291$$
$${}^2J_{FC} = 34$$

$$\overset{116.2 \quad 159.6}{CF_3-CO_2CH_3}$$
$${}^1J_{FC} = 285$$
$${}^2J_{FC} = 42$$

$$\overset{116.7}{F_3C} \overset{O}{\underset{175.2}{\|}} \overset{155.4}{CO_2CH_3}$$
$${}^1J_{FC} = 290$$
$${}^2J_{FC} = 39$$
$${}^3J_{FC} = \sim0$$

$$\overset{HO \quad OH}{\underset{121.7}{F_3C} \underset{90.3}{}} \overset{166.6}{CO_2CH_3}$$
$${}^1J_{FC} = 289$$
$${}^2J_{FC} = 32$$
$${}^3J_{FC} = \sim0$$

$$\underset{126.4}{CF_3}\overset{26.4}{CH_2CH_2CHO}$$
$${}^1J_{FC} = 276$$
$${}^2J_{FC} = 30$$

$$Ph \overset{O}{\underset{42.0}{\|}} \overset{3.80}{\underset{124.0}{CF_3}}$$
$${}^1J_{FC} = 275$$
$${}^2J_{FC} = 28$$

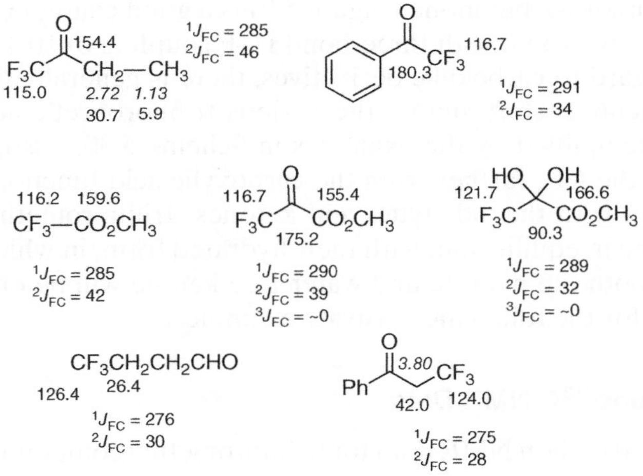

292 and 34 Hz, respectively. In this case, no three-bond F–C coupling is able to be discerned at the CH_2 carbon, which appears as a broad singlet at 30.1 ppm.

Some proton and carbon data for *trifluoroethyl* compounds are given in Scheme 5.32, with the proton and carbon spectra of

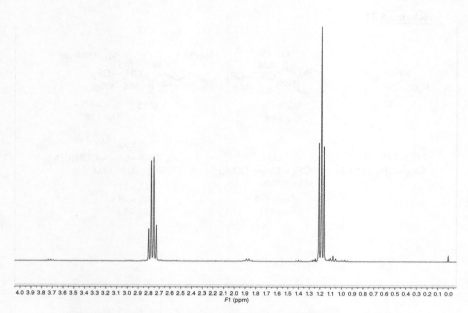

FIGURE 5.10. ¹H NMR spectrum of 1,1,1-trifluoro-2-butanone

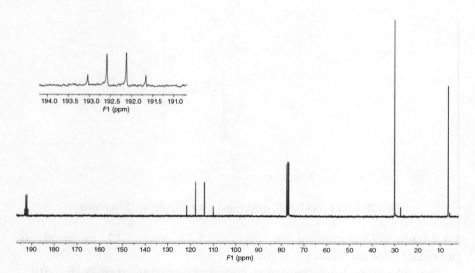

FIGURE 5.11. ¹³C NMR spectrum of 1,1,1-trifluoro-2-butanone

Scheme 5.32

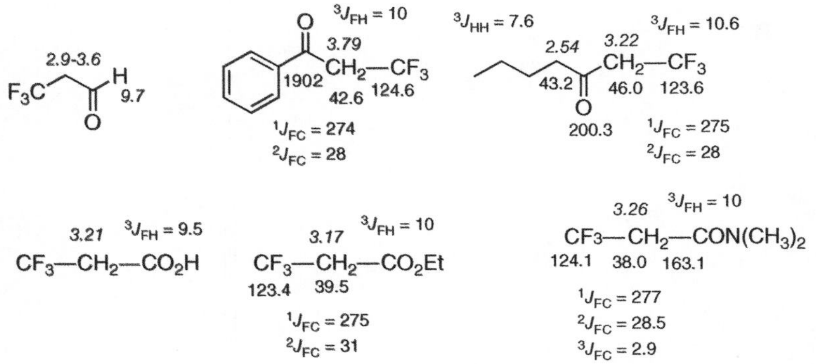

Structures in Scheme 5.32:

F_3C—CH_2—CHO, 2.9–3.6, H 9.7, O

PhC(O)CH$_2$—CF$_3$: $^3J_{FH} = 10$, 3.79, 1902, CH_2—CF_3, 42.6 124.6, $^1J_{FC} = 274$, $^2J_{FC} = 28$

$^3J_{HH} = 7.6$, 2.54, 3.22, $^3J_{FH} = 10.6$, CH_2—CF_3, 43.2, O, 46.0 123.6, 200.3, $^1J_{FC} = 275$, $^2J_{FC} = 28$

CF_3—CH_2—CO_2H: 3.21, $^3J_{FH} = 9.5$

CF_3—CH_2—CO_2Et: 3.17, $^3J_{FH} = 10$, 123.4 39.5, $^1J_{FC} = 275$, $^2J_{FC} = 31$

CF_3—CH_2—$CON(CH_3)_2$: 3.26, $^3J_{FH} = 10$, 124.1 38.0 163.1, $^1J_{FC} = 277$, $^2J_{FC} = 28.5$, $^3J_{FC} = 2.9$

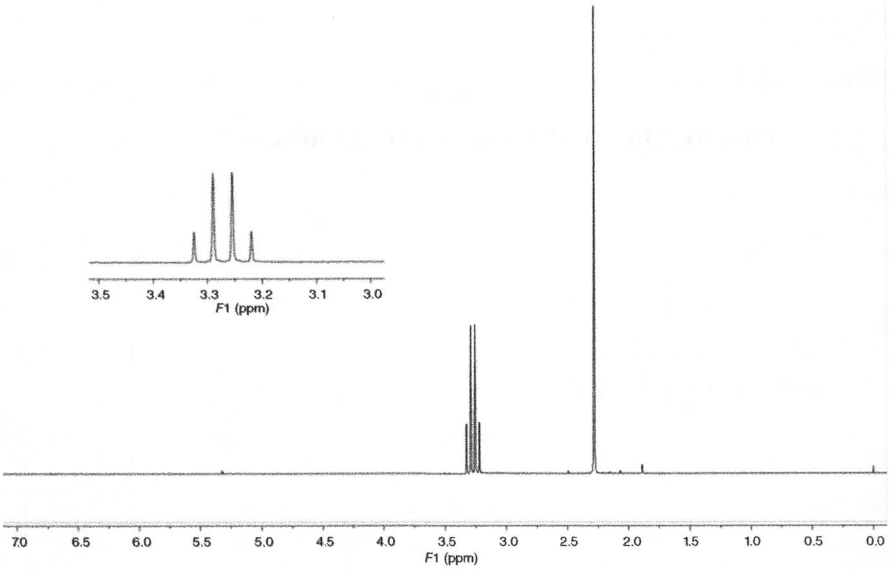

FIGURE 5.12. ^{1}H NMR spectrum of 4,4,4-trifluoro-2-butanone

4,4,4-trifluoro-2-butanone in Figures 5.12 and 5.13, respectively, providing characteristic examples.

The above proton spectrum shows a quartet at δ 3.18 with a three-bond HF coupling of 10.5 Hz for the CH$_2$ group.

The ^{13}C NMR spectrum exhibits a quartet at 123.8 ppm with one-bond F–C coupling of 278 Hz, another quartet at 46.9 ppm, with two-bond F–C coupling of 28 Hz, and a small quartet at 198.2 ppm (for the carbonyl) with still observable three-bond coupling of 2.5 Hz.

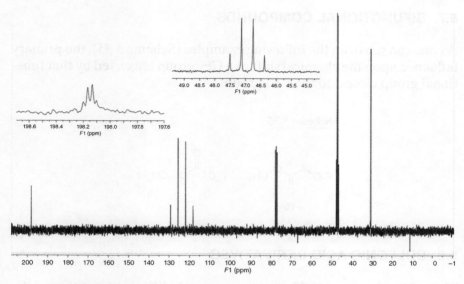

FIGURE 5.13. ^{13}C NMR spectrum of 4,4,4-trifluoro-2-butanone

5.6. NITRILES

Unlike the effect of a carbonyl function, a nitrile group attached to a CF$_3$ group will *deshield* the CF$_3$ fluorines, with a diminished deshielding effect being exerted as the CN is placed farther away (Scheme 5.33).

Scheme 5.33

$$CF_3CN \qquad CF_3CH_2CN \qquad \overset{\displaystyle Br}{\underset{\displaystyle |}{CF_3CHCN}}$$

$$-60 \qquad\qquad -65 \qquad\qquad\quad -69$$

5.6.1. ^{13}C NMR Data for Nitriles

Little data of this type are available (Scheme 5.34).

Scheme 5.34

$$CF_3CN \qquad CF_3CH_2CN \qquad \overset{\displaystyle Br}{\underset{\displaystyle |}{CF_3CHCN}}$$

130.8 128.9 $\qquad$ 1.4 $\qquad\qquad$ 4.9

$^1J_{FC} = 266$ $\qquad\quad$ $^3J_{FH} = 9.2$

$^2J_{FC} = 56$

5.7. BIFUNCTIONAL COMPOUNDS

As one can see from the following examples (Scheme 5.35), the primary influence upon the chemical shift of a CF_3 group is exerted by that functional group closest to the CF_3 group.

Scheme 5.35

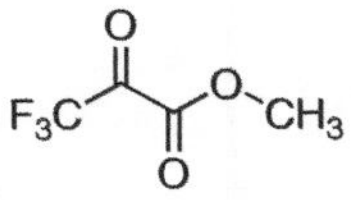

−76 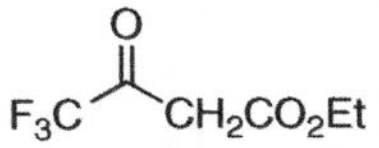   −75

5.8. SULFONIC ACID DERIVATIVES

The influence of an SO_3H group is not much different from that of the CO_2H group. All of the derivatives of triflic acid have very similar chemical shifts (Scheme 5.36).

Scheme 5.36

CF_3SO_3H −80

$CF_3SO_3CH_3$ −75 $(CF_3SO_2)_2O$ −72

CF_3SO_2Cl −76 CF_3SO_3TMS −78

5.9. ALLYLIC AND PROPARGYLIC TRIFLUOROMETHYL GROUPS

Trifluoromethyl groups that are bound to sp^2 carbons of alkenes, arenes, or heterocyclic compounds are slightly deshielded compared to the saturated counterparts, but the influence is relatively small (Scheme 5.37).

Scheme 5.37

$CF_3CH_2CH_3$ $CF_3{-}CH{=}CH_2$

−69   −67   −63

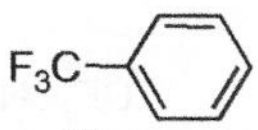

5.9.1. Allylic Trifluoromethyl Groups

The chemical shifts of trifluoromethyl groups directly bound to simple hydrocarbon carbon–carbon double bonds are not significantly different from those of analogous saturated systems (Scheme 5.38). The exact chemical shift for $CF_3CH=C$ compounds varies slightly depending on how many alkyl or aryl groups are bound to the vicinal carbon. Anytime an alkyl or aryl group is cis to the CF_3 group, this leads to deshielding relative to the situation when there is only a trans-substituent. Again, this is consistent with the concept of "steric deshielding" of the CF_3 group by a proximate, in this case, cis-alkyl or aryl group.

Scheme 5.38

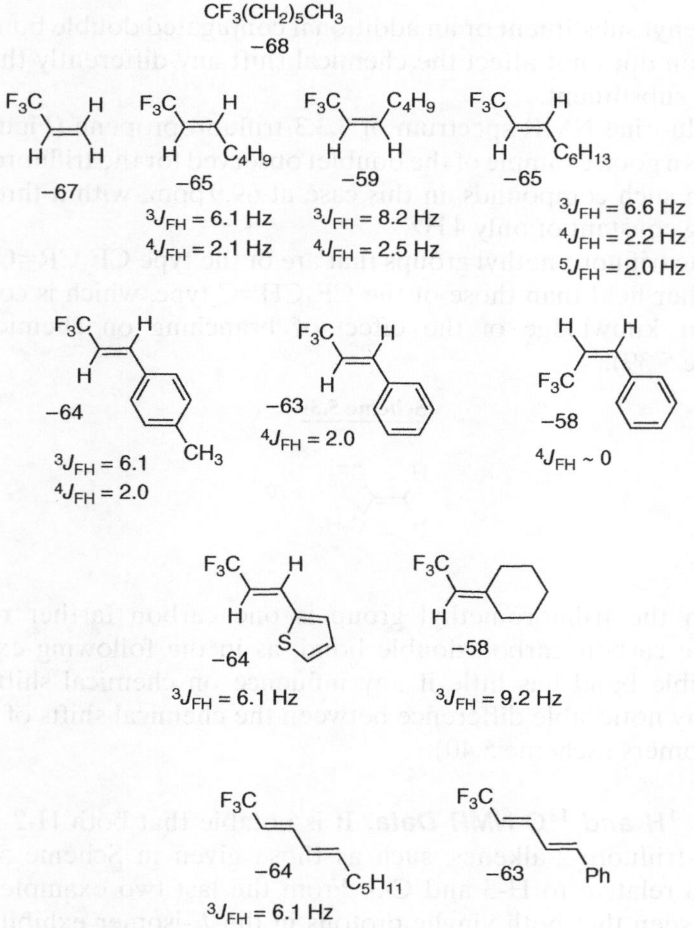

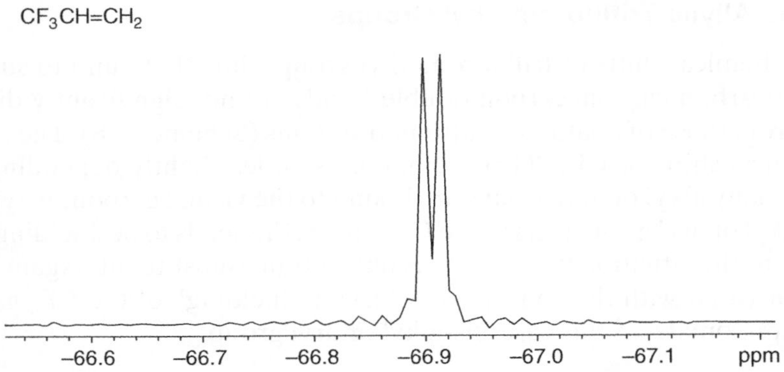

$CF_3CH=CH_2$

FIGURE 5.14. ^{19}F NMR spectrum of 3,3,3-trifluoropropene

A phenyl substituent or an additional conjugated double bond at the β-position does not affect the chemical shift any differently than does an alkyl substituent.

The fluorine NMR spectrum of 3,3,3-trifluoropropene (Figure 5.14) provides a good example of the doublet observed for the trifluoromethyl group in such compounds, in this case at 69.9 ppm, with a three-bond coupling constant of only 4 Hz.

Allylic trifluoromethyl groups that are of the type $CF_3CR=C$ appear at a higher field than those of the $CF_3CH=C$ type, which is consistent with our knowledge of the effect of branching on chemical shift (Scheme 5.39).

Scheme 5.39

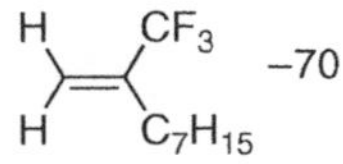

When the trifluoromethyl group is one carbon farther removed from the carbon–carbon double bond, as in the following examples, the double bond has little if any influence on chemical shift, nor is there any noticeable difference between the chemical shifts of cis- and trans-isomers (Scheme 5.40).

5.9.1.1. ^{1}H and ^{13}C NMR Data. It is notable that both H-2 and C-2 in 1,1,1-trifluoro-2-alkenes, such as those given in Scheme 5.41, are shielded relative to H-3 and C-3. From the last two examples, it can also be seen that both vinylic protons in the *E*-isomer exhibit greater

Scheme 5.40

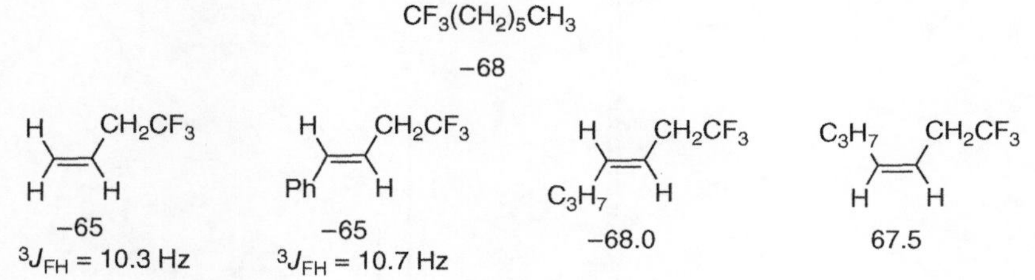
CF3(CH2)5CH3
−68
H CH2CF3
H H
−65
3JFH = 10.3 Hz
H CH2CF3
Ph H
−65
3JFH = 10.7 Hz
H CH2CF3
C3H7 H
−68.0
C3H7 CH2CF3
H H
67.5

Scheme 5.41

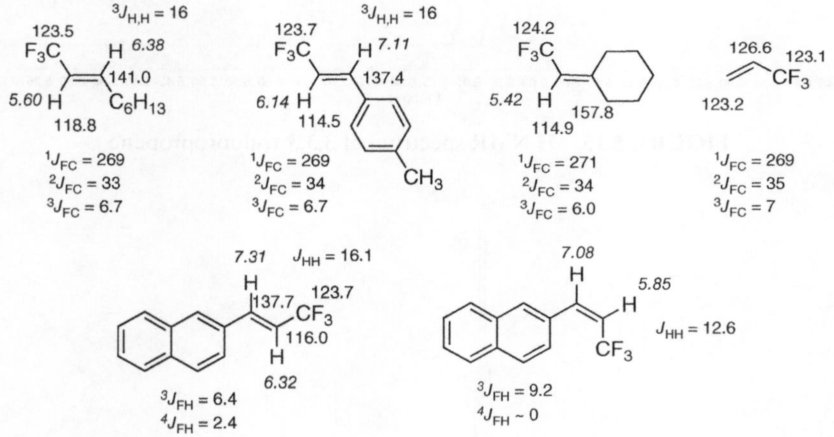
3JH,H = 16
123.5
F3C
H 6.38
141.0
5.60 H
C6H13
118.8
1JFC = 269
2JFC = 33
3JFC = 6.7
3JH,H = 16
123.7
F3C
H 7.11
137.4
6.14 H
114.5
CH3
1JFC = 269
2JFC = 34
3JFC = 6.7
124.2
F3C
5.42 H
157.8
114.9
1JFC = 271
2JFC = 34
3JFC = 6.0
126.6
123.1
CF3
123.2
1JFC = 269
2JFC = 35
3JFC = 7
7.31
JHH = 16.1
H
123.7
137.7
CF3
116.0
H
6.32
3JFH = 6.4
4JFH = 2.4
7.08
H
5.85
H
JHH = 12.6
CF3
3JFH = 9.2
4JFH ~ 0

deshielding than those in the *Z*-isomer. Again we see four-bond F–H coupling only in the *E*-isomer, which has the coupling H and CF_3 cis to each other.

Proton and carbon spectra of 3,3,3-trifluoropropene are provided in Figures 5.15 and 5.16 as specific examples of such spectra. The proton spectrum is more complicated than one would have expected based on a first-order analysis. However, a fluorine-decoupled spectrum becomes first order, as was depicted and discussed in Section 2.8, Figures 2.15 and 2.16.

The carbon spectrum is wonderfully discernable, with all three carbons appearing as quartets in the same general region, with the highly split (270 Hz) trifluoromethyl carbon at 122.8 ppm, the C2 carbon at 126.4 ppm with 34 Hz coupling, and the C1 carbon absorbing at 123.6 ppm with 6.9 Hz coupling.

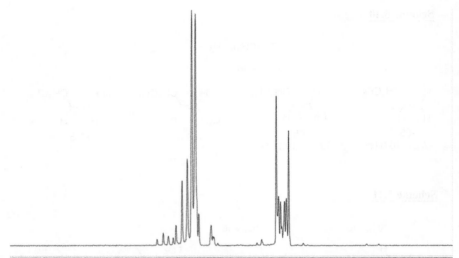

FIGURE 5.15. ^{1}H NMR spectrum of 3,3,3-trifluoropropene

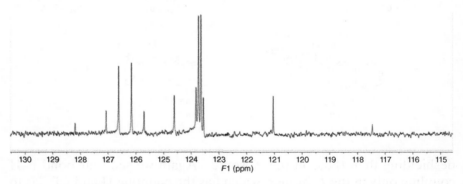

FIGURE 5.16. ^{13}C NMR spectrum of 3,3,3-trifluoropropene

5.9.1.2. Multitrifluoromethyl-Substituted Alkenes.

Spectral data for cis- and trans-1,1,1,4,4,4-hexafluoro-2-butene, and 2-(trifluoromethyl)-3,3,3-trifluoropropene are given in Scheme 5.42 as representative examples of alkenes bearing two CF$_3$ groups. Note the significant shielding of the fluorines of the cis compound versus the trans compound, which has a chemical shift similar to that of 3,3,3-trifluoropropene (−67 ppm).

Scheme 5.42

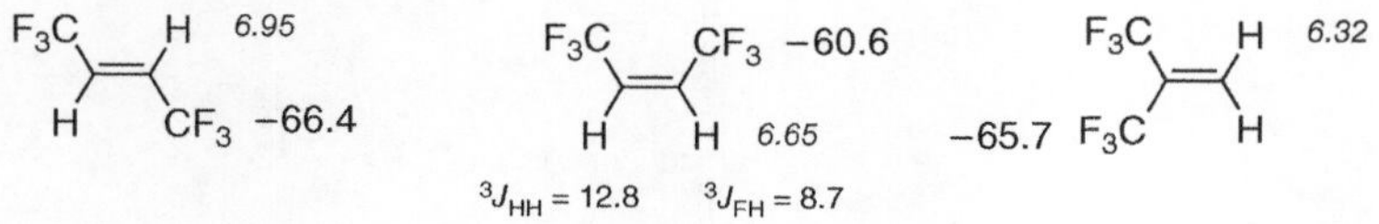

$^3J_{HH} = 12.8$ $^3J_{FH} = 8.7$

5.9.2. α,β-Unsaturated Carbonyl Compounds

The chemical shifts of trifluoromethyl groups at the terminus of α,β-unsaturated carbonyl compounds are not affected by the presence of the carbonyl group, as is indicated by the examples in Scheme 5.43, and as exemplified by the fluorine NMR of 4,4,4-trifluorocrotonic acid, given in Figure 5.17.

Scheme 5.43

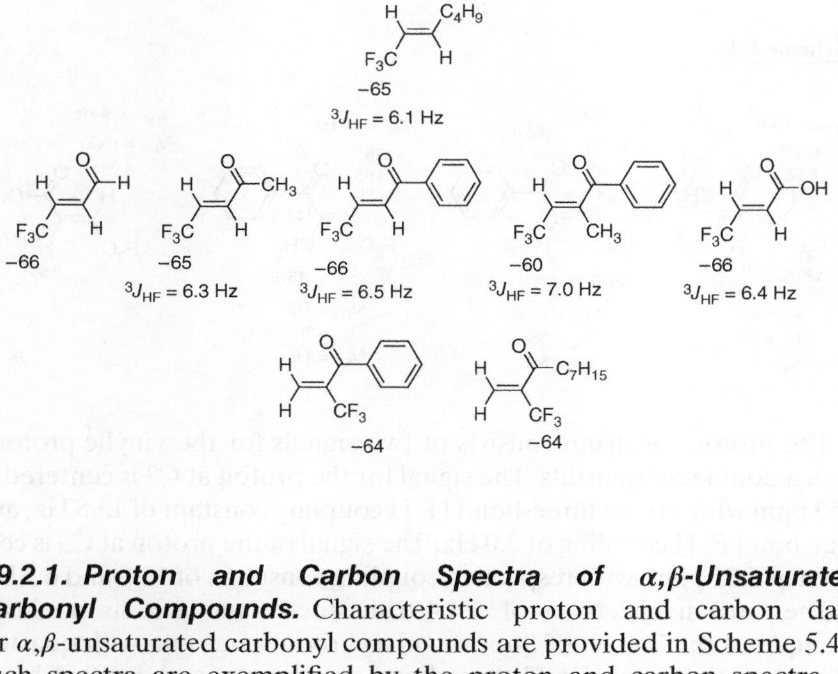

5.9.2.1. Proton and Carbon Spectra of α,β-Unsaturated Carbonyl Compounds. Characteristic proton and carbon data for α,β-unsaturated carbonyl compounds are provided in Scheme 5.44. Such spectra are exemplified by the proton and carbon spectra of 4,4,4-trifluorocrotonic acid in Figures 5.18 and 5.19.

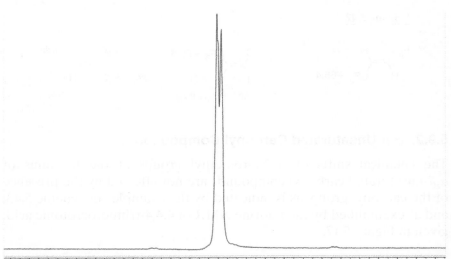

-64.9 -65.0 -65.1 -65.2 -65.3 -65.4 -65.5 -65.6 -65.7 -65.8 -65.9 -66.0 -66.1 -66.2 -66.3 -66.4 -66.5 -66.6 -66.7 -66.8 -66.9 -67.0 -67.1
*F*1 (ppm)

FIGURE 5.17. ^{19}F NMR spectrum of 4,4,4-trifluorocrotonic acid

Scheme 5.44

$^3J_{HH} = 16.1$

129.2   28.4
6.58   2.37

$^3J_{HH} = 15.5$ 188.4

130.6
6.83

$^3J_{HF} = 7.0$ Hz

125.1
6.05

$^3J_{HF} = 6.4$ Hz

$^3J_{HH} = 15.8$

6.93

196.0

122.5   135.0
6.69

$^1J_{FC} = 270$
$^2J_{FC} = 35$
$^3J_{FC} = 5.8$

127.3   131.4
7.55

$^1J_{FC} = 270$
$^2J_{FC} = 35$
$^3J_{FC} = 5.4$

147.3
123.4   15.1

$^1J_{FC} = 272$
$^2J_{FC} = 35$
$^3J_{FC} = 4.8$

6.60

The proton spectrum consists of two signals for the vinylic protons, each a doublet of quartets. The signal for the proton at C2 is centered at 6.53 ppm with a trans three-bond H–H coupling constant of 15.8 Hz, and four-bond F–H coupling of 2.0 Hz. The signal of the proton at C3 is centered at 6.91 ppm, with respective coupling constants of 15.8 and 6.5 Hz.

The carbon spectrum of 4,4,4-trifluorocrotonic acid is a classic example of the impact of fluorine on a carbon spectrum, with quartets due to the one-, two-, and three-bond coupling to carbons 2–4 being clearly seen. The CF$_3$ (C4) carbon is seen at 121.2 ppm with a coupling constant of 271 Hz, the C3 vinylic carbon appearing at 133.8 ppm with a coupling constant of 36 Hz, the other vinylic carbon (C2) appearing at 127.9 ppm with 6.0 Hz coupling, and one can even see some four-bond coupling to the carbonyl carbon that appears at 169.3 ppm.

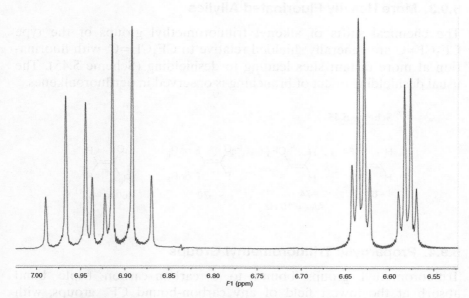

FIGURE 5.18. ^{1}H NMR spectrum of 4,4,4-trifluorocrotonic acid (vinylic area)

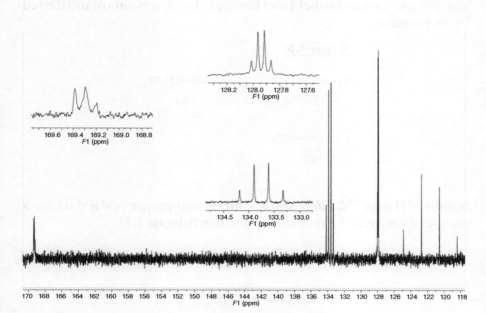

FIGURE 5.19. ^{13}C NMR spectrum of 4,4,4-trifluorocrotonic acid

5.9.3. More Heavily Fluorinated Allylics

The chemical shifts of alkenyl trifluoromethyl groups of the type $CF_3CF=C$ are generally shielded relative to $CF_3CH=C$, with fluorination at more distant sites leading to deshielding (Scheme 5.45). The usual deshielding effect of branching is observed in perfluoroalkenes.

Scheme 5.45

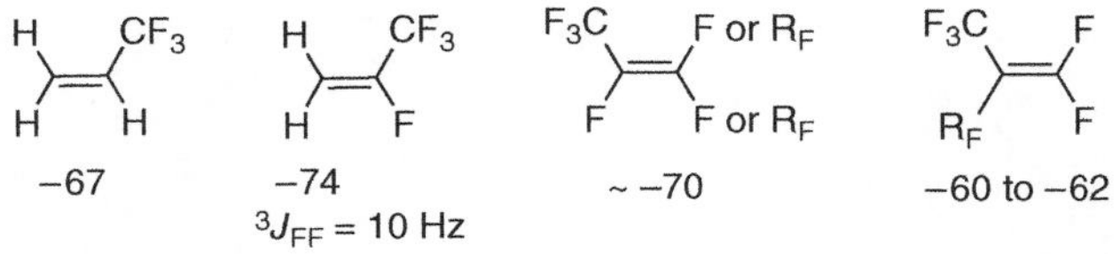

5.9.4. Propargylic Trifluoromethyl Groups

Trifluoromethyl groups bound to a carbon–carbon triple bond absorb at the lowest field of any carbon-bound CF_3 groups, with 3,3,3-trifluoropropyne having a δ_F of −52 (Scheme 5.46). When the CF_3 group is one carbon farther from the triple bond, it is almost unaffected by its presence.

Scheme 5.46

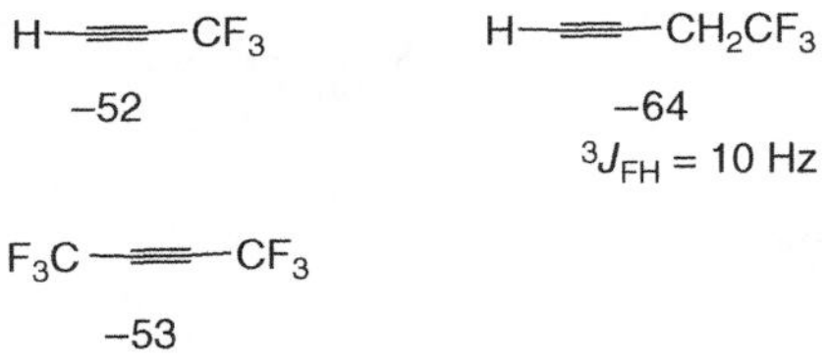

5.9.4.1. *¹H and ¹³C NMR Data.* Carbon and proton NMR data for a couple of alkynyl CF_3 systems are given in Scheme 5.47.

Scheme 5.47

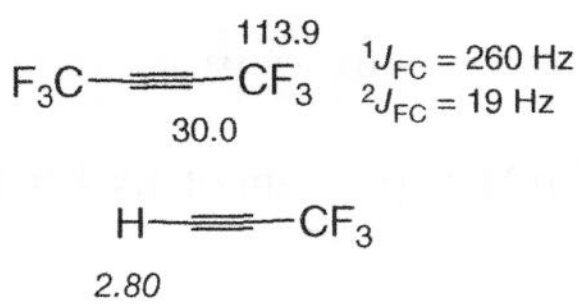

5.10. ARYL-BOUND TRIFLUOROMETHYL GROUPS

Benzylic trifluoromethyl groups generally appear as *a sharp singlet* about 8 ppm downfield of their saturated analog, trifluoromethylcyclohexane, usually in the −63 ppm region (Scheme 5.48), as exemplified by the fluorine spectrum of trifluoromethylbenzene in Figure 5.20.

Scheme 5.48

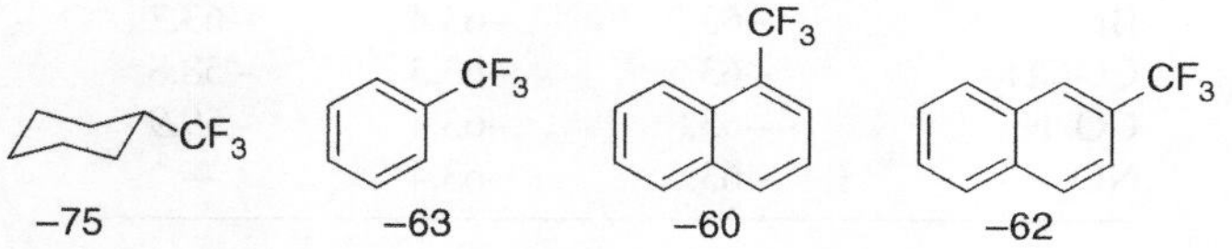

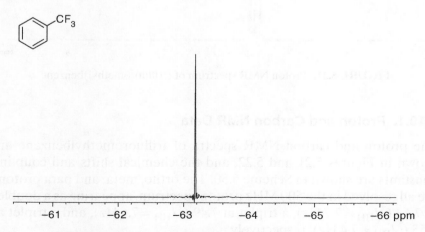

$-75 \qquad -63 \qquad -60 \qquad -62$

The fluorine NMR chemical shift data for disubstituted trifluoromethylbenzenes indicate that meta substitution gives rise to almost no effect, with a range of −63.2 to −63.4 ppm for the chemical shifts of the CF_3 groups (Table 5.4). Ortho substitution has the greatest effect, with the compounds exhibiting a range of −58.8 to −63.2 ppm, with para substituents showing moderate effect, ranging from −61.7 to −63.8 ppm.

When the aryl group is one carbon farther from the CF_3 group, as in (2,2,2-trifluoroethyl)benzene, it has virtually no influence upon the chemical shift of the CF_3 group (Scheme 5.49).

$-61 \qquad -62 \qquad -63 \qquad -64 \qquad -65 \qquad -66 \text{ ppm}$

FIGURE 5.20. ^{19}F NMR spectrum of trifluoromethylbenzene

TABLE 5.4. ^{19}F Chemical Shifts for CF$_3$ Groups in Disubstituted Benzenes

Substituent	Chemical Shifts (δ_F) (CDCl$_3$)		
	Para	meta	ortho
H	−63.2		
NH$_2$	−61.7	−63.4	−63.2
OH	−62.0	−63.3	−61.4
CH$_3$	−62.5	−63.3	−63.6
Cl	−63.1	−63.4	−63.1
Br	−63.3	−63.4	−63.2
COCH$_3$	−63.6	−63.3	−58.8
CO$_2$H	−63.8	−63.4	−59.9
NO$_2$	−63.6	−63.4	—

Scheme 5.49

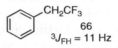

$^3J_{FH} = 11$ Hz

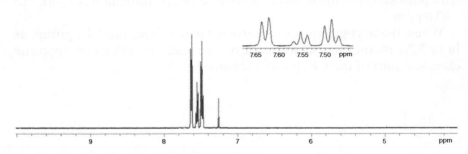

FIGURE 5.21. Proton NMR spectrum of (trifluoromethyl)benzene

5.10.1. Proton and Carbon NMR Data

The proton and carbon NMR spectra of trifluoromethylbenzene are shown in Figures 5.21 and 5.22, and the chemical shifts and coupling constants are shown in Scheme 5.50. The ortho, meta, and para protons are all resolved in the 500 MHz proton spectrum, appearing as a doublet at 7.63 ($^3J_{HH} = 7.7$ Hz), a triplet at 7.49 ($^3J_{HH} = 7.6$ Hz), and a triplet at 7.55 ($^3J_{HH} = 7.4$ Hz), respectively.

In the carbon spectrum, the ortho, meta, and para carbons (hydrogen bound) are clearly seen with diminishing coupling to the CF$_3$ fluorines,

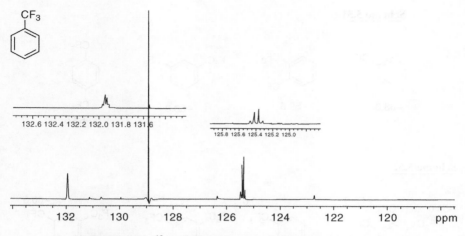

FIGURE 5.22. ^{13}C NMR spectrum of trifluoromethylbenzene

Scheme 5.50

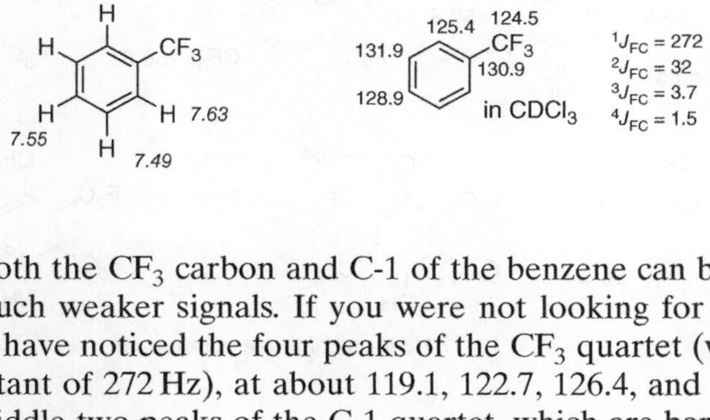

whereas both the CF_3 carbon and C-1 of the benzene can be detected but are much weaker signals. If you were not looking for them, you would not have noticed the four peaks of the CF_3 quartet (with a coupling constant of 272 Hz), at about 119.1, 122.7, 126.4, and 129.9 ppm, and the middle two peaks of the C-1 quartet, which are barely seen at about 130.7 and 131.1 ppm.

5.10.2. Multitrifluoromethylated Benzenes

Fluorine NMR data for the three bis-(trifluoromethyl)benzenes are given in Scheme 5.51, while such data for the tris(trifluoromethyl)-, tetrakis(trifluoromethyl)-, pentakis(trifluoromethyl)-, and hexakis (trifluoromethyl)benzenes are given in Scheme 5.52.[3] In the bis(trifluoromethyl)benzene series, one will notice that adding a second trifluoromethyl group meta or para does not have much effect upon the fluorine chemical shifts, but when vicinal, as in the ortho compound, considerable deshielding is observed.

Scheme 5.51

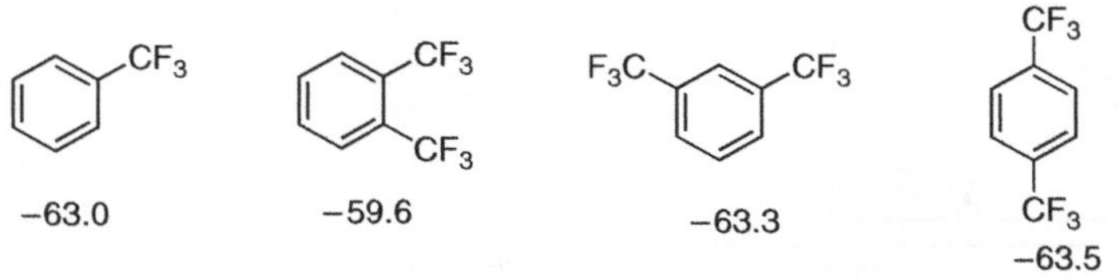

Scheme 5.52

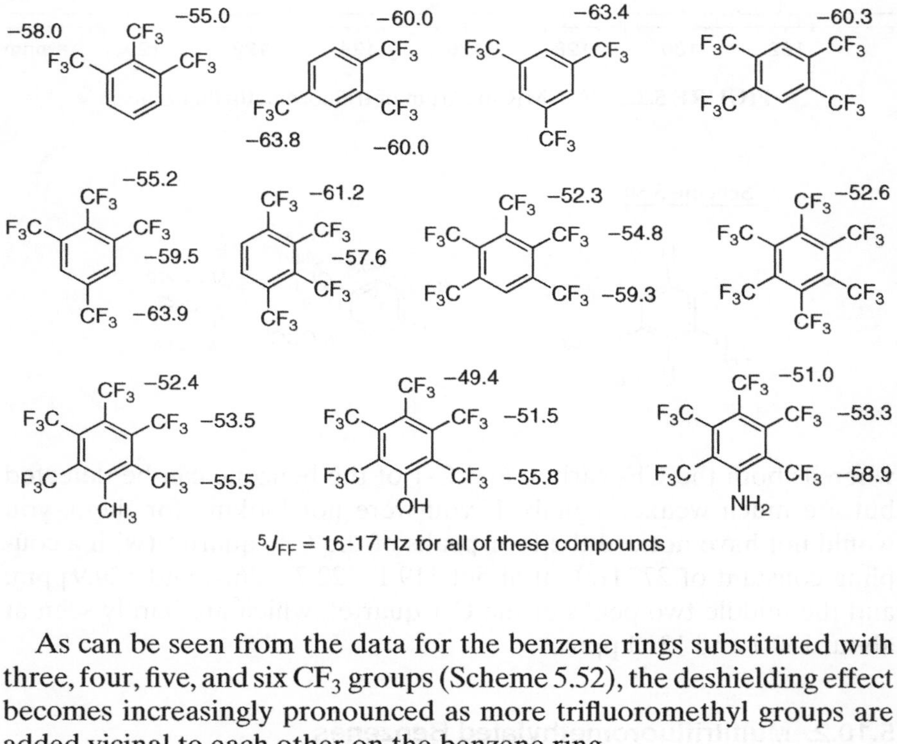

$^5J_{FF}$ = 16-17 Hz for all of these compounds

As can be seen from the data for the benzene rings substituted with three, four, five, and six CF_3 groups (Scheme 5.52), the deshielding effect becomes increasingly pronounced as more trifluoromethyl groups are added vicinal to each other on the benzene ring.

Spectral data for the pentakis(trifluoromethyl)toluene, phenol and aniline compounds are also included in Scheme 5.52.[4] It can be seen that these "donor" groups serve to deshield all of the trifluoromethyl groups but do so most effectively with the CF_3 group that is para to it.

Proton and carbon NMR data for these multi-CF_3-substituted benzenes are provided in Schemes 5.53 and 5.54. Individual one-bond F–C coupling constant data are not provided because virtually all of these coupling constants are in between 272 and 276 Hz.

Scheme 5.53

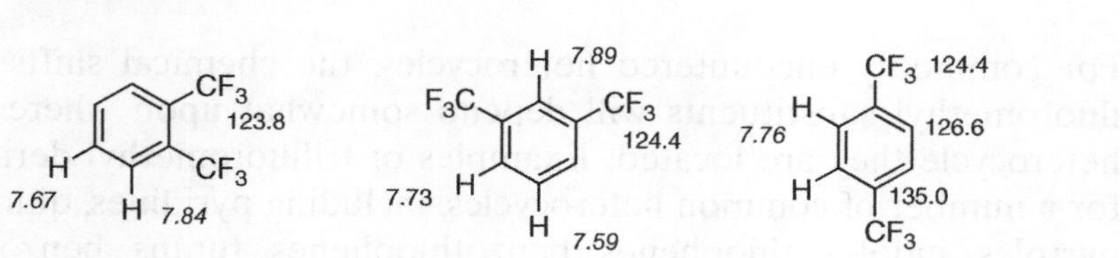

Scheme 5.54

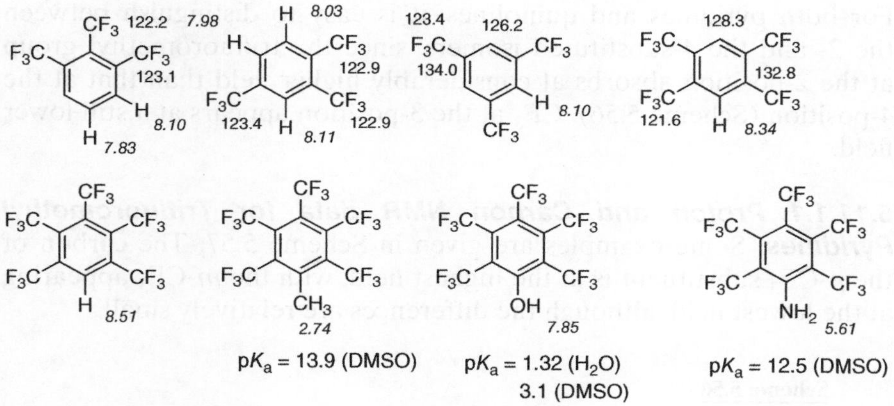

pK_a = 13.9 (DMSO)   pK_a = 1.32 (H_2O)   pK_a = 12.5 (DMSO)
                        3.1 (DMSO)

Similarly, the two-bond F–C coupling constants to the respective aryl carbons are all between 32 and 38 Hz. pK_a data for the pentakis (trifluoromethyl)toluene, phenol and aniline are also provided in the latter scheme.

Spectral data for a few illustrative bis- and tris(trifluoromethyl)phenyl trivalent phosphorous chlorides are provided in Scheme 5.55.

Scheme 5.55

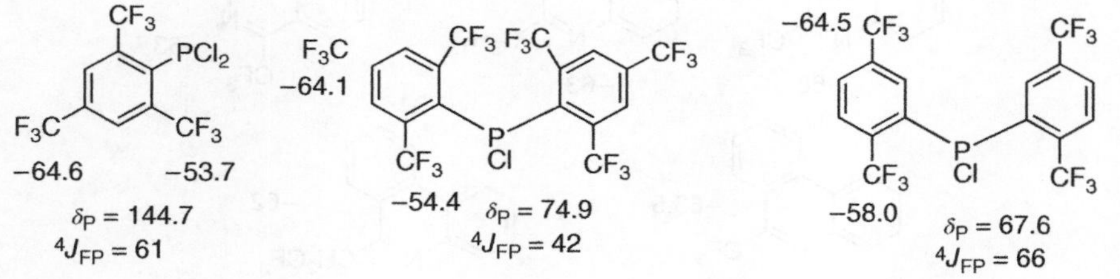

5.11. HETEROARYL-BOUND TRIFLUOROMETHYL GROUPS

For commonly encountered heterocycles, the chemical shifts of trifluoromethyl substituents will depend somewhat upon where in the
heterocycle they are located. Examples of trifluoromethyl derivatives
for a number of common heterocycles, including pyridines, quinolines,
pyrroles, indoles, thiophenes, benzothiophenes, furans, benzofurans,
imidazoles, and uracils are given as follows.

5.11.1. Pyridines, Quinolines, and Isoquinolines

For both pyridines and quinolines, it is easy to distinguish between
the 2- and the 4-substituted isomers, since the trifluoromethyl group
at the 2-position absorbs at considerably higher field than that at the
4-position (Scheme 5.56). CF_3 at the 3-position appears at a still lower
field.

*5.11.1.1. Proton and Carbon NMR data for Trifluoromethyl
Pyridines.* Some examples are given in Scheme 5.57. The carbon of
the p-CF_3 substituent is at the highest field, with the m-CF_3 appearing
at the lowest field, although the differences are relatively small.

Scheme 5.56

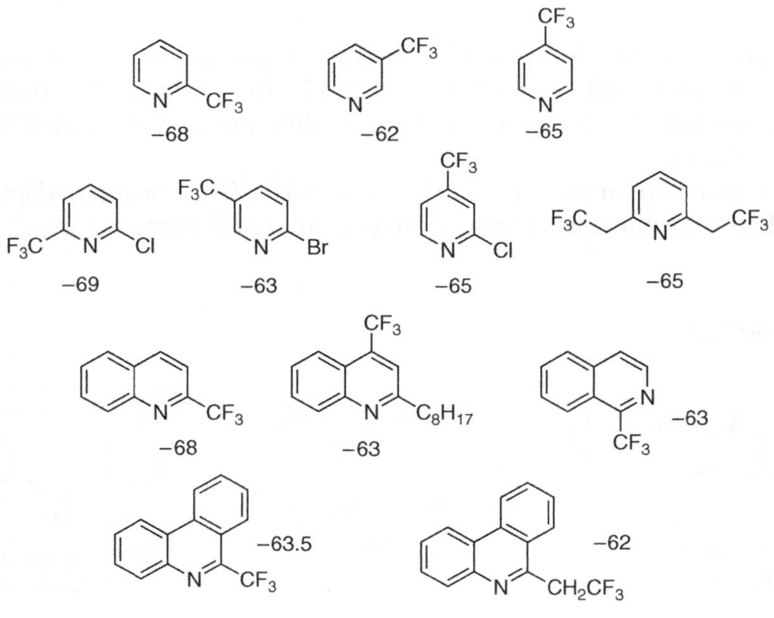

Scheme 5.57

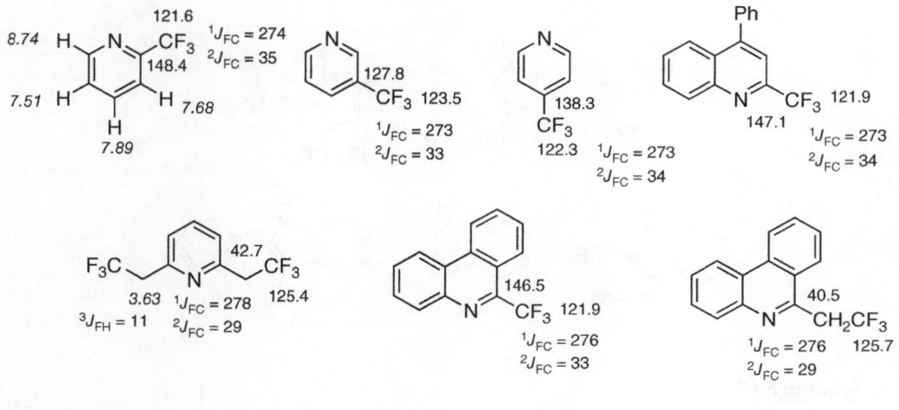

Scheme 5.58

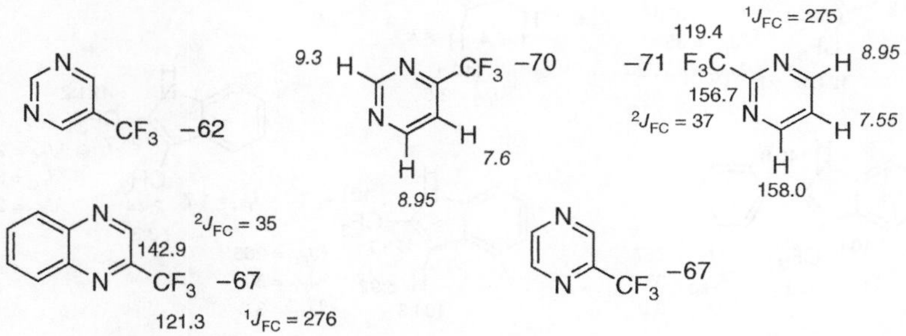

5.11.2. Pyrimidines and Quinoxalines

NMR data for 2-, 4-, and 5-trifluoromethyl pyrimidines and one quinoxaline are given in Scheme 5.58.

5.11.3. Pyrroles and Indoles

Generally, the 2- and 3-trifluoromethyl-substituted pyrroles and indoles are readily distinguished, with the CF_3 groups of the 3-isomers absorbing at a higher field than the 2-isomers (Scheme 5.59). A phenyl or an alkyl vicinal to the CF_3 group gives rise to considerable deshielding.

5.11.3.1. Proton and Carbon NMR Spectra of Pyrroles and Indoles.
Some examples are given in Scheme 5.60 that provide proton and carbon data for pyrroles and indoles bearing a trifluoromethyl group.

Scheme 5.59

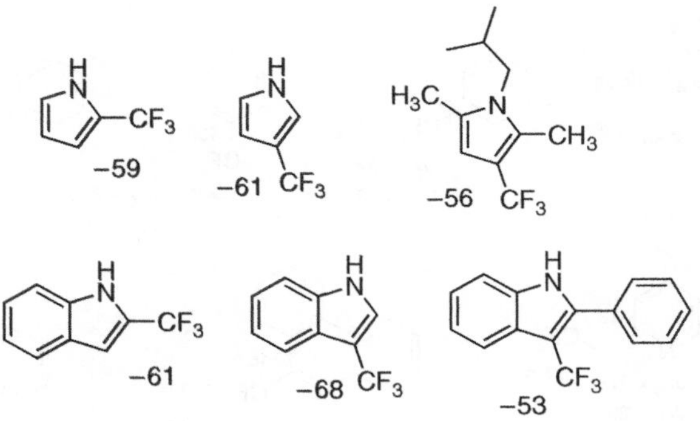

Scheme 5.60

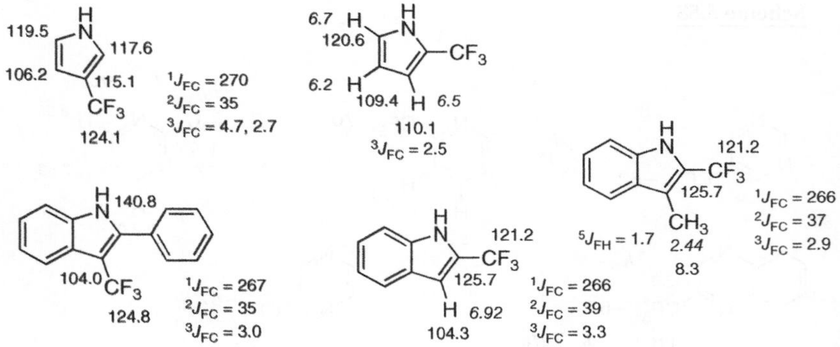

5.11.4. Thiophenes and Benzothiophenes

Similarly, the fluorines of a trifluoromethyl group in the 3-position of thiophenes absorb at a higher field than those of one at the 2-position (Scheme 5.61).

5.11.4.1. Carbon and Proton NMR Spectra of Thiophenes. Some examples are given in Scheme 5.62 that provide proton and carbon data for thiophenes and benzothiophenes that bear a trifluoromethyl group.

5.11.5. Furans

The trifluoromethylfurans and benzofurans have ^{19}F chemical shifts at significantly higher fields than the respective thiophenes or the pyrroles (Scheme 5.63).

5.11.5.1   Proton and Carbon NMR Spectra of Furans. As is the case with furans, vicinal CH_3 groups the proton at the 2- and 3-positions resonates at similarly lower field than do those at the 4- and 4-positions (Scheme 5.61). Vicinal diaryl CF_3 groups are additionally deshielded.

Scheme 5.61

−55   −60   −53

−57   −61

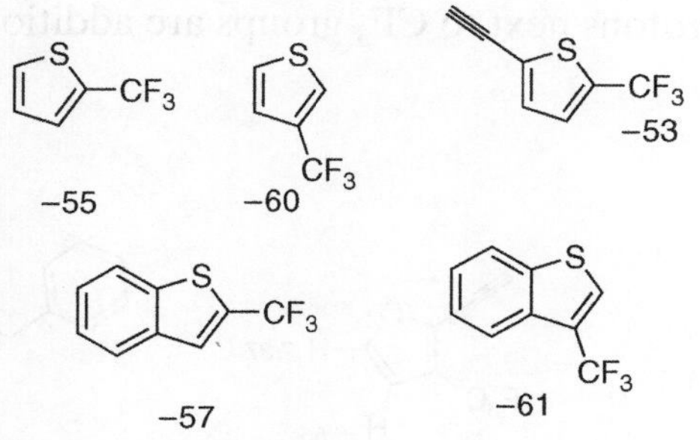

Scheme 5.62

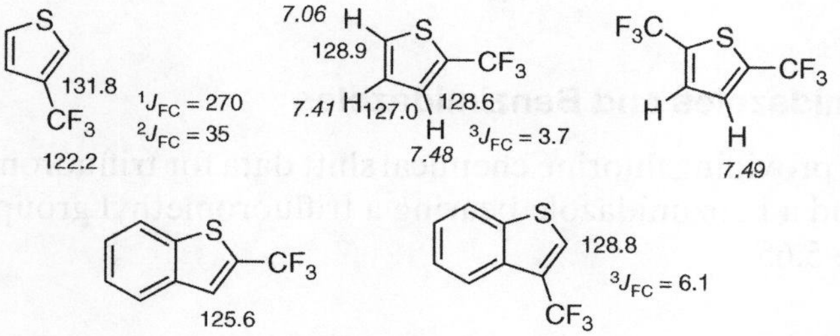

Scheme 5.63

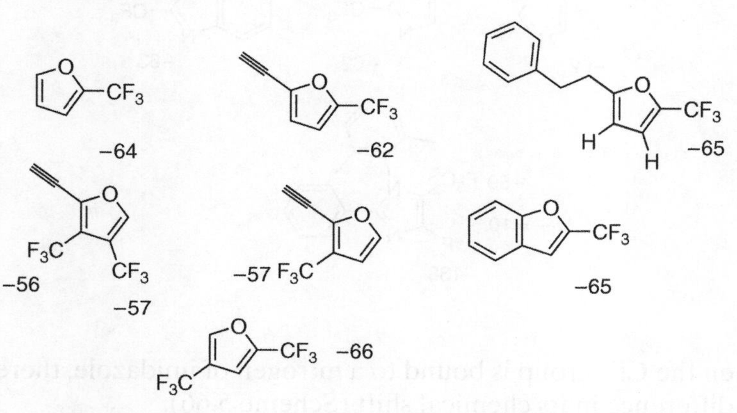

−64   −62   −65

−56   −57   −65

−57

−66

−61

5.11.5.1. Proton and Carbon NMR Spectra of Furans. As is the case with furans without CF_3 groups, the protons at the 2- and 5-positions appear at significantly lower field than do those at the 3- and 4-positions (Scheme 5.64). Protons next to CF_3 groups are additionally deshielded.

Scheme 5.64

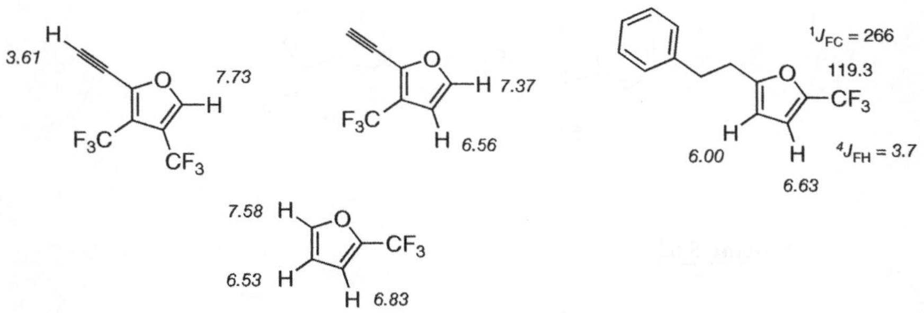

5.11.6. Imidazoles and Benzimidazoles

Examples providing fluorine chemical shift data for trifluoromethyl imidazoles and a benzimidazole bearing a trifluoromethyl group are given in Scheme 5.65.

Scheme 5.65

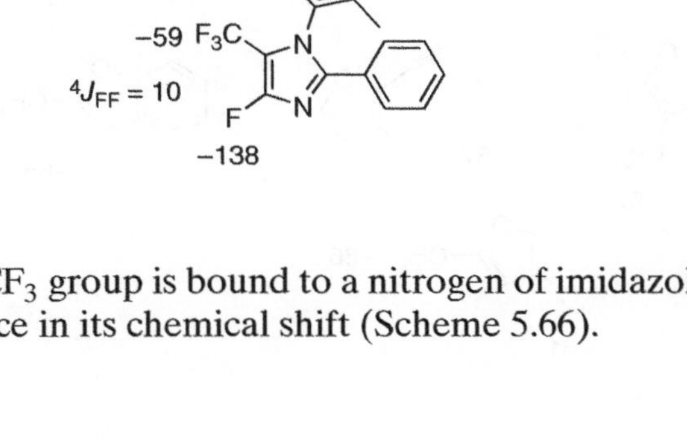

When the CF_3 group is bound to a nitrogen of imidazole, there is not much difference in its chemical shift (Scheme 5.66).

Scheme 5.66

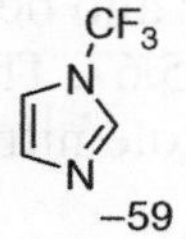

-59

<u>**Scheme 5.67**</u>

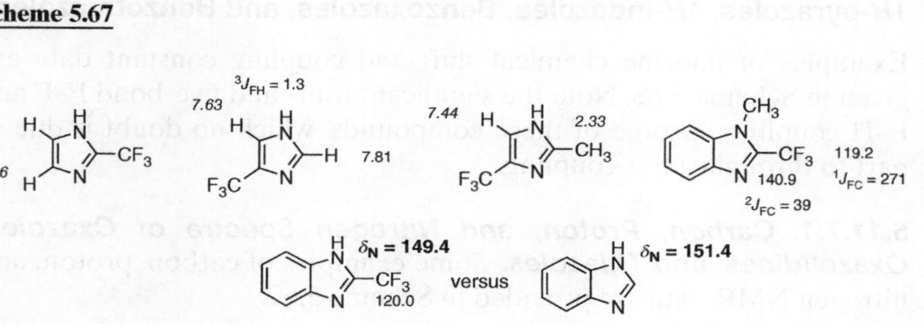

$^3J_{FH} = 1.3$
7.63
7.26
7.44
7.81
2.33

CH$_3$
119.2
140.9
$^1J_{FC} = 271$
$^2J_{FC} = 39$

$\delta_N = 149.4$ versus $\delta_N = 151.4$
120.0

<u>**Scheme 5.68**</u>

-62.1 $^4J_{FF} = 10$ -113

-62.3 $^4J_{HF} = 1.7$

-61.6 $^4J_{FF} = 11.5$ -119

-62

-60

-59

-60

-56

-66.5

-61.6

-59.9 $^4J_{FF} = 15$ -136

-52 $^5J_{FF} = 8.6$ -63

$^5J_{HF} = 2$ -59.4

-67

-70

-62

-62

-62.5

-62.5

-60

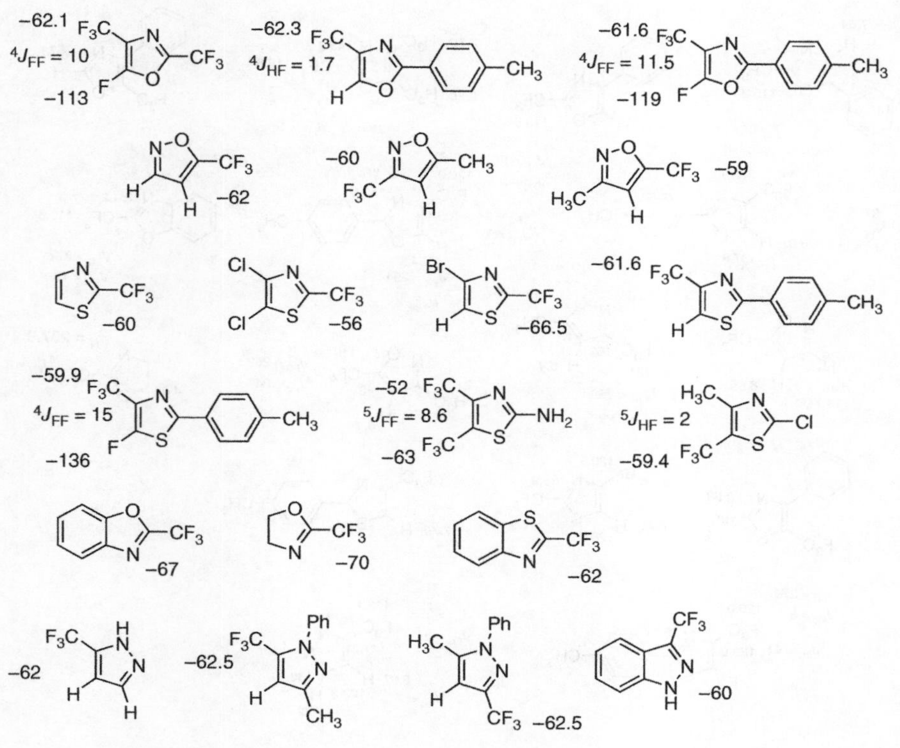

5.11.6.1. Proton Carbon and Nitrogen NMR Spectra of Trifluo-romethylimidazoles and Benzimidazoles.
Available proton, carbon, and nitrogen data for imidazoles and benzimidazoles substituted with a CF_3 group are given in Scheme 5.67. The trifluoromethyl group is seen to slightly shield the benzimidazole nitrogen.

5.11.7. Oxazoles, Isoxazoles, Oxazolidines, Thiazoles, 1*H*-pyrazoles, 1*H*-indazoles, Benzoxazoles, and Benzothiazoles

Examples of fluorine chemical shift and coupling constant data are given in Scheme 5.68. Note the significant four- and five-bond F–F and F–H coupling in some of these compounds, which no doubt is due in part to through-space coupling.

5.11.7.1. Carbon, Proton, and Nitrogen Spectra of Oxazoles, Oxazolidines, and Thiazoles.
Some examples of carbon, proton, and nitrogen NMR data are provided in Scheme 5.69.

Scheme 5.69

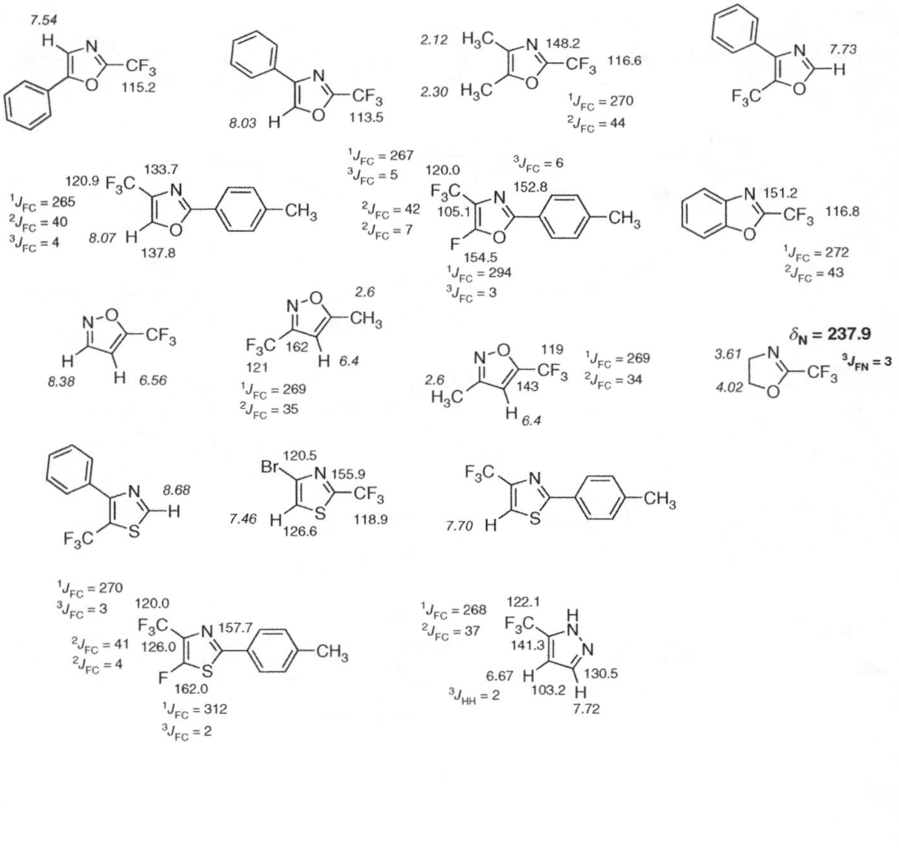

5.11.8. Triazoles and Tetrazoles

An example of each of these trifluoromethyl heterocycles is provided in Scheme 5.70.

Scheme 5.70

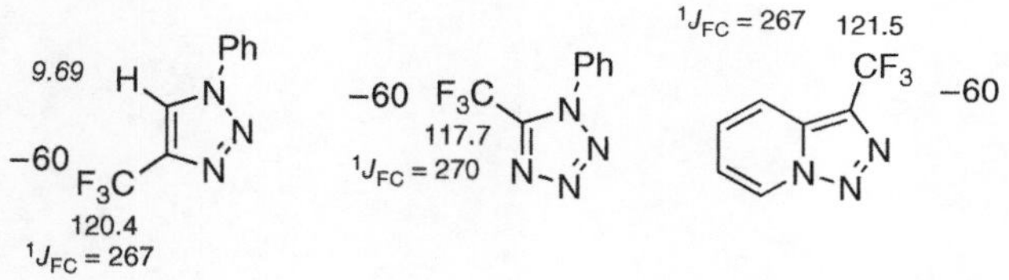

REFERENCES

1. DeMarco, R. A.; Fox, W. B.; Moniz, W. B.; Sojka, S. A. *J. Magn. Res.* **1975**, *18*, 522.
2. Burton, D. J.; Yang, Z.-Y. *Tetrahedron* **1992**, *48*, 189.
3. Takahashi, K.; Yoshino, A.; Hosokawa, K.; Muramatsu, H. *Bull. Chem. Soc. Jpn.* **1985**, *58*, 755.
4. Kutt, A.; Movchun, V.; Rodima, T.; Dansauer, T.; Rusanov, E. B.; Leito, I.; Kaljurand, I.; Koppel, J.; Pihl, V.; Koppel, I.; Ovajannikov, G.; Toom, L.; Mishima, M.; Medebielle, M.; Lork, E.; Roschenthaler, G. V.; Koppel, I. A.; Kolomeitsev, A. A. *J. Org. Chem.* **2008**, *73*, 2607.

5.11.8. Triazoles and Tetrazoles

An example of each of these trifluoromethyl heterocycles is provided in Scheme 5.2l.

Scheme 5.2l

REFERENCES

CHAPTER 6

MORE HIGHLY FLUORINATED GROUPS

6.1. INTRODUCTION

Although most fluorine-containing biologically active pharmaceutical and agrochemical compounds make use of the substituents that have been discussed in the previous three chapters, there are also numerous examples of more highly fluorinated bioactive compounds, the efficacy of which should encourage more examples to be sought.

For example, the 1,2,2 trifluoroethyl group has found use in the fungicide, tetraconazole, **6-1**, whereas the 1,1,2,2-tetrafluoroethyl group is encountered in the benzoylphenylurea insecticide hexaflumuron, **6-2**, which is used to control locust and grasshoppers in sahelian grasslands (Figure 6.1).

Likewise, the pentafluoroethyl substituent plays an important role in the extraordinarily potent antiprogestin properties of compound **6-3**, as well as in the insecticide candidate **6-4** (Figure 6.2).

A variety of fluorinated C_3 substituents have also been found to be useful in facilitating bioactivity, including the 2,2,3,3,3-pentafluoro-propyl group-containing herbicide flupoxam, **6-5**, and the 1,1,2,3,3,3-hexafluoropropyl group-containing insecticide, Lufenuron, **6-6** (Figure 6.3). The hexafluoro-isopropyl group also is found in the

Guide to Fluorine NMR for Organic Chemists, Second Edition. William R. Dolbier, Jr.
© 2016 John Wiley & Sons, Inc. Published 2016 by John Wiley & Sons, Inc.

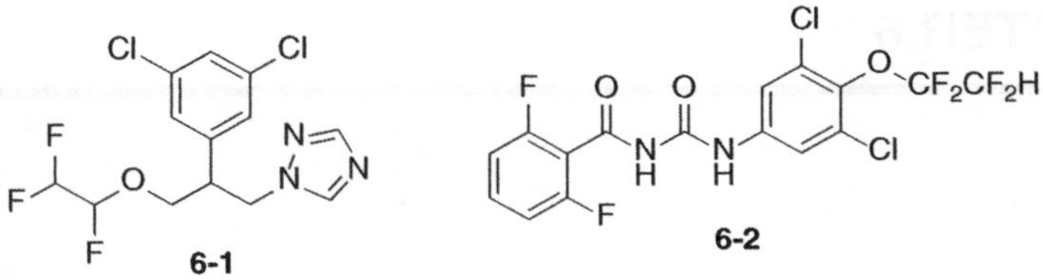

FIGURE 6.1. Examples of bioactive trifluoro- and tetrafluoroethyl compounds

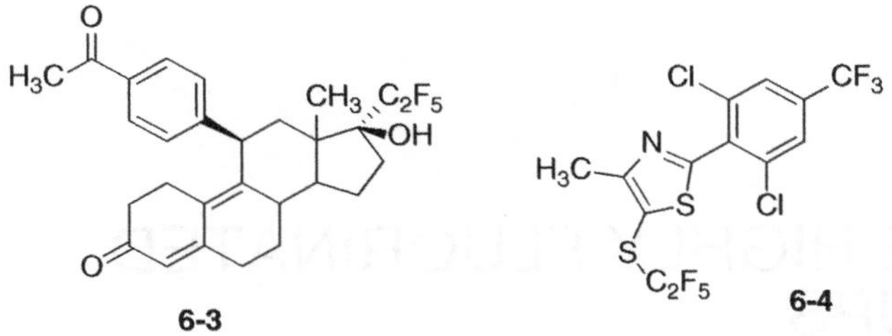

FIGURE 6.2. Examples of bioactive pentafluoroethyl compounds

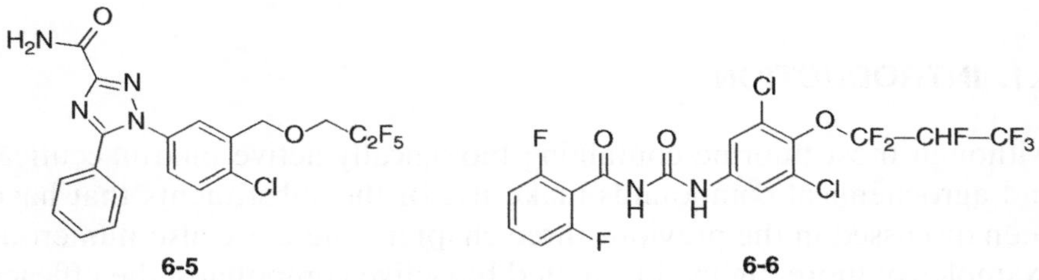

FIGURE 6.3. Examples of bioactive pentafluoro and hexafluoro-*n*-propyl compounds

pyrethroid insecticide, acrinathrin **6-7** and in the fluorinated amino acid, hexafluoroleucine **6-8** (Figure 6.4).

Multifluorinated aromatics also play a significant role within bioactive compounds. Difluoroaromatics have already been discussed in Chapter 3. Bioactive tetrafluorobenzene derivatives are exemplified by the pyrethroid insecticides transfluthrin, **6.9**, and tefluthrin, **6-10**, whereas there are numerous pentafluorobenzene compounds that have proved of interest as potential insecticides, anticancer or antiglaucoma drugs (**6-11**, **6-12**, and **6-13**) (Figure 6.5).

6.2. THE 1,1,2- AND 1,2,2-TRIFLUOROETHYL GROUPS

The common 2,2,2-trifluoroethyl group is discussed in Chapter 5. Less commonly encountered is the 1,1,2-trifluoroethyl group, the ether and

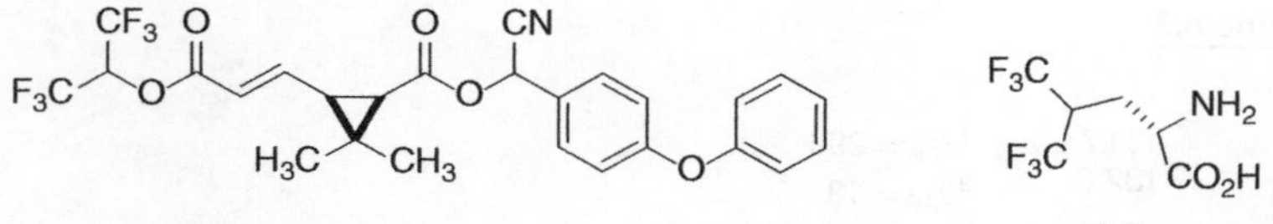

6-7 6-8

FIGURE 6.4. Examples of bioactive hexafluoro isopropyl compounds

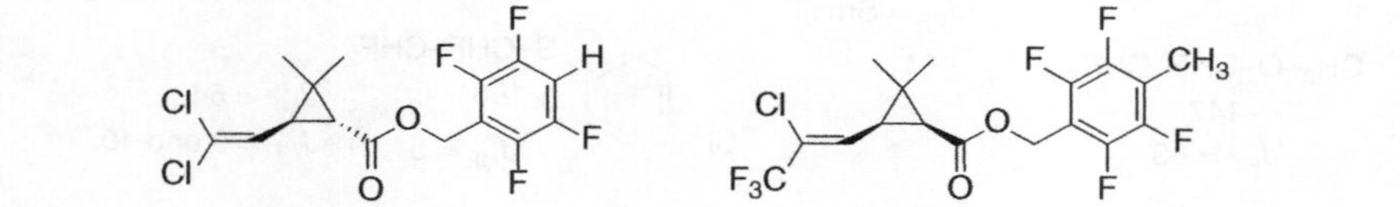

6-9 6-10

Fenfluthrin (6-11) 6-12 6-13

FIGURE 6.5. Examples of bioactive polyfluorobenzene compounds

sulfide of which can be prepared by nucleophilic addition to trifluoroethylene. Still rarer is the more difficult to prepare 1,2,2-trifluoroethyl group. These groups, when seen, are usually ethers or sulfides.

Some explicit examples of 1,1,2-trifluoroethyl ethers and sulfides are provided in Scheme 6.1.

Scheme 6.1

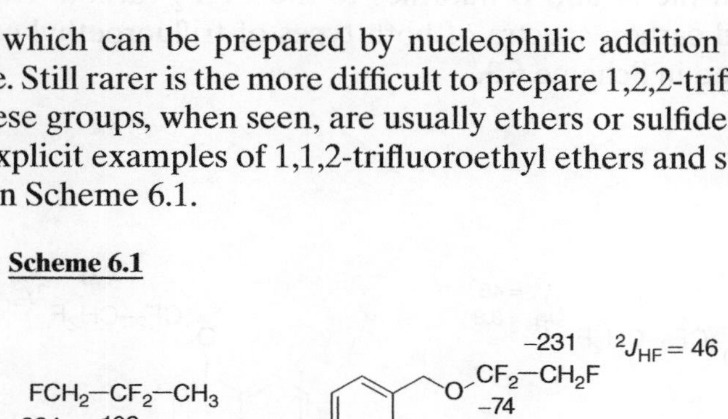

The CHF carbon is chiral in 1,2,2-trifluoroethyl compounds. Thus, the fluorines of the CF_2H group are diastereotopic and appear as an AB system, and each of the fluorines can potentially couple with

Scheme 6.2

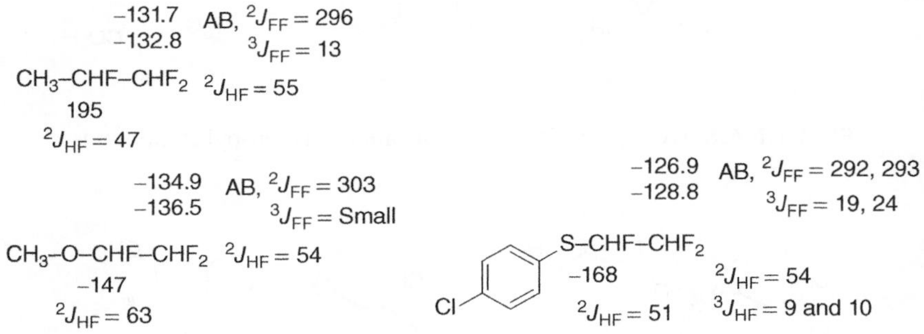

$$-131.7 \quad \text{AB, } ^2J_{FF} = 296$$
$$-132.8 \qquad ^3J_{FF} = 13$$

$$\text{CH}_3\text{–CHF–CHF}_2 \ ^2J_{HF} = 55$$
$$195$$
$$^2J_{HF} = 47$$

$$-134.9 \quad \text{AB, } ^2J_{FF} = 303$$
$$-136.5 \qquad ^3J_{FF} = \text{Small}$$

$$\text{CH}_3\text{–O–CHF–CHF}_2 \ ^2J_{HF} = 54$$
$$-147$$
$$^2J_{HF} = 63$$

$$-126.9 \quad \text{AB, } ^2J_{FF} = 292, 293$$
$$-128.8 \qquad ^3J_{FF} = 19, 24$$

S–CHF–CHF$_2$

$$-168 \qquad ^2J_{HF} = 54$$
$$^2J_{HF} = 51 \qquad ^3J_{HF} = 9 \text{ and } 10$$

different coupling constants to the vicinal H and F. In what is an oft observed phenomenon, the vicinal three-bond F–F coupling in the sulfides is always much greater than that in the ethers. The examples in Scheme 6.2 provide typical data for such compounds.

The proton and carbon NMR spectra of both of these trifluoroethyl systems are marked by the usual large two-bond F–H coupling constants, with the 1,2,2-trifluoro system exhibiting individual coupling constants from the A and B fluorines to the CHF$_2$ carbon. Data for the proton and carbon spectra of both types of trifluoroethyl compounds are provided in Scheme 6.3.

Scheme 6.3

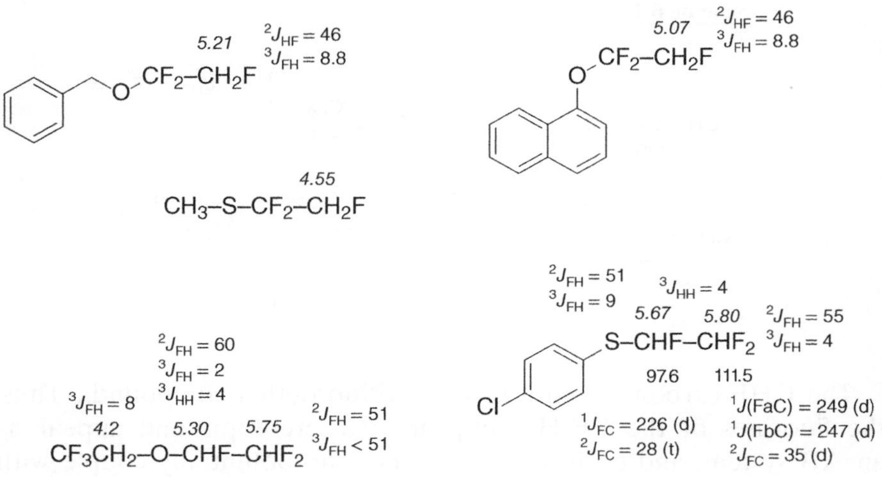

$$5.21 \quad ^2J_{HF} = 46$$
$$^3J_{FH} = 8.8$$

O–CF$_2$–CH$_2$F

$$5.07 \quad ^2J_{HF} = 46$$
$$^3J_{FH} = 8.8$$

O–CF$_2$–CH$_2$F

$$4.55$$
$$\text{CH}_3\text{–S–CF}_2\text{–CH}_2\text{F}$$

$$^2J_{FH} = 60$$
$$^3J_{FH} = 2$$
$$^3J_{FH} = 8 \quad ^3J_{HH} = 4$$
$$4.2 \quad 5.30 \quad 5.75 \quad ^2J_{FH} = 51$$
$$\text{CF}_3\text{CH}_2\text{–O–CHF–CHF}_2 \ ^3J_{FH} < 51$$

$$^2J_{FH} = 51 \quad ^3J_{HH} = 4$$
$$^3J_{FH} = 9$$
$$5.67 \quad 5.80 \quad ^2J_{FH} = 55$$
S–CHF–CHF$_2$ $^3J_{FH} = 4$

$$97.6 \quad 111.5$$

$$^1J_{FC} = 226 \text{ (d)} \qquad ^1J(\text{FaC}) = 249 \text{ (d)}$$
$$^2J_{FC} = 28 \text{ (t)} \qquad ^1J(\text{FbC}) = 247 \text{ (d)}$$
$$^2J_{FC} = 35 \text{ (d)}$$

6.3. THE 1,1,2,2-TETRAFLUOROETHYL AND 2,2,3,3-TETRAFLUOROPROPYL GROUPS

As part of a purely hydrocarbon system, an $RCF_2\text{-}CF_2H$ or an RCF_2CF_2R moieties are rarely encountered. However, Scheme 6.4 provides two examples.

Scheme 6.4

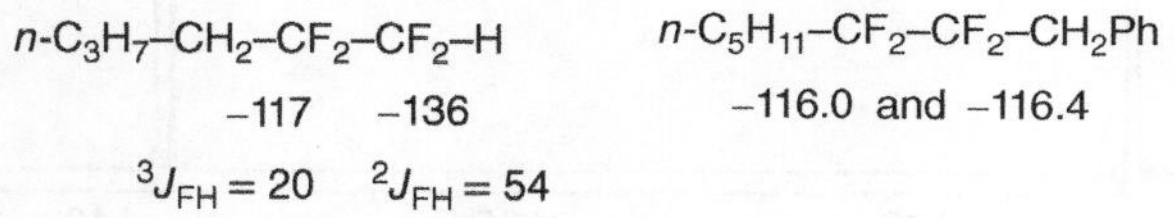

$$n\text{-}C_3H_7\text{-}CH_2\text{-}CF_2\text{-}CF_2\text{-}H \qquad n\text{-}C_5H_{11}\text{-}CF_2\text{-}CF_2\text{-}CH_2Ph$$

$$-117 \quad -136 \qquad\qquad -116.0 \ \text{and} \ -116.4$$

$$^3J_{FH} = 20 \quad ^2J_{FH} = 54$$

The 1,1,2,2-tetrafluoroethyl group, usually appearing as either the ether or the sulfide, is quite commonly encountered, probably because of its relative ease of synthesis from nucleophilic additions to tetrafluoroethylene. The fluorine NMR spectra of such isolated tetrafluoroethyl groups are characterized by the usual large (54 Hz) two-bond H–F coupling constant (Scheme 6.5).

Scheme 6.5

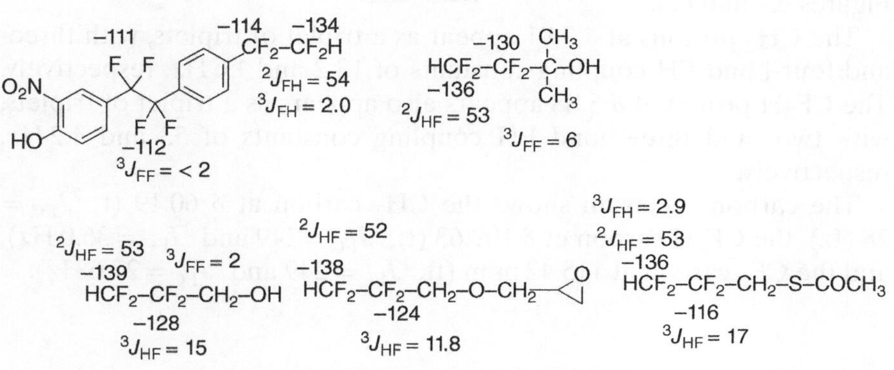

$$CF_3\text{-}CH_2\text{-}O\text{-}CF_2\text{-}CF_2H \qquad\qquad CH_3\text{-}O\text{-}CF_2\text{-}CF_2H$$
$$-76 \qquad -94 \ -138 \qquad\qquad\qquad -95 \ -137$$
$$^3J_{HF} = 8 \qquad\quad ^2J_{HF} = 54 \qquad\qquad ^2J_{HF} = 47$$
$$^5J_{FF} = 2 \qquad\quad ^3J_{FF} = 6 \qquad\qquad\quad ^3J_{FF} = 5.8$$
$$^3J_{HF} = 2.7$$

$$CH_3\text{-}S\text{-}CF_2\text{-}CF_2H$$
$$-94.5 \quad -131.7$$

$$\begin{array}{c} -114 \quad -134 \\ CF_2\text{-}CF_2H \\ ^2J_{FH} = 54 \\ ^3J_{FH} = 2.0 \end{array}$$

$$\begin{array}{c} -130 \ \ CH_3 \\ HCF_2\text{-}CF_2\text{-}C\text{-}OH \\ -136 \quad\ CH_3 \\ ^2J_{HF} = 53 \\ ^3J_{FF} = 6 \end{array}$$

(structure, left)
$$\begin{array}{c} -111 \\ F \quad F \\ O_2N \\ \qquad F \quad F \\ \qquad -112 \\ ^3J_{FF} = <2 \end{array}$$
HO

$$^2J_{HF} = 53 \quad ^3J_{FF} = 2$$
$$-139$$
$$HCF_2\text{-}CF_2\text{-}CH_2\text{-}OH$$
$$-128$$
$$^3J_{HF} = 15$$

$$^2J_{HF} = 52$$
$$-138$$
$$HCF_2\text{-}CF_2\text{-}CH_2\text{-}O\text{-}CH_2\text{-}\triangleleft^O$$
$$-124$$
$$^3J_{HF} = 11.8$$

$$^3J_{FH} = 2.9$$
$$^2J_{HF} = 53$$
$$-136$$
$$HCF_2\text{-}CF_2\text{-}CH_2\text{-}S\text{-}COCH_3$$
$$-116$$
$$^3J_{HF} = 17$$

When the tetrafluoroethyl group is attached to a carbon bearing hydrogen atoms, the three-bond H–F coupling constants are generally > 10 Hz. 2,2,3,3-Tetrafluoropropanol itself is quite inexpensive and

HCF$_2$-CF$_2$-CH$_2$OH

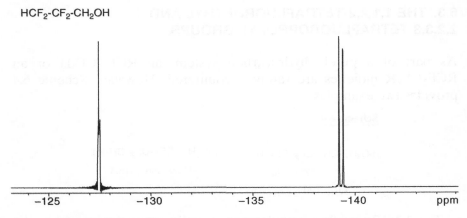

FIGURE 6.6. ^{19}F NMR spectrum of 2,2,3,3-tetrafluoropropanol

building blocks derived from it are quite common. Its fluorine NMR spectrum is given in Figure 6.6 as a good example of such a system. In this case, the 13.2 Hz three-bond F–H coupling constant is best observed in the proton spectrum.

The fluorine NMR exhibits two signals, with the CF$_2$H appearing at δ −139.3 (d, $^2J_{HF}$ = 53 Hz) and the CF$_2$ group appearing at −127.5 ppm as a multiplet.

Proton and carbon NMR spectra of compounds containing the 1,1,2,2-tetrafluoroethyl and the 2,2,3,3-tetrafluoropropyl groups are exemplified in Scheme 6.6, and the proton and carbon NMR spectra of 2,2,3,3-tetrafluoropropanol are provided as a specific example in Figures 6.7 and 6.8.

The CH$_2$ protons at δ 3.91 appear as a triplet of triplets, with three- and four-bond FH coupling constants of 13.2 and 1.6 Hz, respectively. The CF$_2$H proton at δ 5.88 appears also appears as a triplet of triplets, with two- and three-bond FH coupling constants of 53 and 4.2 Hz, respectively.

The carbon spectrum shows the CH$_2$ carbon at δ 60.19 (t, $^2J_{FC}$ = 28 Hz), the CF$_2$H carbon at δ 109.63 (tt, $^1J_{FC}$ = 249 and $^2J_{FC}$ = 36.0 Hz), and the CF$_2$ carbon at 115.42 ppm (tt, $^1J_{FC}$ = 249 and $^2J_{FC}$ = 27.6 Hz).

6.4. THE 1,2,2,2-TETRAFLUOROETHYL GROUP

The commercial anesthetic, desflurane, is a prime example of a bioactive compound containing a 1,2,2,2-tetrafluoroethyl group. In general, this

Scheme 6.6

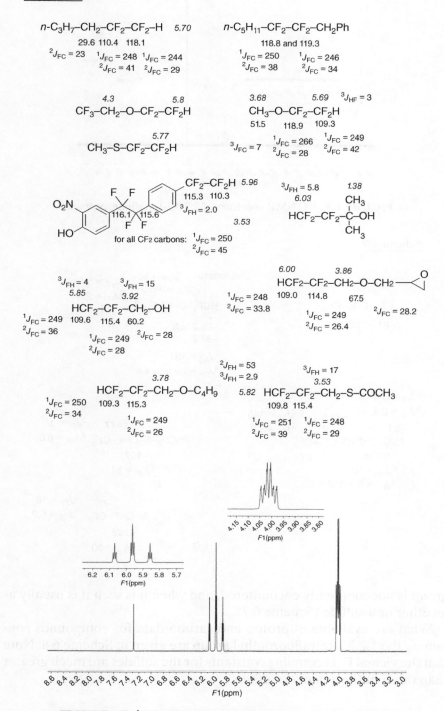

FIGURE 6.7. ^{1}H NMR spectrum of 2,2,3,3-tetrafluoropropanol

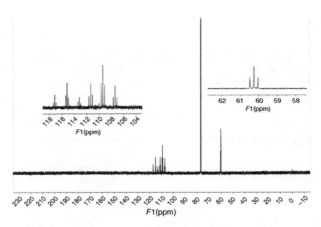

FIGURE 6.8. ^{13}C NMR spectrum of 2,2,3,3-tetrafluoropropanol

Scheme 6.7

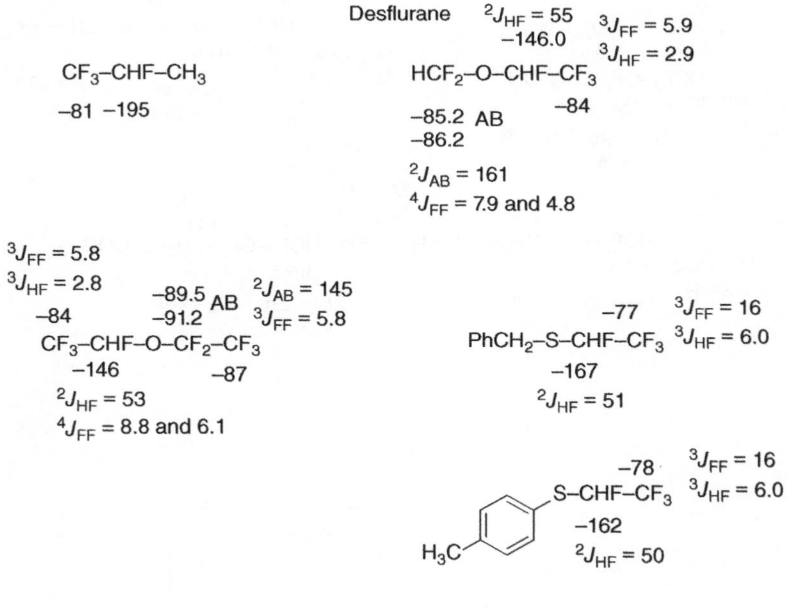

Desflurane $^2J_{HF} = 55$ $^3J_{FF} = 5.9$
                    -146.0
$CF_3-CHF-CH_3$ $HCF_2-O-CHF-CF_3$ $^3J_{HF} = 2.9$

-81 -195 -85.2 AB -84
                            -86.2

                            $^2J_{AB} = 161$
                            $^4J_{FF} = 7.9$ and 4.8

$^3J_{FF} = 5.8$
$^3J_{HF} = 2.8$ -89.5 AB $^2J_{AB} = 145$
     -84 -91.2 $^3J_{FF} = 5.8$ -77 $^3J_{FF} = 16$
     $CF_3-CHF-O-CF_2-CF_3$ $PhCH_2-S-CHF-CF_3$ $^3J_{HF} = 6.0$
          -146 -87 -167
     $^2J_{HF} = 53$ $^2J_{HF} = 51$
     $^4J_{FF} = 8.8$ and 6.1

                                                        -78 $^3J_{FF} = 16$
                                              $S-CHF-CF_3$ $^3J_{HF} = 6.0$
                                                        -162
                                    H_3C $^2J_{HF} = 50$

group is not commonly encountered, and when it is seen it is usually as an ether or a sulfide (Scheme 6.7).

What are available of proton and carbon data for compounds containing the 1,2,2,2-tetrafluoroethyl group are given in Scheme 6.8. Note that the vicinal F–H coupling constants for the sulfides are much greater than those of the analogous ethers.

Scheme 6.8

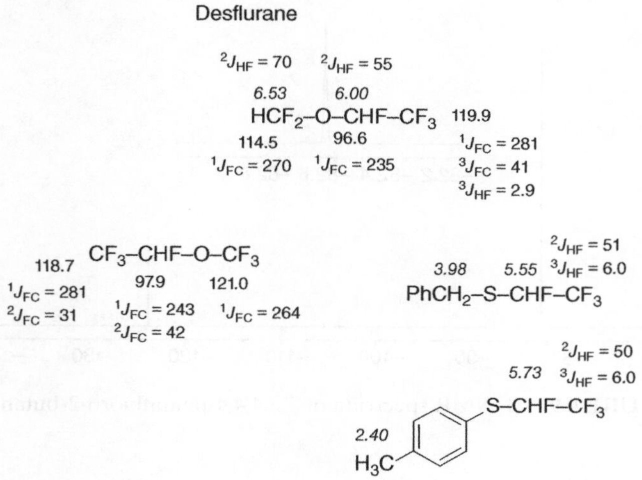

Desflurane

$^2J_{HF} = 70$ $^2J_{HF} = 55$

6.53   6.00

HCF$_2$–O–CHF–CF$_3$ 119.9

114.5   96.6 $^1J_{FC} = 281$

$^1J_{FC} = 270$ $^1J_{FC} = 235$ $^3J_{FC} = 41$

$^3J_{HF} = 2.9$

118.7   CF$_3$–CHF–O–CF$_3$

$^1J_{FC} = 281$ 97.9   121.0

$^2J_{FC} = 31$ $^1J_{FC} = 243$ $^1J_{FC} = 264$

$^2J_{FC} = 42$

$^2J_{HF} = 51$

3.98   5.55 $^3J_{HF} = 6.0$

PhCH$_2$–S–CHF–CF$_3$

$^2J_{HF} = 50$

5.73 $^3J_{HF} = 6.0$

S–CHF–CF$_3$

2.40

H$_3$C

6.5. THE PENTAFLUOROETHYL GROUP

The two signals of an ethyl group, a triplet for the methyl and a quartet for the CH$_2$ group, integrating 3:2, are perhaps the most recognizable in proton NMR. In contrast, the fluorine signals deriving from an isolated pentafluoroethyl group exhibit little vicinal coupling, and they appear effectively as two singlets. Thus, a pentafluoroethyl ketone will exhibit two singlet signals, as is exemplified by the fluorine NMR spectrum of 3,3,4,4,4-pentafluoro-2-butanone in Figure 6.9. As can be seen, the signals deriving from the CF$_3$ and the CF$_2$ groups, appearing at −82.6 and −123.9 ppm, respectively, are singlets in this 282 mHz spectrum.

Similar effectively uncoupled signals are observed for the ester, ethyl pentafluoropropionate (Scheme 6.9), which exhibits singlets at −123 and −84 ppm.

In contrast, in the case of an alkane terminated by a CF$_3$CF$_2$ group, the CF$_3$ group exhibits no apparent coupling, *but its CF$_2$ group experiences strong three-bond HF coupling with the hydrogens on the vicinal carbon* as is seen for the typical compounds in Scheme 6.9. The chemical shift for the CF$_2$ group will be in the range of −120 ppm, with the usual shielding impact of branching being seen. The chemical shift for the CF$_3$ group will be approximately −85 ppm. As seen from the example above and those below, the presence of the adjacent carbonyl group does not have much effect upon the chemical shifts of the pentafluoroethyl group.

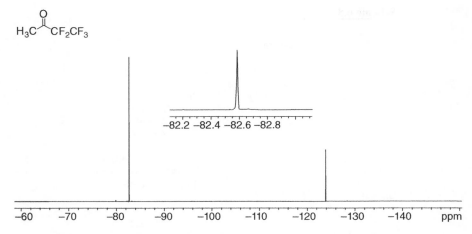

FIGURE 6.9. ^{19}F NMR spectrum of 3,3,4,4,4-pentafluoro-2-butanone

Scheme 6.9

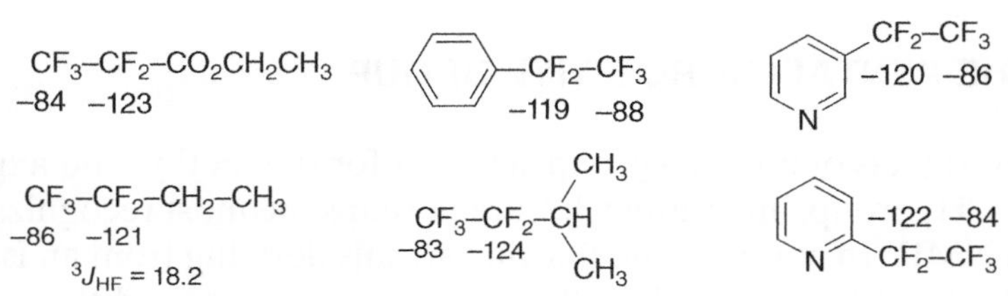

CF_3–CF_2–$CO_2CH_2CH_3$
–84 –123

[benzene ring]–CF_2–CF_3
–119 –88

[pyridine ring]–CF_2–CF_3
–120 –86

CF_3–CF_2–CH_2–CH_3
–86 –121
$^3J_{HF}$ = 18.2

CF_3–CF_2–CH(–CH_3)(CH_3)
–83 –124

[pyridine ring] –122 –84
CF_2–CF_3

Substitution on a benzene or pyridine ring also does not have much effect upon these chemical shifts.

An example of a C_2F_5 substituent bound to a carbon bearing one or more hydrogens is exemplified by the fluorine spectrum of 3,3,4,4,4-pentafluorobutene (Figure 6.10). This spectrum exhibits a singlet for the CF_3 group at −86.0 ppm, but a doublet at −118.2 with three-bond H–F coupling of 7.9 Hz.

The proton spectrum of this C_2F_5-ethylene, similar to that of the CF_3-ethylene described in Chapter 2, is second order in nature (Figure 6.11).

Additional proton and carbon chemical shift and coupling constant data for compounds bearing a C_2F_5 group are provided in Scheme 6.10.

The carbon spectrum of 3,3,4,4,4-pentafluoro-2-butanone provides a good example of the carbon signals of an isolated C_2F_5 group (Figure 6.12)

CF₃CF₂CH=CH₂

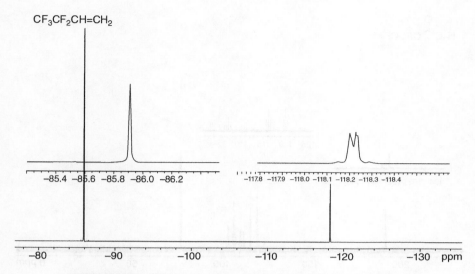

FIGURE 6.10. ¹⁹F NMR spectrum of 3,3,4,4,4-pentafluorobutene

CF₃CF₂CH=CH₂

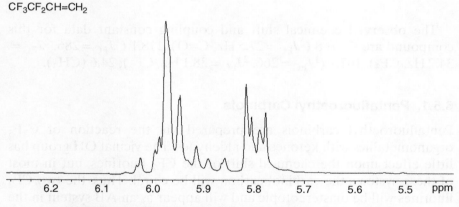

FIGURE 6.11. ¹H NMR spectrum of 3,3,4,4,4-pentafluorobutene

Scheme 6.10

$^3J_{HH} = 7.6$

2.55   1.65
CF₃–CF₂–CH₂–CH₃

$\begin{array}{c} 1.2 \\ CH_3 \end{array}$
CF₃–CF₂–CH 2.35
$\begin{array}{c} \\ CH_3 \end{array}$

2.04–2.22
CF₃–CF₂–CH₂–CH₂–CH₂–OH

$^1J_{FC} = 286$ 119.0 115.9 27.3 23.5 61.3
$^2J_{FC} = 36.5$

$^1J_{FC} = 252$ $^2J_{FC} = 22$
$^2J_{FC} = 36.4$

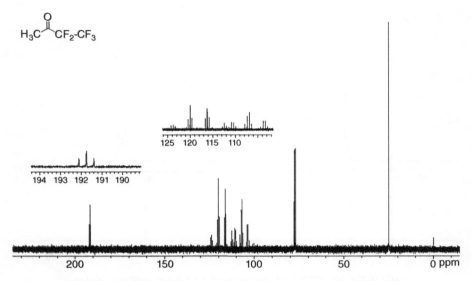

FIGURE 6.12. ^{13}C NMR spectrum of 3,3,4,4,4-pentafluoro-2-butanone

The observed chemical shift and coupling constant data for this compound are δ 191.8 ($^2J_{FC} = 27.5$ Hz, C=O), 118.1 ($^1J_{FC} = 286$, $^2J_{FC} = 34.2$ Hz, CF$_3$), 107.1 ($^1J_{FC} = 266$, $^2J_{FC} = 38.1$ Hz, CF$_2$), 24.6 (CH$_3$).

6.5.1. Pentafluoroethyl Carbinols

Pentafluoroethyl carbinols are prepared via the reaction of C$_2$F$_5$ organometallics with ketones and aldehydes. The vicinal OH group has little effect upon the chemical shift of the CF$_2$ fluorines, but in most of these systems the carbinol carbon will be chiral. Thus, the two CF$_2$ fluorines will be diastereotopic and will appear as an AB system in the fluorine NMR. Two examples of such systems, plus a related lactone are provided in Scheme 6.11, with the carbon NMR data being given in Scheme 6.12. An example of a pentafluoroethyl alkyne is also given. A CF$_2$ group bound to an acetylenic carbon is at an unusually low field for a carbon-bound CF$_2$ group.

6.5.2. Pentafluoroethyl Ethers, Sulfides, and Phosphines

The fluorine NMR spectra of pentafluoroethyl ethers and sulfides are rather nonexceptional, with the fluorine signals for their CF$_2$ groups considerably deshielded and their CF$_3$ groups essentially unaffected, both appearing as singlets ($^3J_{FF} < 2$ Hz for ethers and about 3 Hz for

Scheme 6.11

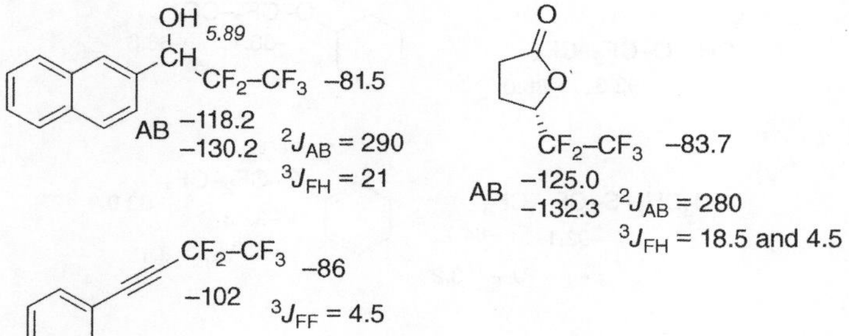

Scheme 6.12

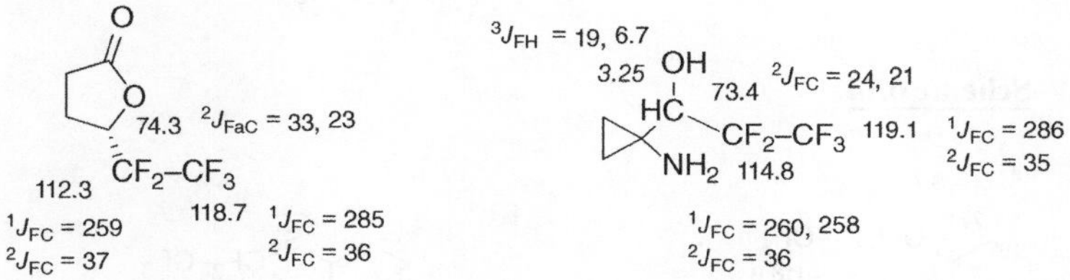

sulfides). Interestingly, the CF_2 signals of the CF_3CF_2 groups bound to oxygen, sulfur, and selenium all have about the same chemical shift, at about -92 ppm. In contrast, the CF_2 group of an analogous phosphine is much more shielded, almost the same as for a carbon-bound CF_2CF_3 group. Examples of ethers, sulfides, a selenide, and a phosphine are given in Scheme 6.13, with some pertinent ^{13}C data being given in Scheme 6.14.

6.5.3. Pentafluoroethyl Organometallics

There are also a number of C_2F_5 organometallics for which ^{19}F spectra are available (Scheme 6.15). It is strange that the CF_2 group that is bound to the metal is not deshielded, whereas CF_3 groups bound to such metals are much deshielded (Section 5.3.4).

6.6. THE 2,2,3,3,3-PENTAFLUOROPROPYL GROUP

Fluorine spectra of compounds containing the 2,2,3,3,3-pentafluoropropyl group resemble very much those of C_2F_5-alkanes, which are

Scheme 6.13

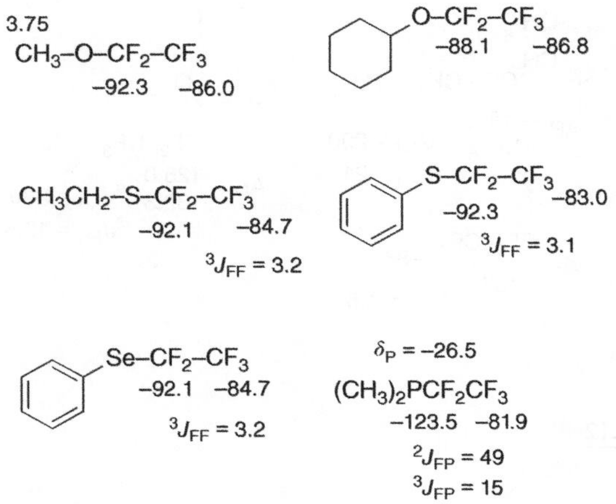

3.75
$CH_3-O-CF_2-CF_3$
−92.3   −86.0

$O-CF_2-CF_3$ (on cyclohexane)
−88.1   −86.8

$CH_3CH_2-S-CF_2-CF_3$
−92.1   −84.7
$^3J_{FF} = 3.2$

$S-CF_2-CF_3$ (on phenyl)
−92.3   −83.0
$^3J_{FF} = 3.1$

$Se-CF_2-CF_3$ (on phenyl)
−92.1   −84.7
$^3J_{FF} = 3.2$

$\delta_P = -26.5$
$(CH_3)_2PCF_2CF_3$
−123.5   −81.9
$^2J_{FP} = 49$
$^3J_{FP} = 15$

Scheme 6.14

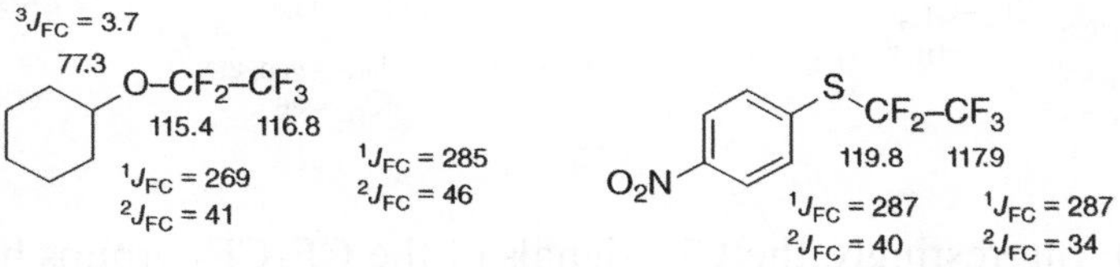

$^3J_{FC} = 3.7$
77.3
$O-CF_2-CF_3$ (on cyclohexane)
115.4   116.8
$^1J_{FC} = 269$ $^1J_{FC} = 285$
$^2J_{FC} = 41$ $^2J_{FC} = 46$

S CF_2-CF_3 (on O_2N-phenyl)
119.8   117.9
$^1J_{FC} = 287$ $^1J_{FC} = 287$
$^2J_{FC} = 40$ $^2J_{FC} = 34$

Scheme 6.15

Pentafluoroethyl organometallics

CF_3-CF_2-R
−121

−84
$CF_3CF_2-Sn(CH_3)_3$
−123

−84
$CF_3CF_2-Pb(CH_3)_3$
−120

−86
$(CF_3CF_2)_2Zn$
−125

−83
$(CF_3CF_2)_2Hg$
−109
$^3J_{FF} = 1.4$ Hz

$(n-C_5F_{11}CF_2)_2Zn$
−126

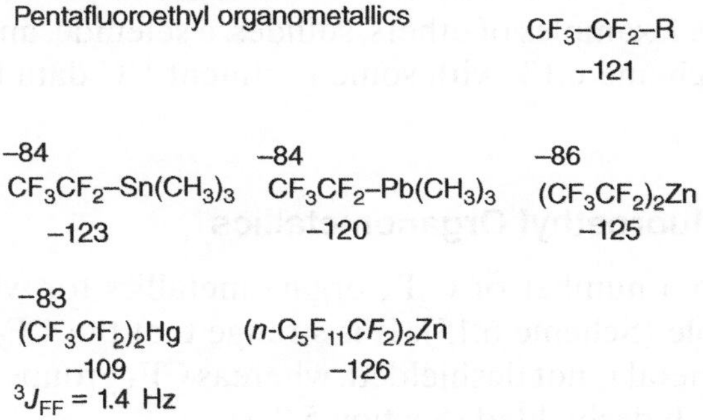

briefly discussed in the previous section, with the CF_2 group appearing as a triplet, coupling strongly to the vicinal CH_2 group. An example of such a compound is 2,2,3,3,3-pentafluoropropylamine, the fluorine spectrum of which is given in Figure 6.13, the CF_2 triplet ($J = 16$ Hz) appearing at −127.4, with the CF_3 appearing as a singlet at −86.5 ppm.

CF$_3$CF$_2$CH$_2$NH$_2$

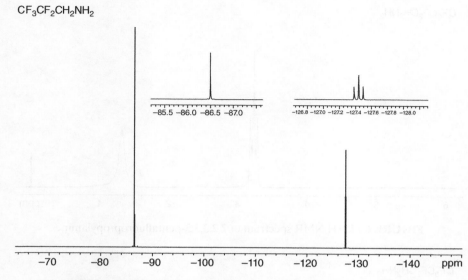

FIGURE 6.13. ^{19}F NMR spectrum of 2,2,3,3,3-pentafluoropropylamine

Scheme 6.16

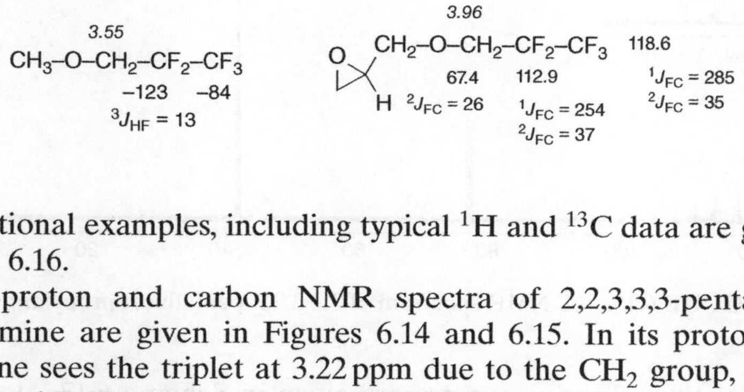

3.55
CH$_3$–O–CH$_2$–CF$_2$–CF$_3$
         –123 –84
      $^3J_{HF}$ = 13

3.96
CH$_2$–O–CH$_2$–CF$_2$–CF$_3$ 118.6
O   67.4   112.9 $^1J_{FC}$ = 285
H $^2J_{FC}$ = 26 $^1J_{FC}$ = 254 $^2J_{FC}$ = 35
                 $^2J_{FC}$ = 37

Additional examples, including typical ^{1}H and ^{13}C data are given in Scheme 6.16.

The proton and carbon NMR spectra of 2,2,3,3,3-pentafluoro-propylamine are given in Figures 6.14 and 6.15. In its proton spectrum, one sees the triplet at 3.22 ppm due to the CH$_2$ group, with its three-bond F–H coupling constant of 15.5 Hz.

In the carbon spectrum, one observes a triplet ($^2J_{FC}$ = 24.8 Hz) at δ 42.6 for the CH$_2$ group, a triplet of quartets ($^1J_{FC}$ = 252 and $^2J_{FC}$ = 36.4 Hz) at 114.2 ppm for the CF$_2$ group, and a quartet of triplets at 119.3 ppm ($^1J_{FC}$ = 286 and $^2J_{FC}$ = 37.3 Hz) for the CF$_3$ group.

6.7. THE 1,1,2,3,3,3-HEXAFLUOROPROPYL GROUP

Ethers and sulfides bearing the CF$_3$CHFCF$_2$—group are readily prepared by addition of the alcohol or thiol to hexafluoropropene. Some

CF$_3$CF$_2$CH$_2$NH$_2$

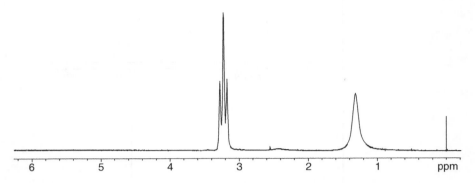

FIGURE 6.14. ^{1}H NMR spectrum of 2,2,3,3,3-pentafluoropropylamine

CF$_3$CF$_2$CH$_2$NH$_2$

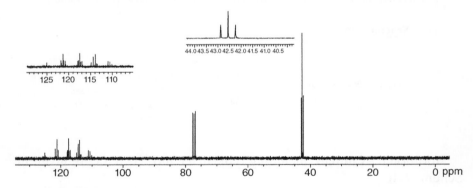

FIGURE 6.15. ^{13}C NMR spectrum of 2,2,3,3,3-pentafluoropropylamine

data from their fluorine spectra are given in Scheme 6.17, along with proton data. With the CHF carbon being chiral, the CF$_2$ group should appear as an AB quartet.

Typical proton and carbon NMR data for various substituted hexafluoropropyl compounds are given in Scheme 6.18.

6.8. 1,1,2,2,3,3-HEXAFLUOROPROPYL SYSTEM

Spectra of a hydrocarbon with a CF$_2$CF$_2$CF$_2$H group and 1,1,2,2,3,3-hexafluorocyclohexane, a compound with three contiguous CF$_2$ groups are provided in Scheme 6.19. The RCF$_2$ group always appears at the

Scheme 6.17

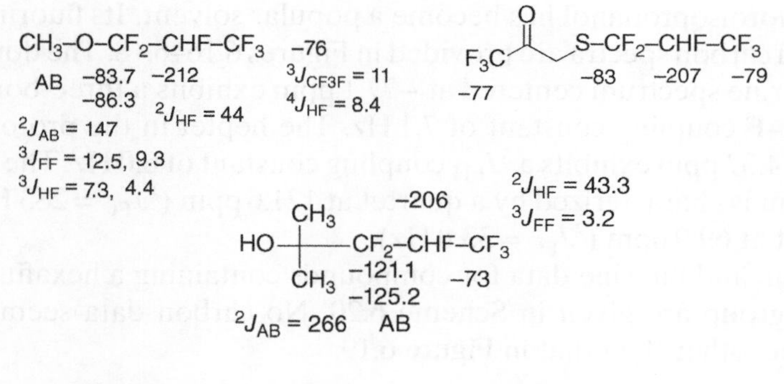

CH₃–O–CF₂–CHF–CF₃   –76

$$\text{CH}_3\text{-O-CF}_2\text{-CHF-CF}_3$$

AB –83.7 –212

   –86.3

$^2J_{AB} = 147$

$^3J_{FF} = 12.5, 9.3$

$^3J_{HF} = 7.3, 4.4$

$^3J_{CF3F} = 11$

$^4J_{HF} = 8.4$

$^2J_{HF} = 44$

F₃C–C(=O)–S–CF₂–CHF–CF₃

   –77   –83 –207 –79

                      CH₃
HO–C(CH₃)–CF₂–CHF–CF₃

              –206   –121.1 –73
                        –125.2

$^2J_{AB} = 266$ AB

$^2J_{HF} = 43.3$

$^3J_{FF} = 3.2$

Scheme 6.18

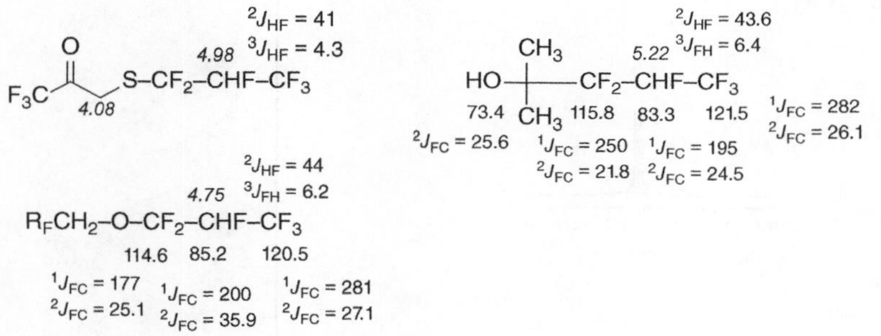

$^2J_{HF} = 41$

4.98 $^3J_{HF} = 4.3$

F₃C–C(=O)–CH₂–S–CF₂–CHF–CF₃

   4.08

$^2J_{HF} = 44$

4.75 $^3J_{FH} = 6.2$

R_FCH₂–O–CF₂–CHF–CF₃

      114.6   85.2   120.5

$^1J_{FC} = 177$ $^1J_{FC} = 200$ $^1J_{FC} = 281$

$^2J_{FC} = 25.1$ $^2J_{FC} = 35.9$ $^2J_{FC} = 27.1$

$^2J_{HF} = 43.6$

CH₃ 5.22 $^3J_{FH} = 6.4$

HO–C(CH₃)–CF₂–CHF–CF₃

73.4 CH₃ 115.8   83.3   121.5

$^2J_{FC} = 25.6$ $^1J_{FC} = 250$ $^1J_{FC} = 195$

          $^2J_{FC} = 21.8$ $^2J_{FC} = 24.5$

$^1J_{FC} = 282$

$^2J_{FC} = 26.1$

Scheme 6.19

–115 –132 –138

CF₂–CF₂–CF₂H

      –139
(perfluorocyclohexane ring) –118

      –118
(cyclohexane ring) –135

(cyclopropane ring) –159

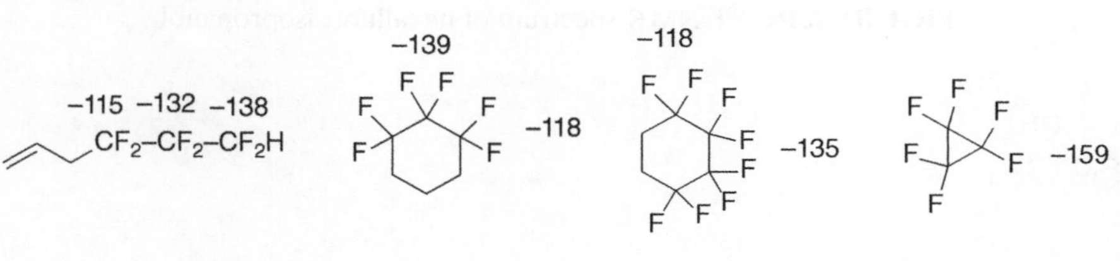

lowest field, with the CF₂H group at the highest. A cyclohexane with four contiguous CF₂ groups has analogous chemical shifts to that of the hexafluoro compound. Contrast the −135 ppm chemical shift for the internal fluorines of this strain-free compound, in comparison to the −159 ppm chemical shift of hexafluorocyclopropane – another example of the unique shielding effect of the cyclopropane ring on any fluorine substituents. In fact, the CF₂ group of hexafluorocyclopropane exhibits the largest (most shielded) chemical shift reported for a CF₂ group.

6.9. THE HEXAFLUORO-ISOPROPYL GROUP

Hexafluoroisopropanol has become a popular solvent. Its fluorine, proton, and carbon spectra are provided in Figures 6.16–6.18. The doublet in the fluorine spectrum centered at −77.1 ppm exhibits a three-bond coupling H–F coupling constant of 7.1 Hz. The heptet in the proton spectrum at 4.37 ppm exhibits a $^3J_{FH}$ coupling constant of 6.0 Hz. The carbon spectrum is characterized by a quartet at 121.6 ppm ($^1J_{FC} = 283$ Hz) and a heptet at 69.9 ppm ($^2J_{FC} = 33.8$ Hz).

Proton and fluorine data for compounds containing a hexafluoroisopropyl group are given in Scheme 6.20. No carbon data seems to be available, other than that in Figure 6.19.

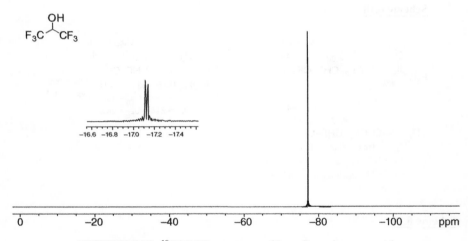

FIGURE 6.16. ^{19}F NMR spectrum of hexafluoroisopropanol

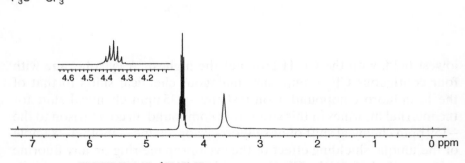

FIGURE 6.17. ^{1}H NMR spectrum of hexafluoroisopropanol

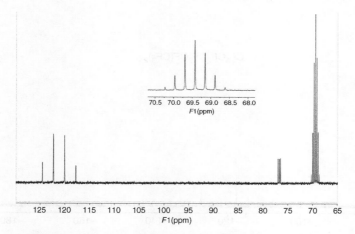

FIGURE 6.18. ^{13}C NMR spectrum for hexafluoroisopropanol

Scheme 6.20

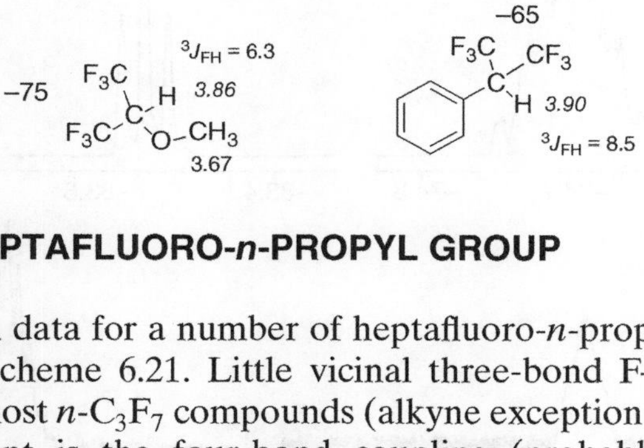

−75   F$_3$C $^3J_{FH}$ = 6.3
       $\overset{|}{C}$ H 3.86
       F$_3$C   O—CH$_3$
                    3.67

                    −65
       F$_3$C   CF$_3$
          $\overset{|}{C}$
             H 3.90
                    $^3J_{FH}$ = 8.5

6.10. THE HEPTAFLUORO-*n*-PROPYL GROUP

Fluorine NMR data for a number of heptafluoro-*n*-propyl compounds are given in Scheme 6.21. Little vicinal three-bond F–F coupling is observed for most *n*-C$_3$F$_7$ compounds (alkyne exception below); rather more prominent is the four-bond coupling (probably significantly *through-space*).

6.11. THE HEPTAFLUORO-*iso*-PROPYL GROUP

One recurring difference between the heptafluoro-isopropyl ethers and sulfides is the significant ~9 Hz vicinal coupling observed for the sulfur compounds, but not for the ethers (Scheme 6.22).

6.12. THE NONAFLUORO-*n*-BUTYL GROUP

As was the case for the *n*-C$_3$F$_7$ group, the most prominent F–F coupling in the *n*-C$_4$F$_9$ group is usually the four-bond coupling (Scheme 6.23).

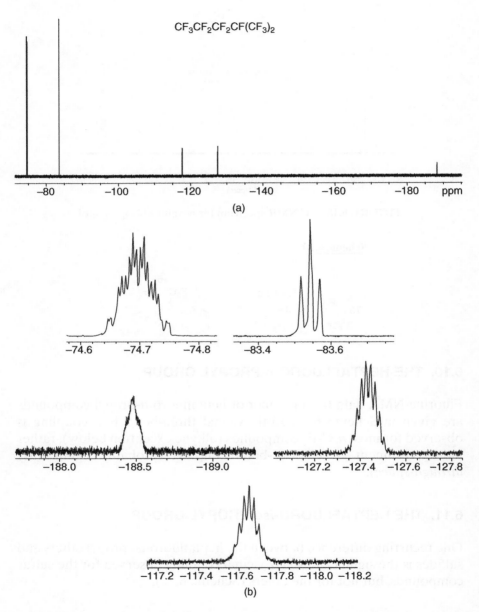

FIGURE 6.19. (a) ^{19}F NMR spectrum of perfluoro-2-methylpentane. (b) Expansions of individual peaks in the ^{19}F NMR spectrum of perfluoro-2-methylpentane

Scheme 6.21

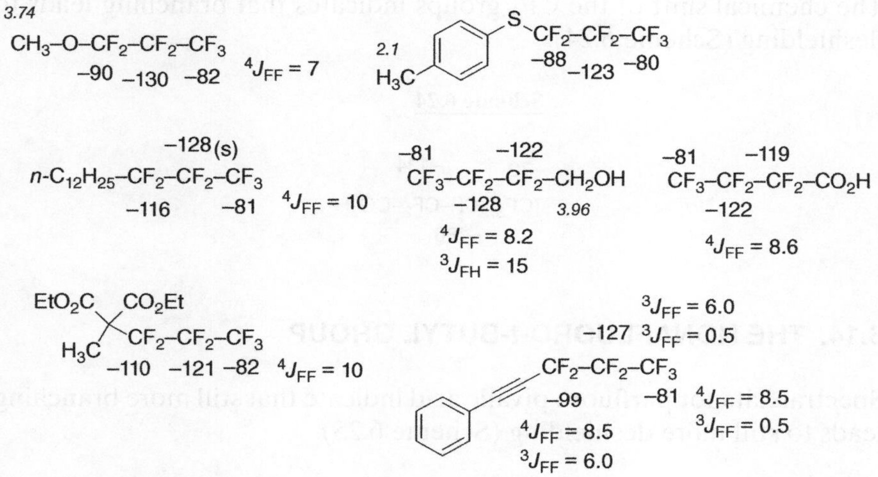

3.74

$$CH_3-O-CF_2-CF_2-CF_3$$
$$_{-90}\ _{-130}\ _{-82}\ \ ^4J_{FF} = 7$$

2.1

$$H_3C-\langle\!\!\!\bigcirc\!\!\!\rangle-S-CF_2-CF_2-CF_3$$
$$_{-88}\ _{-123}\ _{-80}$$

$$_{-128\,(s)}$$
$$n\text{-}C_{12}H_{25}-CF_2-CF_2-CF_3$$
$$_{-116}\qquad_{-81}\ \ ^4J_{FF}=10$$

$$_{-81}\qquad_{-122}$$
$$CF_3-CF_2-CF_2-CH_2OH$$
$$_{-128}\qquad_{3.96}$$
$$^4J_{FF}=8.2$$
$$^3J_{FH}=15$$

$$_{-81}\qquad_{-119}$$
$$CF_3-CF_2-CF_2-CO_2H$$
$$_{-122}$$
$$^4J_{FF}=8.6$$

$$EtO_2C\quad CO_2Et$$
$$_{H_3C}\!\diagdown\!\diagup\!CF_2-CF_2-CF_3$$
$$_{-110}\ _{-121}\ _{-82}\ \ ^4J_{FF}=10$$

$$^3J_{FF}=6.0$$
$$_{-127}\ ^3J_{FF}=0.5$$
$$CF_2-CF_2-CF_3$$
$$_{-99}\qquad_{-81}\ \ ^4J_{FF}=8.5$$
$$\langle\!\!\!\bigcirc\!\!\!\rangle-C\!\equiv\!C\qquad\qquad^3J_{FF}=0.5$$
$$^4J_{FF}=8.5$$
$$^3J_{FF}=6.0$$

Scheme 6.22

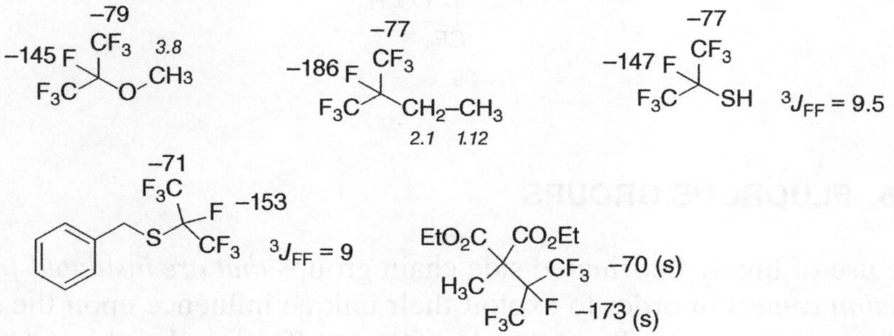

$$_{-79}$$
$$_{-145}\,F\diagdown\!C F_3\ \ _{3.8}$$
$$F_3C\diagup\!O-CH3$$

$$_{-77}$$
$$_{-186}\,F\diagdown\!CF_3$$
$$F_3C\diagup\!CH_2-CH_3$$
$$_{2.1\ \ 1.12}$$

$$_{-77}$$
$$_{-147}\,F\diagdown\!CF_3$$
$$F_3C\diagup\!SH\ \ ^3J_{FF}=9.5$$

$$_{-71}$$
$$_{F_3C}\diagdown$$
$$\langle\!\!\!\bigcirc\!\!\!\rangle-CH_2-S\!\!-\!\!C\!\!-\!\!F\ _{-153}$$
$$CF_3\ \ ^3J_{FF}=9$$

$$EtO_2C\quad CO_2Et$$
$$_{H_3C}\!\diagdown\!\diagup\!CF_3\ _{-70\,(s)}$$
$$F_3C\!\diagup\!F\ _{-173\,(s)}$$

Scheme 6.23

$$_{-118}\qquad_{-78}\ ^4J_{FF}=9$$
$$\langle\!\!\!\bigcirc\!\!\!\rangle-CH_2-S\!-\!CF_2-CF_2-CF_2-CF_3$$
$$_{-85}\qquad_{-123}$$

1.11 2.03
$$_{-123/-125}$$
$$CH_3-CH_2-CF_2-CF_2-CF_2-CF_3$$
$$_{-116}\qquad\qquad_{-81}$$

$$_{-81}\qquad_{-123}$$
$$CF_3-CF_2-CF_2-CF_2-CO_2H$$
$$_{-119}\qquad_{-126}$$

$$_{-81}\qquad_{-114}$$
$$CF_3-CF_2-CF_2-I$$
$$_{-126}\qquad_{-60}$$

$$_{-123}\qquad_{-81}$$
$$\langle\!\!\!\bigcirc\!\!\!\rangle-CF_2-CF_2-CF_2-CF_3$$
$$_{-111}\qquad_{-126}$$

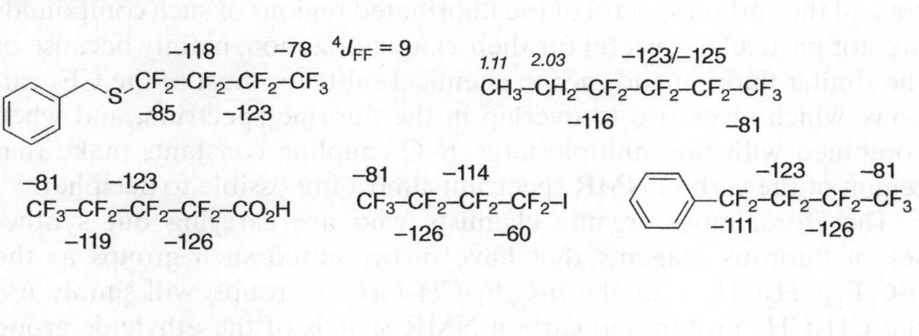

6.13. THE NONAFLUORO-*iso*-BUTYL GROUP

The chemical shift of the CF_3 groups indicates that branching leads to deshielding (Scheme 6.24).

Scheme 6.24

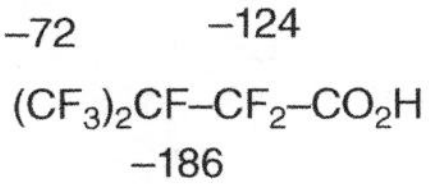

$$
\begin{array}{cc}
-72 & -124 \\
\end{array}
$$
$$(CF_3)_2CF{-}CF_2{-}CO_2H$$
$$-186$$

6.14. THE NONAFLUORO-*t*-BUTYL GROUP

Spectral data for perfluoro pivalic acid indicate that still more branching leads to still more deshielding (Scheme 6.25).

Scheme 6.25

$$
\begin{array}{c}
CF_3 \\
\mid \\
F_3C{-}\!\!-\!\!-CO_2H \\
\mid \\
CF_3 \\
-69
\end{array}
$$

6.15. FLUOROUS GROUPS

The use of highly fluorinated side chain groups *that are insulated from reaction centers* in order to exploit their unique influence upon the solubility properties of the molecule without affecting the chemistry of the functional groups has spawned a new subfield of fluorine chemistry known as Fluorous Chemistry. For the most part, both the fluorine spectra and the carbon spectra of the fluorinated regions of such compounds are not particularly useful for their characterization, mainly because of the similar fluorine and carbon chemical shifts for most of the CF_2 carbons, which gives rise to overlap in the fluorine spectrum, and when combined with the multiple large F–C coupling constants make that region of the carbon NMR spectrum almost impossible to decipher.

Therefore, most organic chemists who are carrying out syntheses of fluorous reagents that have incorporated such groups as the $n\text{-}C_6F_{13}CH_2CH_2{-}$ or the $n\text{-}C_8F_{17}CH_2CH_2{-}$ groups, will simply use the CH_2CH_2 proton and carbon NMR signals of the ethylenic group

Scheme 6.26

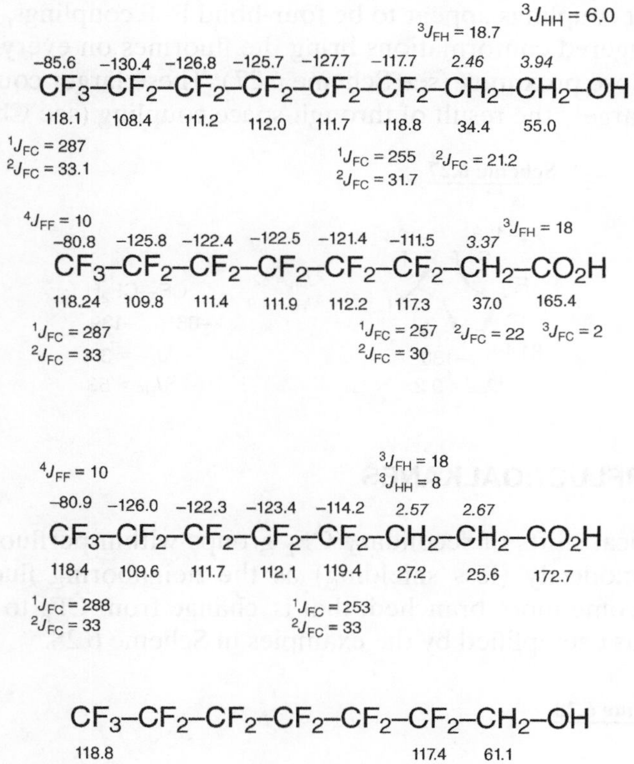

$^3J_{HH} = 6.0$

$^3J_{FH} = 18.7$

−85.6   −130.4   −126.8   −125.7   −127.7   −117.7   2.46   3.94

$$CF_3-CF_2-CF_2-CF_2-CF_2-CF_2-CH_2-CH_2-OH$$

118.1   108.4   111.2   112.0   111.7   118.8   34.4   55.0

$^1J_{FC} = 287$

$^2J_{FC} = 33.1$

$^1J_{FC} = 255$ $^2J_{FC} = 21.2$

$^2J_{FC} = 31.7$

$^4J_{FF} = 10$

$^3J_{FH} = 18$

−80.8   −125.8   −122.4   −122.5   −121.4   −111.5   3.37

$$CF_3-CF_2-CF_2-CF_2-CF_2-CF_2-CH_2-CO_2H$$

118.24   109.8   111.4   111.9   112.2   117.3   37.0   165.4

$^1J_{FC} = 287$

$^2J_{FC} = 33$

$^1J_{FC} = 257$ $^2J_{FC} = 22$ $^3J_{FC} = 2$

$^2J_{FC} = 30$

$^4J_{FF} = 10$

$^3J_{FH} = 18$

$^3J_{HH} = 8$

−80.9   −126.0   −122.3   −123.4   −114.2   2.57   2.67

$$CF_3-CF_2-CF_2-CF_2-CF_2-CH_2-CH_2-CO_2H$$

118.4   109.6   111.7   112.1   119.4   27.2   25.6   172.7

$^1J_{FC} = 288$

$^2J_{FC} = 33$

$^1J_{FC} = 253$

$^2J_{FC} = 33$

$$CF_3-CF_2-CF_2-CF_2-CF_2-CF_2-CH_2-OH$$

118.8   117.4   61.1

that bridges the fluorinated segment to the nonfluorinated segment in characterizing their compounds. Scheme 6.26 provides such typical proton NMR data along with the very difficult to obtain fluorine and carbon NMR data for the tridecafluoro-*n*-octyl alcohol precursor that is often used in the synthesis of fluorous compounds. Some data for the analogous system with a carboxylic acid function and with only a single CH$_2$ linkage are provided for comparison purposes.

6.16. 1-HYDRO-PERFLUOROALKANES

When a primary CF$_2$H group is attached to any perfluoroalkyl group, the respective chemical shifts of the CF$_2$H fluorine, hydrogen, and carbon are not greatly affected by the length of the perfluoro group; thus for R$_F$CF$_2$H, the chemical shifts for the CF$_2$H groups for R$_F$ = CF$_3$,

C_2F_5, and n-C_3F_7 are −142.0, −139.0, and −137.6 ppm, respectively. Vicinal couplings, both F–F and F–H, can be quite small in such systems. The largest couplings appear to be four-bond F–F couplings, where the zigzag staggered conformations bring the fluorines on every third carbon into close proximity (see Scheme 6.27). These larger couplings are probably largely the result of through-space coupling (see Chapter 2).[1]

Scheme 6.27

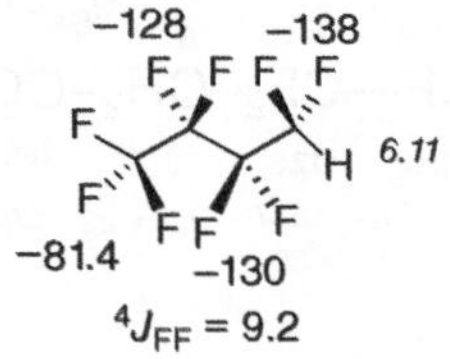

$^2J_{FH}$ = 52
$^3J_{FH}$ = 4.9

CF_3–CF_2H
−86 −139
$^3J_{FF}$ = 3.4
$^2J_{HF}$ = 53

6.17. PERFLUOROALKANES

The chemical shifts of secondary CF_2 groups within perfluorocarbons decrease modestly (less shielding) as the neighboring fluorocarbon groups become more branched, that is, change from CF_3 to CF_2R_F to $CF(R_F)_2$, as exemplified by the examples in Scheme 6.28.

Scheme 6.28

$CF_3\boldsymbol{CF_2}CF_3$	$CF_3CF_2\boldsymbol{CF_2}CF_3$	$CF_3CF_2\boldsymbol{CF_2}CF_2CF_3$	$(CF_3)_2CF\boldsymbol{CF_2}CF_3$
−132	−127	−123	−119
$^3J_{FF}$ = 7.3		$^4J_{FF}$ = 10.2	

As indicated earlier, vicinal F–F coupling constants in perfluoroalkanes are often very small, virtually negligible in comparison to longer range couplings, as is the case for perfluoro-n-pentane above where the four-bond F–F coupling is 10.2 Hz.

Trifluoromethyl groups in unbranched perfluoroalkanes are the most deshielded, with chemical shifts around −81 ppm, whereas branching near the CF_3 will move the signal to lower field, as can be seen in Scheme 6.29.

Scheme 6.29

CF_3–CF_3	CF_3–CF_2R_F	CF_3–$CF(R_f)_2$	CF_3–$C(R_F)_3$
−89	~ −81	~ −73	~ −63

Scheme 6.30

$$CF_3CF_2CF_2CF_2CF_2CF_2CF_2CF_3$$
$$a\ \ b\ \ c\ \ d$$

δ_F (a) −81.1, (b) −126.1, (c) −122.5, (d) −121.7   in cyclohexane

$$CF_3CF_2CF_2CF_2CF_2CF_3$$
$$a\ \ b\ \ c$$

δ_F (a) −81.2, (b) −125.9, (c) −122.4   in CFCl$_3$

$$(CF_3)_2CFCF_2CF_2CF_3$$
$$a\ \ b\ \ c\ \ d\ \ e$$

δ_F (a) −72.0, (b) −185.8, (c) −115.0, (d) −124.8, (e) −80.8

all in CFCl$_3$

$$(CF_3CF_2)_2CFCF_3 \qquad\qquad (CF_3)_2CFCF(CF_3)_2$$
$$a\ \ b\ \ c\ \ d \qquad\qquad\qquad a\ \ \ b$$

δ_F (a) −80.3, (b) −116.7, (c) −184.4, (d) −71.0 δ_F (a) −70.4, (b) −178.8

$$(CF_3)_3CCF_2CF_3$$
$$a\ \ b\ \ c$$

δ_F (a) −62.0, (b) −108.7, (c) −78.6

Scheme 6.30 provides chemical shifts for all of the fluorines in a representative group of perfluorocarbons. The various environments exhibited should allow one to estimate the chemical shift for almost any fluorine in a perfluorocarbon system.

The fluorine chemical shifts of four-, five- and six-membered ring perfluoroalicyclics are quite consistently in the range of −133 to −134 ppm, but as usual, fluorines on a cyclopropane ring appear at a much higher field than those of other fluorinated alicyclics, perfluorocyclopropane having a chemical shift of −159 ppm.

Examining perfluoro-2-methylpentane more carefully, the chemical shifts of the fluorines of individual carbons are given in Scheme 6.31, with the actual spectrum provided in Figure 6.19a and b.

Substantial long-range (four-bond) couplings with the CF$_3$ groups can be observed, whereas little three-bond coupling is evident.

The ^{13}C NMR spectrum of perfluoro-2-methylhexane (Figure 6.20) both exemplifies the difficulty in analyzing such spectra because

Scheme 6.31

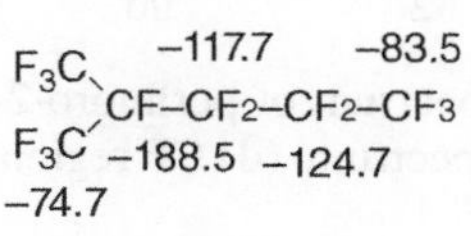

$$\begin{array}{c} F_3C \\ \diagdown \\ F_3C \diagup \\ \end{array}$$

F$_3$C   −117.7   −83.5
　　　＼
　　　　CF−CF$_2$−CF$_2$−CF$_3$
　　　／
F$_3$C   −188.5 −124.7
−74.7

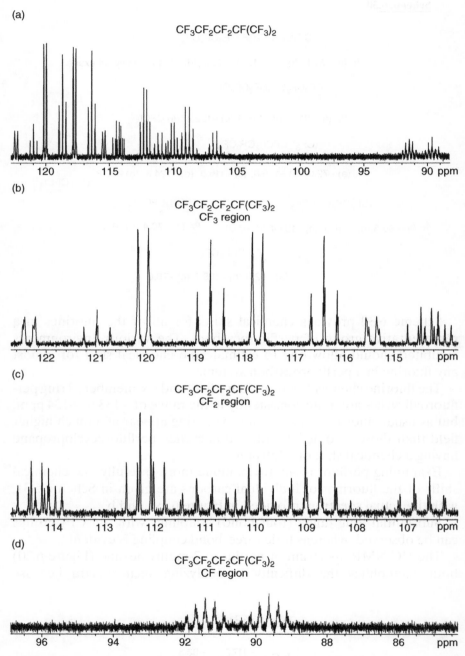

FIGURE 6.20. (a) ^{13}C NMR spectrum of perfluoro-2-methylpentane. (b) CF_3 region of spectrum. (c) CF_2 region of spectrum. (d) CF region of spectrum

TABLE 6.1. Fluorine δ-Values for Perfluoro-*n*-Alkyl Halides

	CF$_3$(CF$_2$)$_5$CF$_2$CF$_2$X	
X	δ_F(CF$_2$X)	δ_F(2-CF$_2$)
F	−81.8	−126.7
Cl	−68.6	−120.6
Br	−63.8	−117.7
I	−58.8	−113.5

of overlapping multiplets, but also shows that for relatively small molecules with some symmetry, the spectra, with careful analysis, can be fully characterized.

6.18. PERFLUORO-*n*-ALKYL HALIDES

The examples in Table 6.1 indicate how the CF$_2$ chemical shifts vary for typical examples of perfluoro-*n*-alkyl halides.

6.19. PERFLUOROALKYL AMINES, ETHERS, AND CARBOXYLIC ACID DERIVATIVES

The examples given in Scheme 6.32 provide insight regarding CF$_2$ chemical shifts of perfluoroalkyl groups bound to oxygen, nitrogen, and

Scheme 6.32

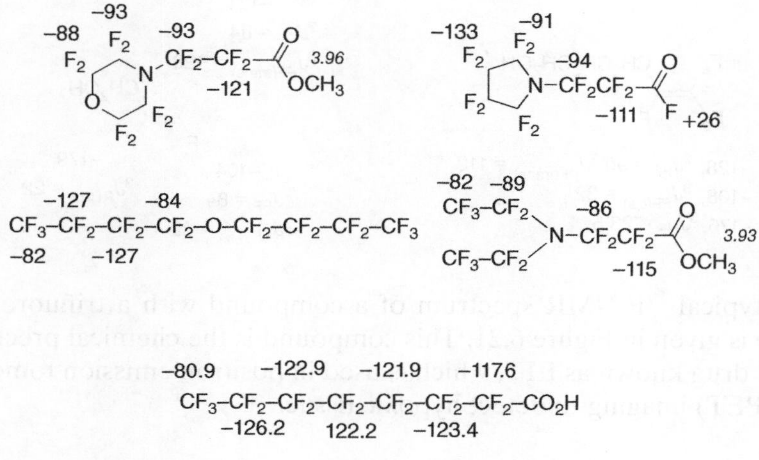

carbonyl groups,[2] whereas the chemical shift data for perfluorooctanoic acid is a classic example of how assignments of a group of signals for CF_2 groups with similar chemical shifts can be accomplished by use of F–F COSY and NOESY experiments,[1] and the knowledge that the largest coupling constants observed between fluorines in a perfluoroalkyl chain are four-bond couplings.

6.20. POLYFLUOROALKENES

6.20.1. Trifluorovinyl Groups

The fluorine substituent at the 2-position of a trifluorovinyl group is much more highly shielded than the other two fluorines, and its presence gives rise to an enhanced "split" of the diastereotopic fluorines at the 1-position and enhanced coupling constants, both geminal and vicinal.

Trifluorovinyl groups have characteristic chemical shifts and coupling constants that are exemplified as follows (Scheme 6.33).

Scheme 6.33

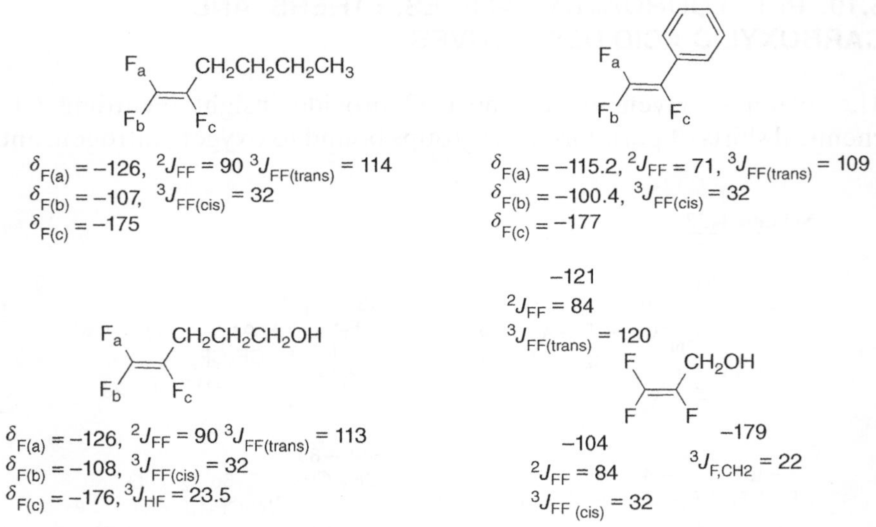
Fa CH2CH2CH2CH3
Fb Fc
δF(a) = −126, 2JFF = 90 3JFF(trans) = 114
δF(b) = −107, 3JFF(cis) = 32
δF(c) = −175
Fa
Fb Fc
δF(a) = −115.2, 2JFF = 71, 3JFF(trans) = 109
δF(b) = −100.4, 3JFF(cis) = 32
δF(c) = −177
Fa CH2CH2CH2OH
Fb Fc
δF(a) = −126, 2JFF = 90 3JFF(trans) = 113
δF(b) = −108, 3JFF(cis) = 32
δF(c) = −176, 3JHF = 23.5
−121
2JFF = 84
3JFF(trans) = 120
F CH2OH
F F
−104
2JFF = 84
3JFF (cis) = 32
−179
3JF,CH2 = 22

A typical ^{19}F NMR spectrum of a compound with a trifluorovinyl group is given in Figure 6.21. This compound is the chemical precursor of the drug known as EF5, which is used in positron emission tomography (PET) imaging to detect hypoxic tissue.

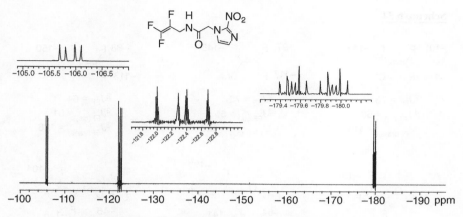

FIGURE 6.21. ^{19}F NMR spectrum 2,3,3-trifluoroallyl amide

The fluorine NMR data for the EF5 precursor are δ −105.9 (dd, $^2J_{FF} = 82$, $^3J_{FF(cis)} = 33\,\text{Hz}$), −122.3 (ddt, $^2J_{FF} = 81$, $^3J_{FF(trans)} = 114$, $^4J_{HF} = 3.7\,\text{Hz}$), −179.7 (ddt, $^3J_{FF(trans)} = 114$, $^3J_{FF(cis)} = 32$, $^3J_{HF} = 21.7\,\text{Hz}$).

6.20.1.1. Trifluorovinyl Halides and Ethers.

Trifluorovinyl halides are quite commonly encountered reagents, while trifluorovinyl ethers have increased interest as novel monomers.[3] There is a nice recent paper dealing with the NMR spectra of some trifluorovinyl ethers.[4] Fluorine data for a couple of examples of halides, ethers, and thioethers are given in Scheme 6.34.

6.20.1.2. Carbon and Proton NMR Spectra of Trifluorovinyl Compounds.

Scheme 6.35 provides the carbon and proton NMR data for a few trifluorovinyl compounds.

A specific example of the C–F regions of a carbon spectrum of a trifluorovinyl compound, that of the EF5 precursor is provided in Figure 6.22a and b. Note that the carbon bearing one fluorine at 127.56 ppm is a doublet of doublets of doublets (8 peaks), with a large (236 Hz) one-bond F–C coupling constant and then into two smaller doublets with two-bond F–C coupling constants of 52 and 15.7 Hz. (One of the eight peaks is obscured by the intense signal deriving from one of the imidazole C–H carbons.)

The CF_2 carbon at 154.16 ppm is essentially split into a triplet of doublets with an even larger (~281 Hz) one-bond F–C coupling constant, although the coupling constants for the Z and the E fluorines

Scheme 6.34

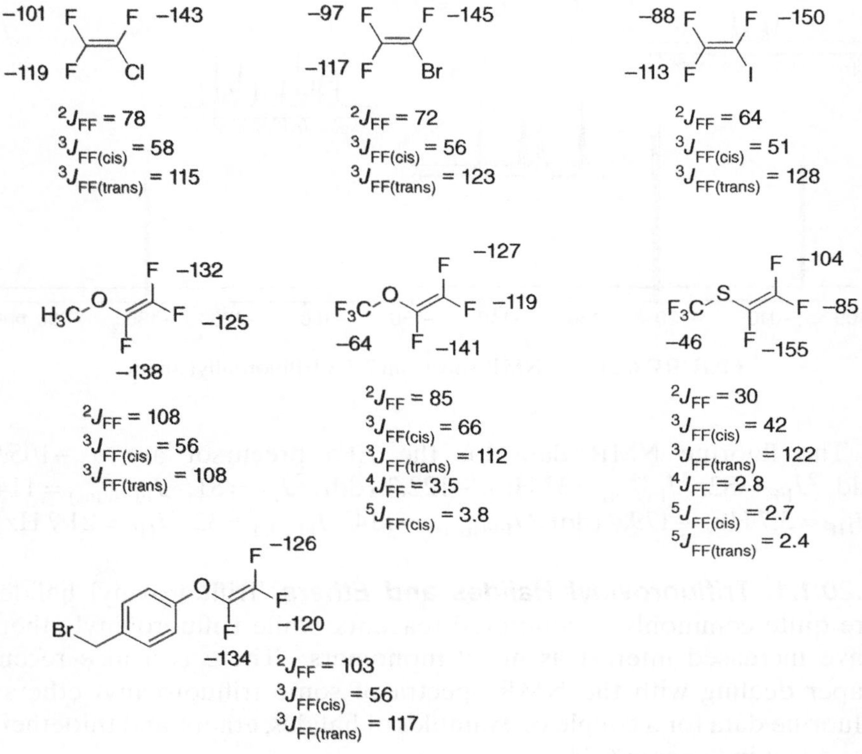

-101 F   F -143

-119 F   Cl

$^2J_{FF} = 78$
$^3J_{FF(cis)} = 58$
$^3J_{FF(trans)} = 115$

-97 F   F -145

-117 F   Br

$^2J_{FF} = 72$
$^3J_{FF(cis)} = 56$
$^3J_{FF(trans)} = 123$

-88 F   F -150

-113 F   I

$^2J_{FF} = 64$
$^3J_{FF(cis)} = 51$
$^3J_{FF(trans)} = 128$

F -132

H_3C–O   F -125

F

-138

$^2J_{FF} = 108$
$^3J_{FF(cis)} = 56$
$^3J_{FF(trans)} = 108$

-127

F_3C–O   F -119

-64 F -141

$^2J_{FF} = 85$
$^3J_{FF(cis)} = 66$
$^3J_{FF(trans)} = 112$
$^4J_{FF} = 3.5$
$^5J_{FF(cis)} = 3.8$

F -104

F_3C–S   F -85

-46 F -155

$^2J_{FF} = 30$
$^3J_{FF(cis)} = 42$
$^3J_{FF(trans)} = 122$
$^4J_{FF} = 2.8$
$^5J_{FF(cis)} = 2.7$
$^5J_{FF(trans)} = 2.4$

F -126

Br   O   F -120

-134 $^2J_{FF} = 103$
$^3J_{FF(cis)} = 56$
$^3J_{FF(trans)} = 117$

Scheme 6.35

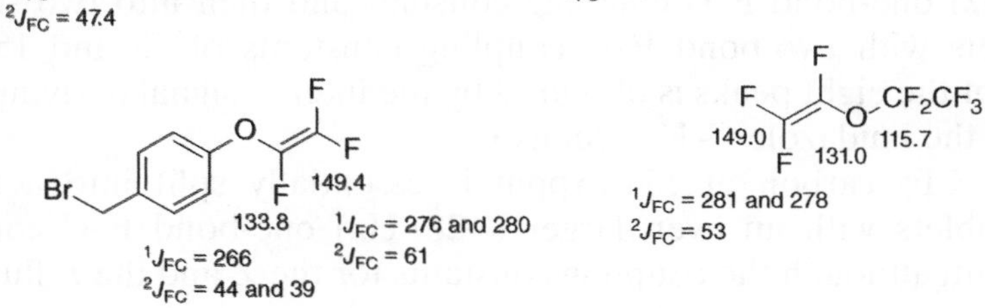

$^1J_{FC} = 234$
$^2J_{FC} = 53.6$ and 16.9
128.7

28.1
F 1.64

F   OH

2.23 3.50 $^3J_{HH} = 6.0$
21.9 60.9

153.3 $^3J_{FH} = 23.5$
$^1J_{FC} = 285$ and 272
$^2J_{FC} = 47.4$

F

154.1 F 128.8

F   F

$^1J_{FC} = 290$ and 282 $^1J_{FC} = 227$
$^2J_{FC} = 50$ $^2J_{FC} = 46$ and 20

F

Br   O   F

133.8 F 149.4

$^1J_{FC} = 266$ $^1J_{FC} = 276$ and 280
$^2J_{FC} = 44$ and 39 $^2J_{FC} = 61$

F   O CF_2CF_3

149.0 F 131.0 115.7

$^1J_{FC} = 281$ and 278
$^2J_{FC} = 53$

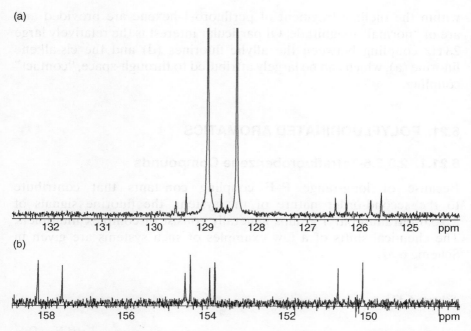

FIGURE 6.22. (a) CF part of the $CF_2=CF-$ section of the ^{13}C NMR spectrum of the EF5 precursor. (b) CF_2 part of the $CF_2=CF-$ section of the ^{13}C NMR spectrum of the EF5 precursor

are slightly different, along with the doublet deriving from the smaller 45.6 Hz two-bond F–C coupling. The four peak central part of the "triplet" is typical for systems such as this, where two diastereotopic fluorines have slightly different couplings to the same carbon.

6.20.2. Perfluoroalkenes

The chemical shifts for *all fluorines* in two representative examples of perfluoro-1-alkenes are given in Scheme 6.36. F–F coupling constants

Scheme 6.36

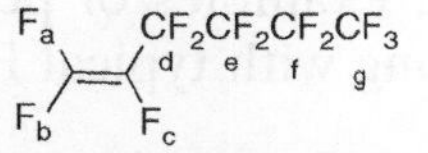

$$\underset{F_b}{\overset{F_a}{>}}C=C\underset{F_c}{\overset{CF_2CF_2CF_3}{<}}_{d\ \ e\ \ f\ \ g}$$

$\delta_F =$ (a) −108.1, (b) −91.7, (c) −193.4, (d) −120.5, (e) −127.0, (f) −129.0, (g) −83.7

$^2J_{ab} = 52;\ ^3J_{ac} = 117;\ ^3J_{bc} = 40$
$^3J_{cd} = 14,\ ^4J_{ad} = 28;\ ^4J_{bd} = 6$

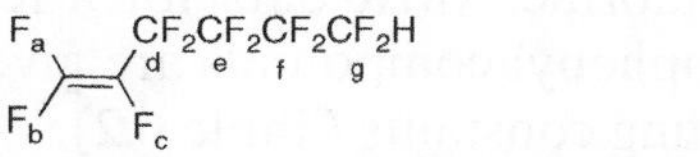

$$\underset{F_b}{\overset{F_a}{>}}C=C\underset{F_c}{\overset{CF_2CF_2CF_2CF_2H}{<}}_{d\ \ e\ \ f\ \ g}$$

$\delta_F =$ (a) −105.1, (b) −87.8, (c) −188.7, (d) −118.5, (e) −125.5, (f) −129.8, (g) −137.0 ($^2J_{FH} = 50$ Hz) within the olefinic fragment of perfluoro-1-hexene are provided and are of "normal" magnitude. Of particular interest is the relatively large 28 Hz coupling between the allylic fluorines (d) and the cis-alkene fluorine (a), which can be largely attributed to through-space, "contact" coupling.

6.21. POLYFLUORINATED AROMATICS

6.21.1. 2,3,5,6-Tetrafluorobenzene Compounds

Because of long-range F–F coupling constants that contribute to the second-order nature of the system, the fluorine signals of 2,3,5,6-tetrafluoroaryl groups generally consist of complex multiplets. The chemical shifts of a few examples of such systems are given in Scheme 6.37.

Scheme 6.37

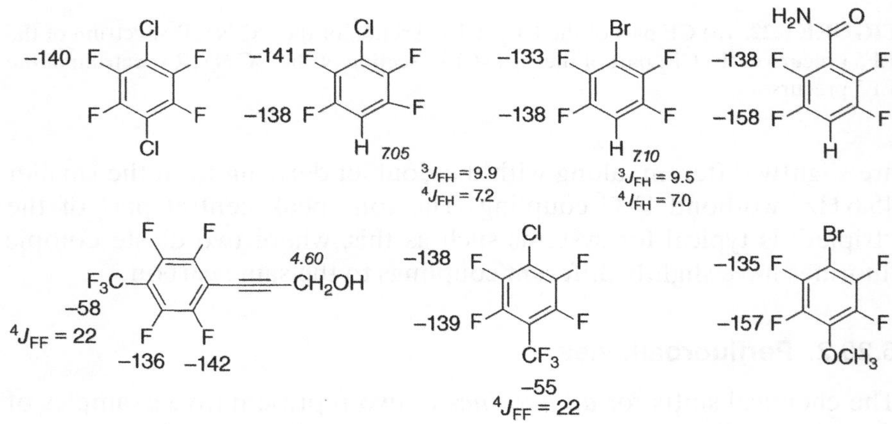

6.21.2. The Pentafluorophenyl Group

Fluorine NMR data for a few representative examples of pentafluorophenyl compounds are given as follows, along with typical F–F coupling constants (Table 6.2).[5]

It can be seen that electron-donating groups generally shield, and electron-withdrawing substituents generally deshield all of the fluorine

TABLE 6.2. Typical Chemical Shifts for Pentafluorobenzenes

	X	δ_F values		
		ortho	meta	para
	OH	−164.4	−165.8	−171.2
	CH_3	−143.9	−164.4	−159.1
	H	−138.7	−162.6	−154.3
	CN	−132.0	−158.9	−143.2

For $X = CH_3$, $^3J_{2,3} = 20.4$, $^3J_{3,4} = 18,9$, and $^3J_{2,5} = 8.6\,Hz$

atoms, but that the ortho and para fluorines are affected most greatly, and the shielding substituents exert the greatest influence.

6.22. POLYFLUOROHETEROCYCLICS

6.22.1. Polyfluoropyridines

Most of the multifluoro-substituted pyridines were prepared more than 30 years ago in the Birmingham fluorine group using CoF_3 technology. Scheme 6.38 provides the multitude of fluorine and proton chemical shift data that were accumulated at that time. It will be seen that, all other things being equal, fluorines at the 2-position are most deshielded and fluorines at the 3-position are the most shielded. Scheme 6.39 provides a few examples of substituted tetrafluoropyridines.

6.22.2. Polyfluorofurans

Again, the family of di-, tri-, and tetrafluorofurans is prepared via CoF_3 chemistry, and their fluorine and proton spectra reported, as given in Scheme 6.40. All things being equal, fluorines at the 2-position are more deshielded than those at the 3-position.

6.22.3. Polyfluorothiophenes

It appears that only tetrafluorothiophene has been reported with its fluorine NMR spectrum (Scheme 6.41). Interestingly, for this heterocycle, fluorine at the 3-position appears to be slightly more deshielded

Scheme 6.38

Difluoropyridines:

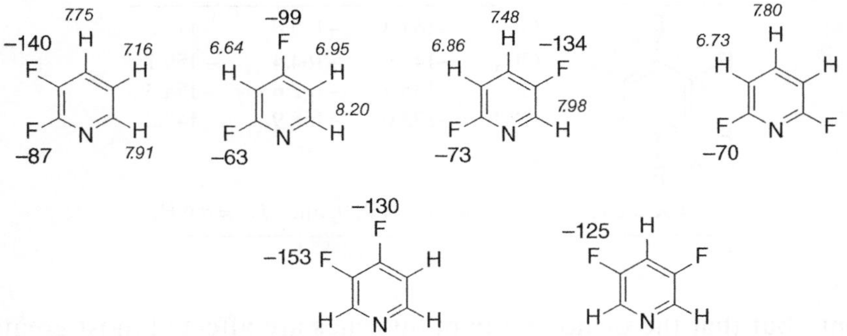

Trifluoropyridines:

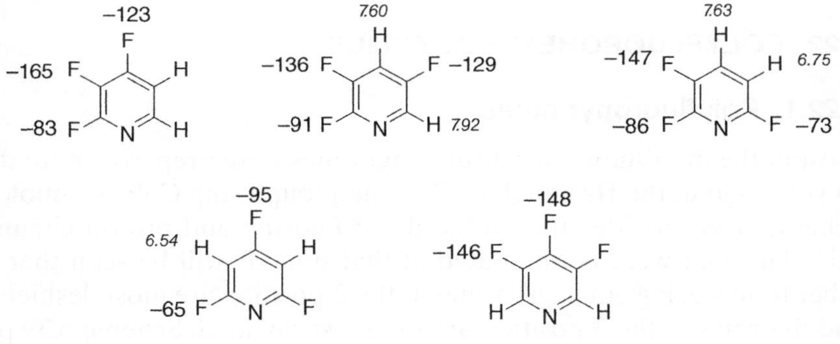

Tetrafluoropyridines:

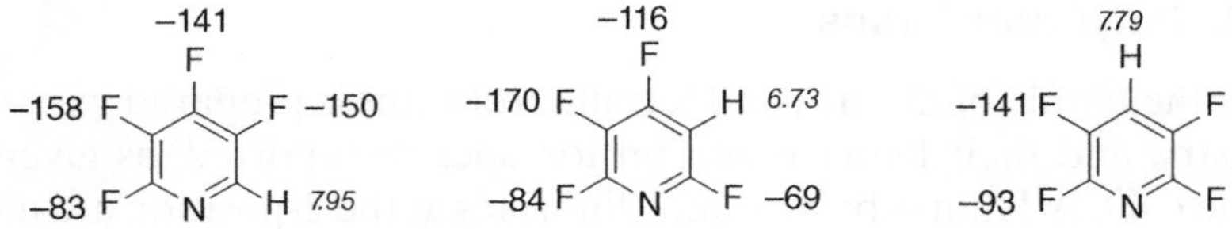

Pentafluoropyridine:

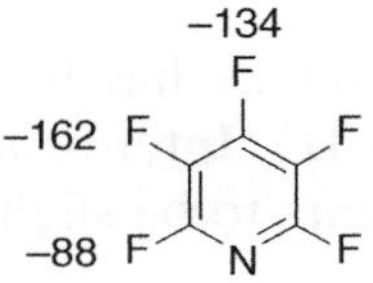

Scheme 6.39

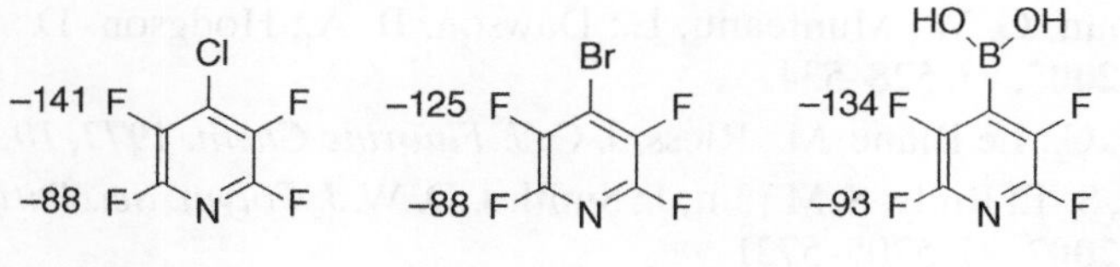

Scheme 6.40

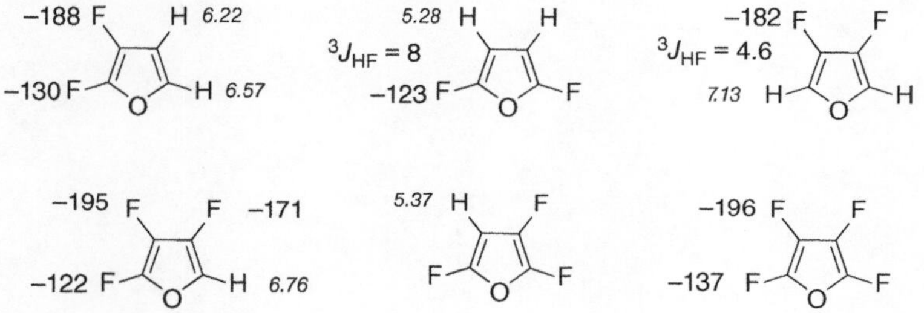

Scheme 6.41

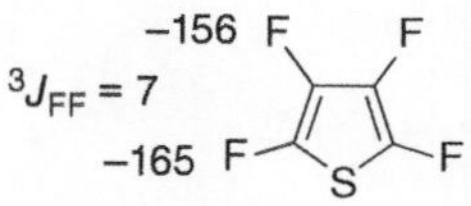

than that at the 2-position. That is also the case for the monofluorothiophenes, as reported in Chapter 3.

6.22.4. Polyfluoropyrimidines

Fluorine NMR data for 2,5-difluoro-, 2,4,5-trifluoro-, and 2,4,5,6-tetrafluoropyrimidines are given in Scheme 6.42.

Scheme 6.42

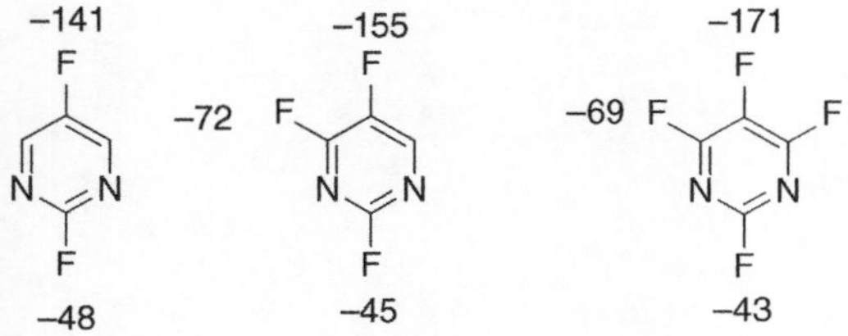

REFERENCES

1. Buchanan, G. W.; Munteanu, E.; Dawson, B. A.; Hodgson, D. *Magn. Res. Chem.* **2005**, *43*, 528–534.
2. Santini, G.; Le Blanc, M.; Riess, J. G. *J. Fluorine Chem.* **1977**, *10*, 363–373.
3. Iacono, S. T.; Budy, S. M.; Jin, J.; Smith J. D. W. *J. Polym. Sci. Part A: Polym. Chem.* **2007**, *45*, 5705–5721.
4. Brey, W. S. *J. Fluorine Chem.* **2005**, *126*, 389–399.
5. Hogben, M. G.; Graham, W. A. G. *J. Am. Chem. Soc.* **1969**, *91*, 283–291.

CHAPTER 7

COMPOUNDS AND SUBSTITUENTS WITH FLUORINE DIRECTLY BOUND TO A HETEROATOM

7.1. INTRODUCTION

This chapter deals with compounds that have one or more fluorine atoms bound directly to a heteroatom, such as boron, silicon, nitrogen, phosphorous, oxygen, and sulfur. Many such compounds, especially those with N—F, O—F, or S—F bonds, will be recognized by organic chemists as being highly reactive fluorinating reagents, and thus it is only rarely that such compounds will be encountered in a characterization situation. Nevertheless, they are included in this survey of fluorine NMR for two reasons: first so that they will be recognized when present in mixtures, and second, because knowledge of such chemical shifts can provide insight into factors that govern trends in fluorine chemical shifts.

This chapter also includes a discussion of compounds bearing the SF_5 substituent, which is a thermally and hydrolytically stable substituent that has attracted much attention in the last decade with respect to its potential impact on biological activity of compounds of pharmacological and agrochemical interest.

Guide to Fluorine NMR for Organic Chemists, Second Edition. William R. Dolbier, Jr.
© 2016 John Wiley & Sons, Inc. Published 2016 by John Wiley & Sons, Inc.

TABLE 7.1. Fluorine Chemical Shifts of Second-Period Fluorides

	F_2	FOF	NF_3	CF_4	BF_3	BeF_2
δ_F	+423	+250	+143	−65	−126	−171

TABLE 7.2. Fluorine Chemical Shifts of Third-Period Fluorides

	ClF	SF_2	PF_3	SiF_4
δ_F	−437	−167	−32	−160

First let us look at some trends. The binary fluorides of the second period exhibit a consistent increase in shielding, as reflected by the significant decrease in their ^{19}F chemical shifts (more negative) as the difference in electronegativity of the bound atom and fluorine increases and thus as the negative charge on the respective fluorine atoms increases (Table 7.1).

Such electronegativity arguments would predict that, because the third-period elements are less electronegative than their second-period counterparts, a similar trend should exist for the binary fluorides of the third period, with the fluorines being more shielded. However, no similar consistent trend toward high-field absorption exists for this series (Table 7.2). The expected trend holds true for SiF_4 and PF_3, the fluorines of which appear 95 and 175 ppm upfield from those of CF_4 and NF_3, respectively. However, the fluorines of SF_2 and ClF are much more shielded than one would have expected, absorbing even farther upfield than SiF_4.

Electronegativity arguments obviously cannot explain the lack of trend in chemical shift for the third-period binary fluorides. The unexpectedly large shielding exhibited for SF_2 and ClF has been attributed to a $\pi^* \rightarrow \sigma^*$ excitation caused by the external magnetic field.[1]

The net result of such effects is the quite amazing difference of *more than 850 ppm* between the chemical shifts of F_2 (+423 ppm) and ClF (−437 ppm), all of which are not of great consequence to organic chemists, who would rarely if ever find themselves in a situation requiring knowledge of the chemical shifts of the compounds in Tables 7.1 and 7.2.

However, there *are* compounds of boron, silicon, nitrogen, phosphorous, sulfur, and oxygen that are of potential interest to organic chemists, and most of the remainder of this chapter is devoted to such compounds.

7.2. BORON FLUORIDES

Boron being more electropositive than carbon, the fluorines of BF_3 (−126 ppm) are more highly shielded than those of CF_4 (−65 ppm). When this strong Lewis acid interacts with a Lewis base, as in $BF_3 \cdot OEt_2$ (−154 ppm), the fluorines will become even more highly shielded. The fluorines in BF_4^- are also highly shielded, appearing at ∼−150 ppm. Data for this and other fluoroborates are given in Scheme 7.1.

Scheme 7.1

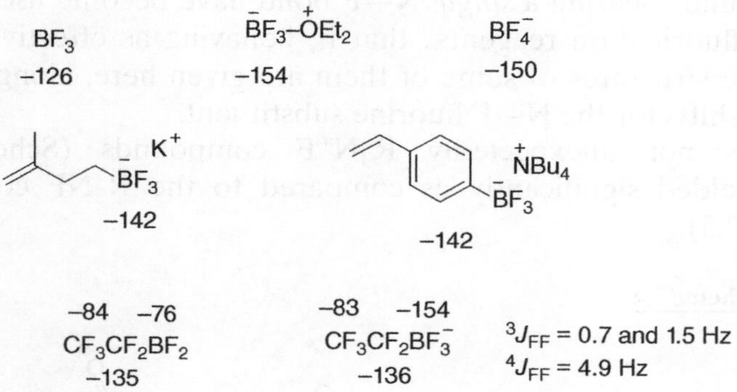

BF_3
−126

$\bar{B}F_3-\overset{+}{O}Et_2$
−154

BF_4^-
−150

K+
−BF_3
−142

+
NBu_4
−BF_3
−142

−84 −76
$CF_3CF_2BF_2$
−135

−83 −154
$CF_3CF_2BF_3^-$
−136

$^3J_{FF}$ = 0.7 and 1.5 Hz
$^4J_{FF}$ = 4.9 Hz

7.3. FLUOROSILANES

Fluorosilanes are all more highly shielded than their carbon counterparts (Scheme 7.2).

Scheme 7.2

SiF_4	CH_3SiF_3	$(CH_3)_2SiF_2$	$(CH_3)_3SiF$
−160	−131	−128	−154

CF_4	CH_3CF_3	$(CH_3)_2CF_2$	$(CH_3)_3CF$
−65	−65	−85	−131

7.4. NITROGEN FLUORIDES

All nitrogen fluorides are very reactive, often dangerous species, generally powerful oxidizers. Compounds such as NF_3 (δ_F = +143) and N_2F_4 would virtually never be encountered in an organic synthesis laboratory, and no NMR data has even been reported for the latter compound. NF_3

has a pyramidal structure, with F—N—F bond angles of ~102°. Some data do exist for NF_4^+ salts; an example is given in Scheme 7.3.

Scheme 7.3

$$NF_4^+ \ PF_6^-$$
$$-217 \quad {}^1J_{FN} = 230$$

7.4.1. Electrophilic Fluorinating Agents

More pertinent to the interests of fluoroorganic chemists, a number of compounds bearing a *single* N—F bond have become useful "electrophilic fluorination reagents", that is, behaving as effective sources of F⁺. The structures of some of them are given here, along with the chemical shifts for the N—F fluorine substituent.

Perhaps not unexpectedly R_3N^+F compounds (Scheme 7.4) are deshielded significantly as compared to the R_2NF compounds (Scheme 7.5).

Scheme 7.4

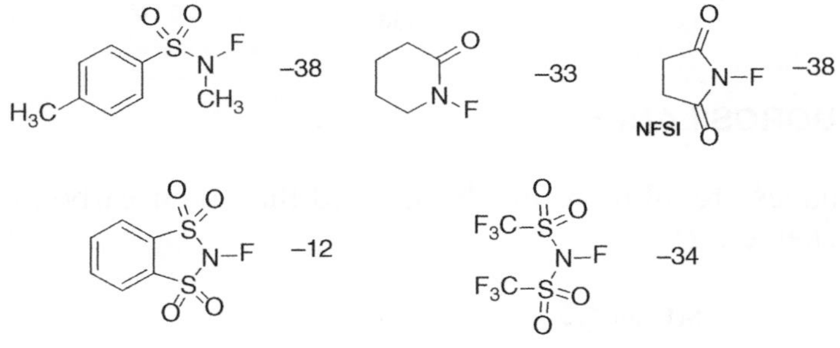

Scheme 7.5

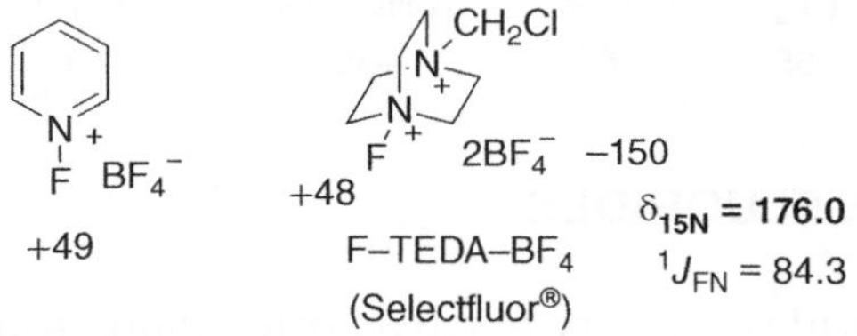

NMR spectra of some alkyl nitrogen difluorides have recently been reported. Their characteristic spectra are given in Scheme 7.6. Note the huge (560 Hz) two-bond F–F coupling through nitrogen that is reported

Scheme 7.6

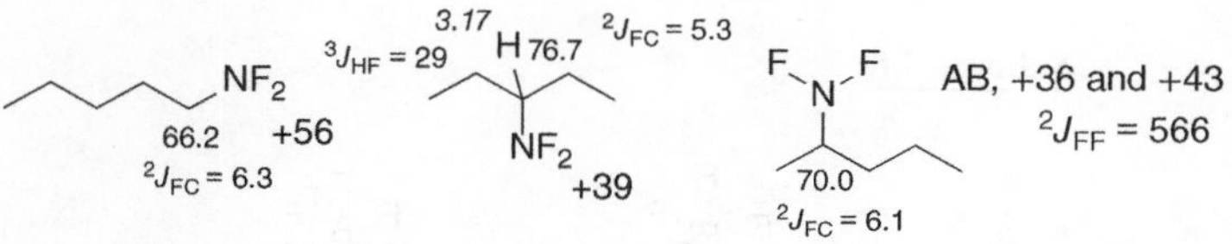

$$^{3}J_{HF} = 29 \quad \overset{3.17}{H} \; 76.7 \quad ^{2}J_{FC} = 5.3$$

NF$_2$ 66.2   +56 $^{2}J_{FC} = 6.3$

NF$_2$ +39

F—N—F   AB, +36 and +43 $^{2}J_{FF} = 566$ 70.0 $^{2}J_{FC} = 6.1$

for the AB system. By comparison (see Section 2.3.2), the largest gemi-nal coupling constant mentioned in this book for two fluorines bound to carbon is 290 Hz. The difference may derive from the pyramidal nature of the NF$_2$ group, which probably would have a smaller F—N—F angle relative to the F—C—F angle.

7.5. PHOSPHOROUS FLUORIDES

There are a large number of types of compounds that contain phospho-rous fluorine bonds, including some of the most toxic compounds known to humankind (nerve gases such as Sarin).

7.5.1. Phosphorous (III) Fluorides

PF$_3$ and other phosphorous (III) fluorides have pyramidal struc-tures, with the F—P—F bond angle for PF$_3$ being ~98°. Thus, all P—F fluorines in such compounds should be magnetically equivalent (Scheme 7.7).

Scheme 7.7

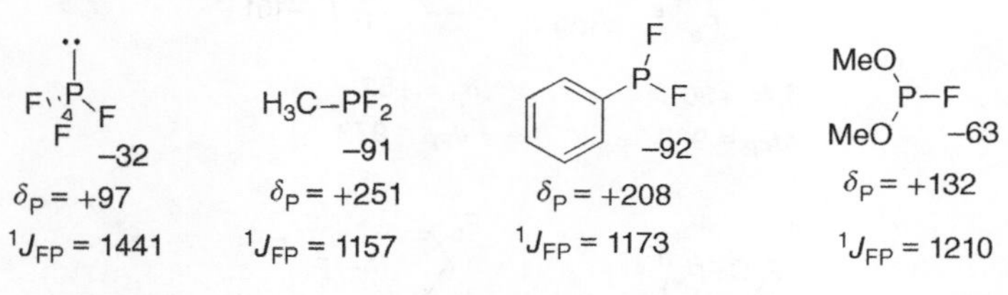

F—P—F, F   −32 $\delta_P = +97$ $^{1}J_{FP} = 1441$

H$_3$C—PF$_2$ −91 $\delta_P = +251$ $^{1}J_{FP} = 1157$

[phenyl]—P(F)—F   −92 $\delta_P = +208$ $^{1}J_{FP} = 1173$

MeO—P(OMe)—F   −63 $\delta_P = +132$ $^{1}J_{FP} = 1210$

Note that the change in the coupling constant is consistent with the relative amount of s-character in the phosphorous orbital bound to the fluorines.

7.5.2. Phosphorous (V) Fluorides

Phosphorous (V) fluorides have some interesting structural and dynamic features that influence their ^{19}F NMR spectra. All such com-pounds appear to have a trigonal bipyramid structure, which features two axial and three equatorial sites. Hexafluorophosphate, PF_6^-, is of course octahedral with all sites equivalent (Scheme 7.8).

Scheme 7.8

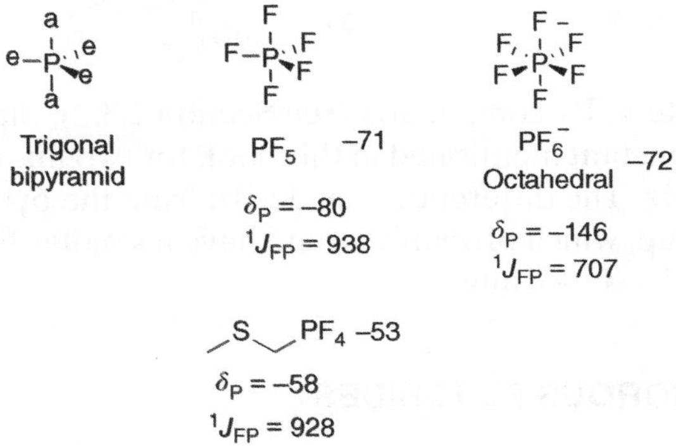

Trigonal bipyramid

PF_5 -71
$\delta_P = -80$
$^1J_{FP} = 938$

PF_6^- Octahedral -72
$\delta_P = -146$
$^1J_{FP} = 707$

$\text{S}\diagup\diagdown PF_4$ -53
$\delta_P = -58$
$^1J_{FP} = 928$

In spite of bearing nonequivalent fluorine substituents, both PF_5 and all compounds of the type $R\text{-}PF_4$ exhibit only a single signal in their fluorine NMR spectra. The observed magnetic equivalence of the fluorines in such compounds is believed to derive from a rapid intramolecular, pseudorotational exchange process that is too rapid, even at $-80\,°C$ to allow distinction of the axial and equatorial fluorine atoms (Scheme 7.9).

Scheme 7.9 R-PF$_4$ compounds

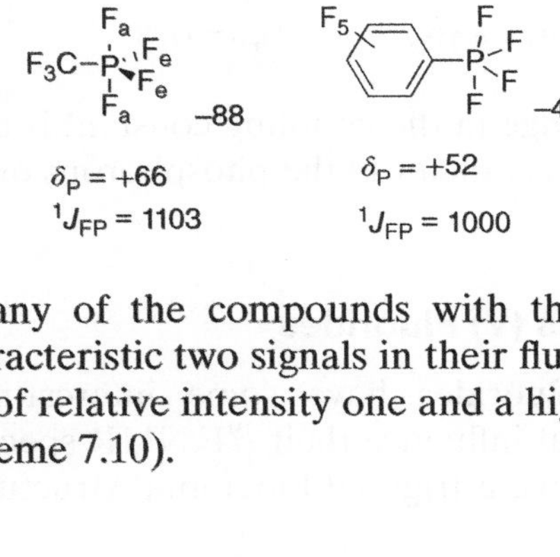

$H_3C\text{-}PF_aF_eF_eF_a$ -109
$\delta_P = +30$
$^1J_{FP} = 965$

$C_6H_5\text{-}PF_4$ -101
$\delta_P = +52$
$^1J_{FP} = 973$

$F_3C\text{-}PF_aF_eF_eF_a$ -88
$\delta_P = +66$
$^1J_{FP} = 1103$

$F_5C_6\text{-}PF_4$ -41
$\delta_P = +52$
$^1J_{FP} = 1000$

In contrast, many of the compounds with the general structure R_2PF_3 have a characteristic two signals in their fluorine NMR spectra, a low-field triplet of relative intensity one and a higher field doublet of intensity two (Scheme 7.10).

Scheme 7.10 R_2PF_3 compounds

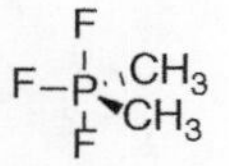

$$F-\underset{\underset{F}{|}}{\overset{\overset{F}{|}}{P}}\overset{CH_3}{\underset{CH_3}{\diagdown}}$$

$\delta_{Fe} = -68$ (t, 1F)
$\delta_{Fa} = -152$ (d, 2F)
$\delta_P = -8$
$^1J_{FeP} = 975$
$^1J_{FaP} = 787$
$^3J_{FF} = 26$

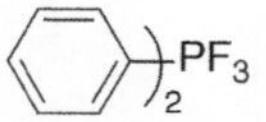

$$\text{(phenyl)}_2 PF_3$$

$\delta_{Fe} = -76$ (t, 1F)
$\delta_{Fa} = -122$ (d, 2F)
$\delta_P = +35$
$^1J_{FeP} = 966$
$^1J_{FaP} = 837$

$(CF_3)_2PF_3$
-85

$\delta_P = +51$
$^1J_{FP} = 1260$

In order to correctly predict which ligands occupy which sites in such compounds, one must recognize that, as a general rule, fluorines will always prefer the axial site in a trigonal bipyramidal system, perhaps because of fluorine's small size, but probably also because of its preference to bind to orbitals with as little s-character as possible. The orbitals used by P to make its axial bonds have less s-character than those used to make its equatorial bonds. This is reflected by the larger F–P coupling constants to the equatorial fluorine substituents.

One must assume that the example with the electronegative CF_3 groups that only exhibits one signal in its fluorine NMR spectrum must be undergoing rapid intramolecular exchange of its axial and equatorial fluorines.

All compounds of the type R_3PF_2 will have both fluorines occupying the axial sites and thus their fluorine NMR spectra consist of but one doublet (Scheme 7.11). Figure 7.1 provides the fluorine spectrum of Ph_3PF_2 as an example.

Scheme 7.11 R_3PF_2 compounds

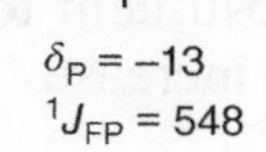

$$H_3C-\underset{\underset{F}{|}}{\overset{\overset{F}{|}}{P}}\overset{CH_3}{\underset{CH_3}{\diagdown}} \quad -6$$

$\delta_P = -13$
$^1J_{FP} = 548$

$Me_3PF_2 \quad -5$
$\delta_P = -16$
$^1J_{FP} = 542$

$Ph_3PF_2 \quad -41$
$\delta_P = -54$
$^1J_{FP} = 664$

$(EtO)_3PF_2 \quad -58$
$\delta_P = -75$
$^1J_{FP} = 723$

Spectra of analogous triphenylarsenic and antimony difluorides have also been reported (Scheme 7.12), these having fluorines progressively more shielded.

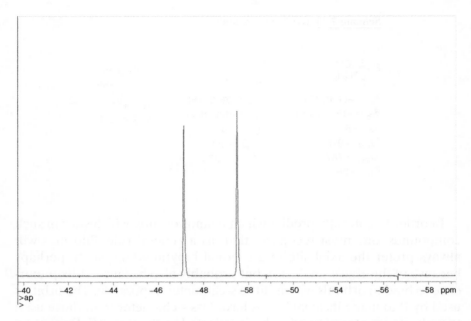

FIGURE 7.1. ^{19}F NMR spectrum of triphenylphosphine difluoride

Scheme 7.12

Ph$_3$AsF$_2$ Ph$_3$SbF$_2$

−92   −151

7.5.3. Phosphorous (V) Oxyfluorides

Fluorine and phosphorous NMR data are given in Scheme 7.13 for a number of phosphorous oxyfluorides, including phosphorous data for the deadly nerve agent, Sarin (no one has yet been bold enough to obtain its fluorine spectrum).

Regarding the phosphorous chemical shifts, it can be seen that replacing an alkyl substituent with a methoxy group gives rise to shielding of the phosphorous and going from one fluorine substituent to two in a similar series also leads to shielding and generally increases the F–P coupling constant.

7.5.4. Cyclophosphazenes

Fluorine and phosphorous NMR data are given in Scheme 7.14 for two cyclophosphazenes.

Scheme 7.13

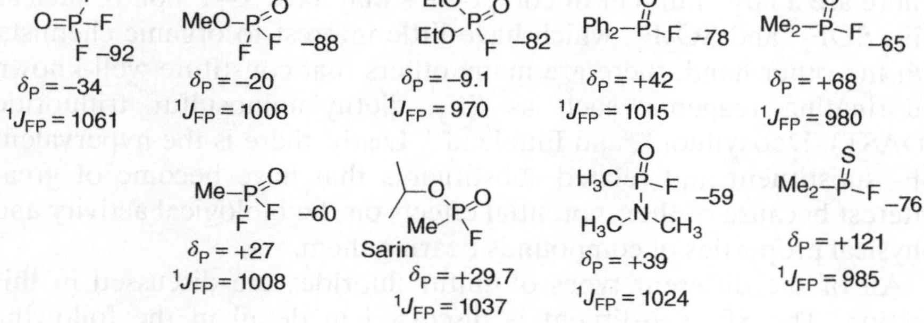

$O=\overset{F}{\underset{F}{P}}-F$ -92

$\delta_P = -34$
$^1J_{FP} = 1061$

$MeO-\overset{O}{\underset{F}{P}}-F$ -88

$\delta_P = -20$
$^1J_{FP} = 1008$

$\overset{EtO}{\underset{EtO}{}}\overset{O}{P}\overset{}{\underset{F}{}}F$ -82

$\delta_P = -9.1$
$^1J_{FP} = 970$

$Ph_2-\overset{O}{P}-F$ -78

$\delta_P = +42$
$^1J_{FP} = 1015$

$Me_2-\overset{O}{P}-F$ -65

$\delta_P = +68$
$^1J_{FP} = 980$

$Me-\overset{O}{\underset{F}{P}}-F$ -60

$\delta_P = +27$
$^1J_{FP} = 1008$

$\overset{}{\underset{Me}{}}\overset{O}{\underset{}{P}}\overset{O}{\underset{F}{}}$ Sarin
$\delta_P = +29.7$
$^1J_{FP} = 1037$

$\overset{H_3C-\overset{O}{P}-F}{\underset{H_3C}{\overset{|}{N}}}$ -59
CH_3
$\delta_P = +39$
$^1J_{FP} = 1024$

$Me_2-\overset{S}{P}-F$ -76

$\delta_P = +121$
$^1J_{FP} = 985$

Scheme 7.14

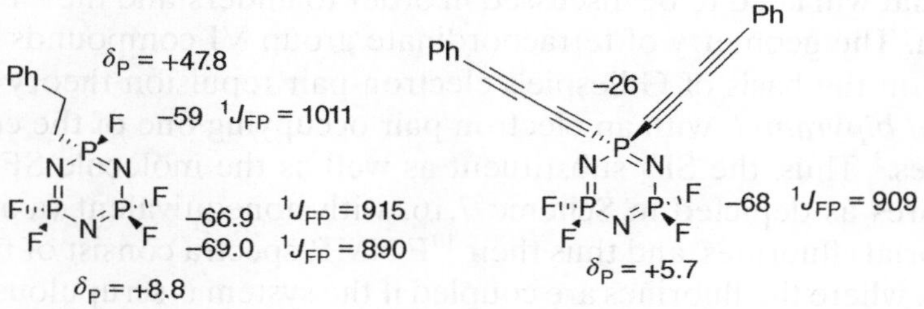

$\delta_P = +47.8$
Ph
-59 $^1J_{FP} = 1011$

-66.9 $^1J_{FP} = 915$
-69.0 $^1J_{FP} = 890$
$\delta_P = +8.8$

Ph -26
Ph

-68 $^1J_{FP} = 909$
$\delta_P = +5.7$

7.6. OXYGEN FLUORIDES (HYPOFLUORITES)

Similar to the nitrogen fluorides, compounds having an O–F bond are also very strong oxidants and, when utilized as such, effective sources of electrophilic fluorine. Because the O–F bond is so weak, the chemistry of such compounds approaches that of F_2 itself, and such chemistry can be free radical or electrophilic in nature, depending on the conditions and the reactants.

There are two types of hypofluorite compounds that have been prepared, the perfluoroalkyl hypofluorites, such as CF_3OF, and the acyl hypofluorites, such as CH_3CO_2F. The fluorines of such compounds are very deshielded and thus have significantly positive chemical shifts (Scheme 7.15). Also shown is the powerful electrophilic fluorinating agent, perchloryl fluoride.

Scheme 7.15

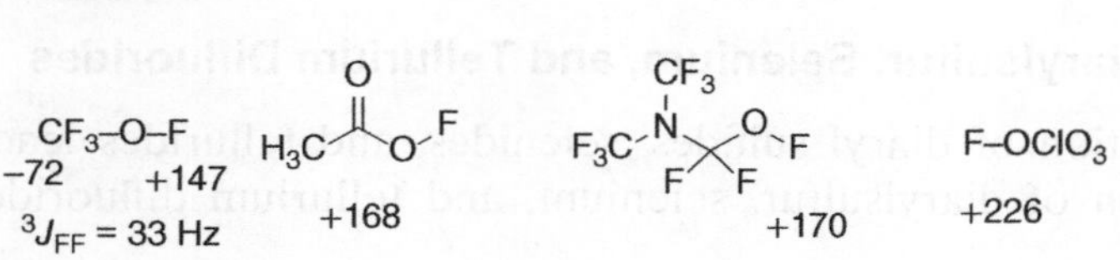

CF_3-O-F
-72 $+147$
$^3J_{FF} = 33$ Hz

$\overset{O}{\underset{}{}}H_3C-\overset{O}{C}-O-F$
$+168$

$\overset{CF_3}{\underset{F_3C-N}{}}\overset{}{\underset{F}{C}}\overset{O-F}{\underset{F}{}}$
$+170$

$F-OClO_3$
$+226$

7.7. SULFUR FLUORIDES

There are a large number of compounds that have S—F bonds, such as SF_2, SOF_2, and SO_2F_2, which have little interest to organic chemists. On the other hand, there are many others that constitute well-known fluorinating reagents, such as SF_4, diethylaminosulfur trifluoride (DAST), Deoxyfluor®, and Fluolead®. Lastly, there is the hypervalent SF_5 substituent and related substituents that have become of great interest because of their potential effects on the biological activity and physical properties of compounds bearing them.

All of the different types of sulfur fluorides are discussed in this section. The SF_5 substituent is discussed in detail in the following section.

There are some unique structural aspects of some of the sulfur fluorides that will need to be discussed in order to understand the ^{19}F NMR spectra. The geometry of tetracoordinate group VI compounds is predicted on the basis of Gillespie's electron-pair repulsion theory[2] to be *trigonal bipyramid*, with an electron pair occupying one of the equatorial sites.[3] Thus, the SF_3 substituent as well as the molecule SF_4 have structures as depicted in Scheme 7.16, with nonequivalent (axial and equatorial) fluorines, and thus their ^{19}F NMR spectra consist of two ^{19}F signals, where the fluorines are coupled if the system is scrupulously dry.

Scheme 7.16

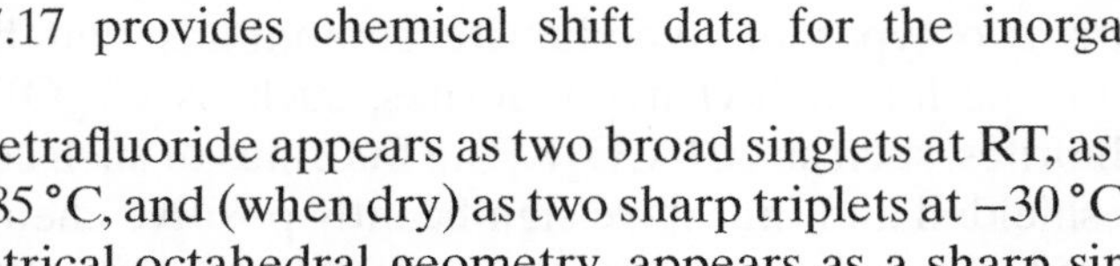

$$R{-}SF_3 \quad = \quad \underset{\text{ab}_2\text{ system}}{R{-}\overset{F}{\underset{F}{S}}{\cdot}{\cdot}F} \qquad\qquad SF_4 \quad = \quad \underset{\text{a}_2\text{b}_2\text{ system}}{{:}{-}\overset{F}{\underset{F}{S}}{\cdot}{<}^{F}_{F}}$$

7.7.1. Inorganic Sulfur, Selenium, and Tellurium Fluorides

Scheme 7.17 provides chemical shift data for the inorganic sulfur fluorides.

Sulfur tetrafluoride appears as two broad singlets at RT, as one broad singlet at 85 °C, and (when dry) as two sharp triplets at −30 °C. SF_6, with its symmetrical octahedral geometry, appears as a sharp singlet at all temperatures. The activation energy for pseudorotation of SF_4, which interconverts its axial and equatorial fluorines, is ~12 kcal/mol.[4]

7.7.2. Diarylsulfur, Selenium, and Tellurium Difluorides

Fluorination of diaryl sulfides, selenides, and tellurides leads to the formation of diarylsulfur, selenium, and tellurium difluorides, all of

Scheme 7.17

F–S–F	SF$_4$	SF$_6$
–167	+97 (Axial)	+56
	+37 (Equatorial)	

$^2J_{FF} = 76$ Hz (seen at – 30 °C)

SOF$_2$	SO$_2$F$_2$
+77	+34

Scheme 7.18

Ph$_2$SF$_2$	Ph$_2$SeF$_2$	Ph$_2$TeF$_2$
+6.8	–67	–127

which exhibit only one fluorine signal (singlet) in their ^{19}F NMR spectra (Scheme 7.18), and with progressively more shielded fluorines going down the group. Presumably, both fluorines occupy the axial sites of the trigonal bipyramidal structure.

7.7.3. Aryl and Alkyl SF$_3$ Compounds

Scheme 7.19 provides fluorine NMR data for some organic SF$_3$ compounds, each of which exhibits peaks of vastly different chemical shift for their axial and their equatorial fluorines. Aryl-SF$_3$ compounds have attracted recent interest as deoxofluorinating reagents, in particular 4-(t-butyl)-2,6-dimethylphenylsulfur trifluoride, which is known as Fluolead. The fluorine NMR spectrum of p-bromophenyl-SF$_3$ is given in Figure 7.2 as an example of this type of compound.

Scheme 7.19

Ph–SF$_3$	p-Br–Ph–SF$_3$	t-Bu–SF$_3$	CH$_3$–SF$_3$	CF$_3$–SF$_3$
$\delta_{Fa} = +58$ (d, 2F)	$\delta_{Fa} = +63$ (d, 2F)	$\delta_{Fa} = +33$ (d, 2F)	$\delta_{Fa} = +61$ (d, 2F)	$\delta_{Fa} = +49$ (d, 2F)
$\delta_{Fe} = -40$ (t, 1F)	$\delta_{Fe} = -36$ (t, 1F)	$\delta_{Fe} = -64$ (t, 1F)	$\delta_{Fe} = -53$ (t, 1F)	$\delta_{Fe} = -48$ (t, 1F)
$^2J_{FF} = 59$ Hz	$^2J_{FF} = 59$ Hz	$^2J_{FF} = 53$ Hz	$^2J_{FF} = 75$ Hz	$^2J_{FF} = 63$ Hz

7.7.4. Dialkylaminosulfur Trifluorides

Dialkylaminosulfur trifluorides are widely used as a safe substitute for SF$_4$ in the replacement of oxygen by fluorine in many types of

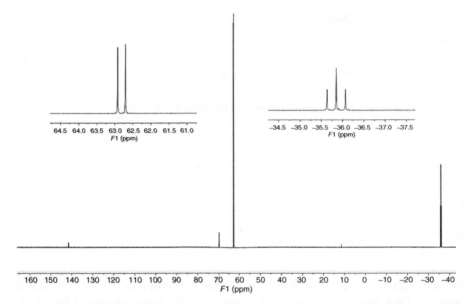

FIGURE 7.2. ^{19}F spectrum of the ab$_2$ system of p-bromophenyl-SF$_3$

organic compounds, for example, converting alcohols into fluorides, or aldehydes and ketones into *gem*-difluorides. The most familiar of these reagents is DAST, while an increasingly popular one (considered safer to use than DAST) is bis(2-methoxyethyl)aminosulfur trifluoride (Deoxo-Fluor Reagent®) (Scheme 7.20).

The equatorial fluorines of these nitrogen-bound SF$_3$ groups are greatly deshielded compared to those in their carbon-bound counterparts in Scheme 7.19. Note also that the four-bond coupling of both the axial and the equatorial fluorines to hydrogen are able to be observed in these compounds. In all likelihood, this coupling is "through-space."

An example of another deoxofluorination reagent, an *N,N*-disubstituted aminodifluorosulfinium tetrafluoroborate salt, is given in Scheme 7.21, along with a bis(dialkylamino)sulfur difluoride, which can be formed in reactions of SF$_4$ with secondary amines.

7.7.5. Hypervalent Sulfur Fluorides

Hexacoordinate, hypervalent sulfur fluorides have an octahedral geometry that is symmetrical for SF$_6$, the fluorines of which appear as a sharp singlet at +50 ppm, as do those of SeF$_6$ (+54 ppm) and TeF$_6$ (−56 ppm) (Scheme 7.22). NMR data for other molecular hexafluorides are also available in a recent review.[5]

Scheme 7.20

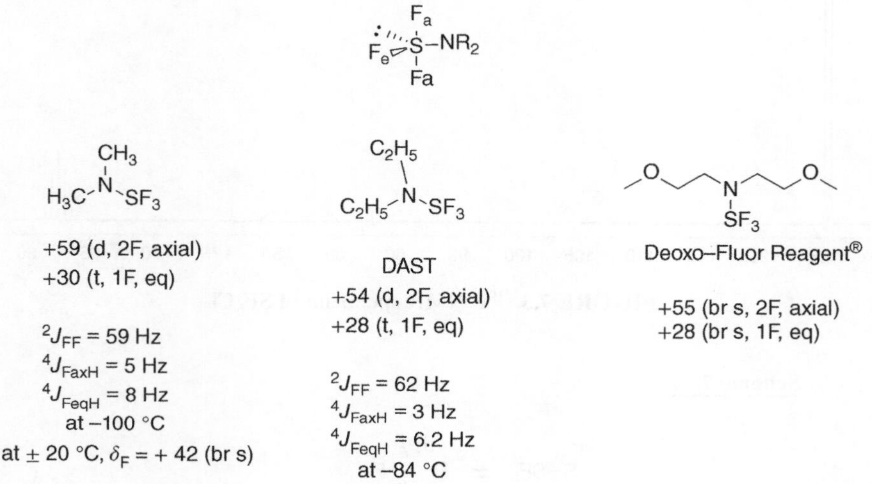

$$\ddot{F_e} \overset{\cdots}{\underset{\overset{|}{Fa}}{\overset{F_a}{\diagup}}} S-NR_2$$

$$H_3C-\underset{\underset{CH_3}{|}}{N}-SF_3$$

+59 (d, 2F, axial)
+30 (t, 1F, eq)

$^2J_{FF}$ = 59 Hz
$^4J_{FaxH}$ = 5 Hz
$^4J_{FeqH}$ = 8 Hz
at −100 °C
at ± 20 °C, δ_F = + 42 (br s)

$$C_2H_5-\underset{\underset{C_2H_5}{|}}{N}-SF_3$$

DAST
+54 (d, 2F, axial)
+28 (t, 1F, eq)

$^2J_{FF}$ = 62 Hz
$^4J_{FaxH}$ = 3 Hz
$^4J_{FeqH}$ = 6.2 Hz
at −84 °C

$$\diagup O \diagdown \diagup N \diagdown \diagup O \diagup$$
$$\underset{SF_3}{|}$$

Deoxo–Fluor Reagent®

+55 (br s, 2F, axial)
+28 (br s, 1F, eq)

Scheme 7.21

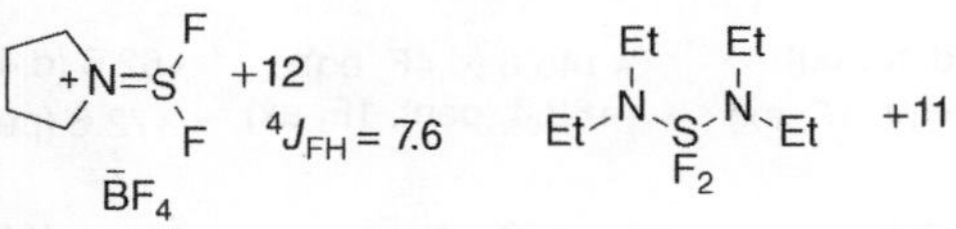

$$\overset{F}{\underset{\underset{\bar{B}F_4}{F}}{N=S}} \quad +12$$
$^4J_{FH}$ = 7.6

$$Et-\underset{Et}{N}-\underset{\underset{F_2}{S}}{}-\underset{Et}{N}-Et \quad +11$$

Scheme 7.22

+50   SF$_6$
δ_{33S} = 177
$^2J_{FS}$ = 252

SeF$_6$
+54

TeF$_6$
−56

On the other hand, compounds of the structure R-SF$_5$ have nonequivalent (axial and equatorial, AB$_4$ system) fluorines. The fluorine NMR spectrum of SF$_5$Cl exemplifies this AB$_4$ system (Figure 7.3), with a pentet representing the axial fluorine at +62.3 ppm and a doublet representing the four equatorial fluorines at 125.8 ppm ($^2J_{FF}$ = 151 Hz).

Compounds with the general structure R-SF$_4$-X have one additional complication, being able to exist as either cis- or trans-isomers, the former having three types of fluorines, and the latter only one (Scheme 7.23). The fluorine NMR spectrum of *trans-p*-tolyl-SF$_5$Cl is provided in Figure 7.4.

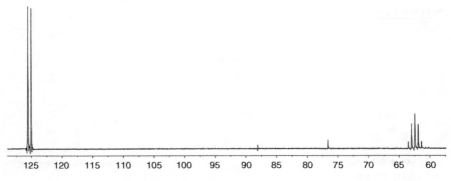

FIGURE 7.3. ^{19}F NMR spectrum of SF$_5$Cl

Scheme 7.23

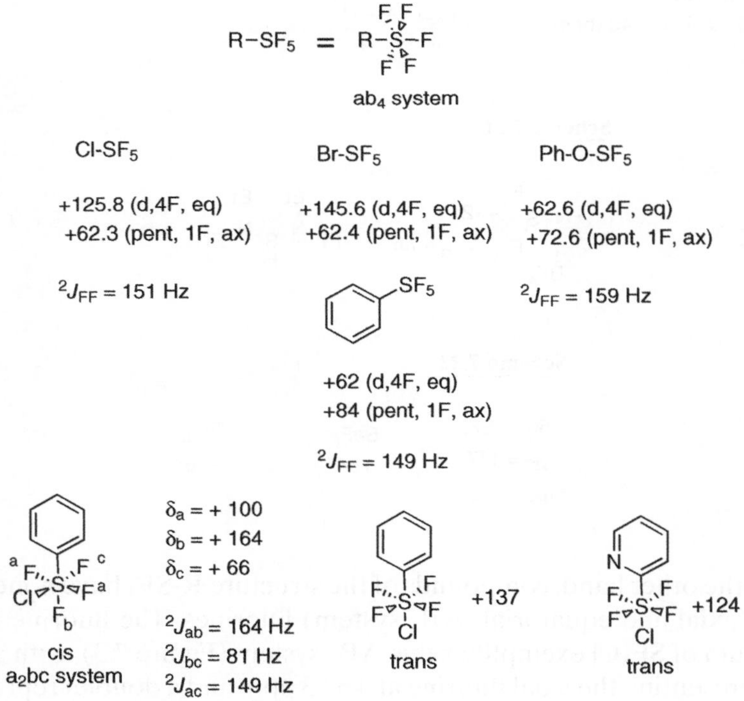

$$R{-}SF_5 \;=\; R{-}\underset{\underset{F}{\overset{\cdot}{|}}}{\overset{\overset{F}{\overset{F}{\diagup}}}{S}}{-}F$$

ab$_4$ system

Cl-SF$_5$

+125.8 (d,4F, eq)
+62.3 (pent, 1F, ax)

$^2J_{FF}$ = 151 Hz

Br-SF$_5$

+145.6 (d,4F, eq)
+62.4 (pent, 1F, ax)

Ph-O-SF$_5$

+62.6 (d,4F, eq)
+72.6 (pent, 1F, ax)

$^2J_{FF}$ = 159 Hz

SF$_5$

+62 (d,4F, eq)
+84 (pent, 1F, ax)

$^2J_{FF}$ = 149 Hz

δ_a = + 100
δ_b = + 164
δ_c = + 66

$^2J_{ab}$ = 164 Hz
$^2J_{bc}$ = 81 Hz
$^2J_{ac}$ = 149 Hz

cis
a$_2$bc system

+137

trans

+124

trans

The four equatorial fluorines of *p*-tolyl-SF$_4$Cl appear as a singlet at +137.7 ppm. Since sulfur has a significant (4.3%) 34-isotope, there appears a small signal just upfield that is due to the isotope effect on the fluorine chemical shift of the S-34 component.

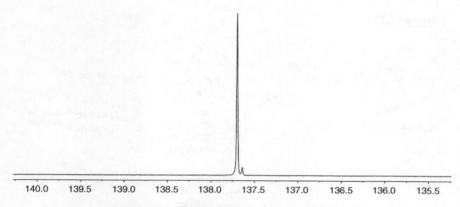

FIGURE 7.4. ^{19}F NMR spectrum of *p*-tolyl-SF$_4$Cl

When the SF$_4$Cl group is attached to the 2-position of pyridine, its fluorines are significantly shielded (+124.2 ppm) compared to the respective aryl-SF$_4$Cl fluorines (Scheme 7.23).

The free radical addition of aryl-SF$_4$Cl compounds to alkynes provides adducts with the SF$_4$ group in the trans orientation, as seen from the data given in Scheme 7.24.

Scheme 7.24

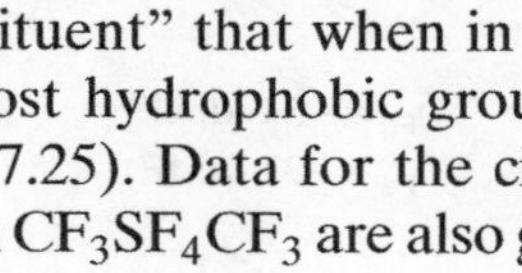

$$Ph\text{--}F_4S\underset{^3J_{FH}=8.9}{\overset{+68}{\diagup}}\overset{(CH_2)_3CH_3}{\underset{H\ \ 6.87}{\diagdown}Cl}$$

When a trifluoromethyl group replaces the Cl in the above system, one has created a "substituent" that when in the trans form holds the current record as the most hydrophobic group in existence, with a π value of +2.13 (Scheme 7.25). Data for the cis and trans forms of the more symmetrical system CF$_3$SF$_4$CF$_3$ are also given in the same scheme.

7.7.6. Related Hypervalent Selenium and Tellurium Fluorides

For comparative purposes and in order to enhance our understanding of trends in chemical shifts, typical examples of hypervalent selenium and tellurium fluorides are provided in Scheme 7.26. Unfortunately, it appears that neither the SeF$_5$ nor the TeSF$_5$ groups have the chemical stability that is exhibited by the SF$_5$ group.

Scheme 7.25

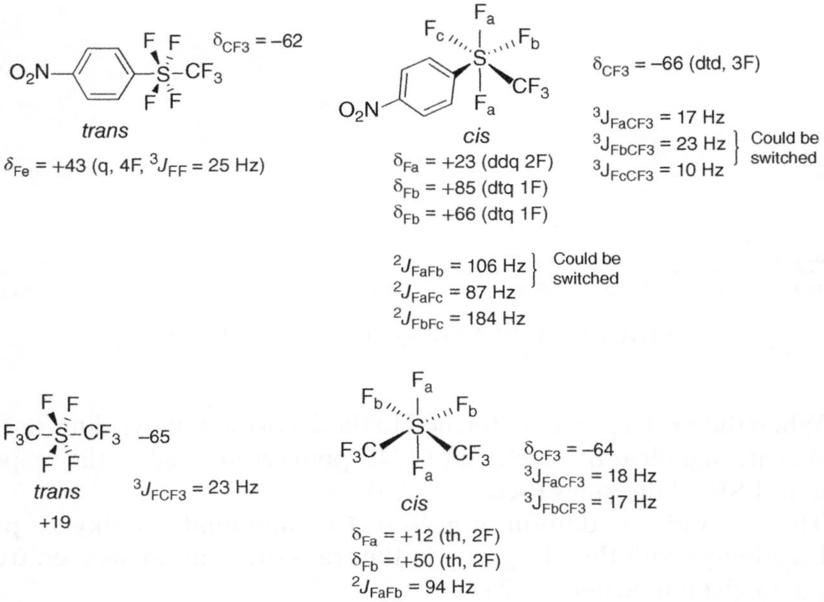

trans

$\delta_{Fe} = +43$ (q, 4F, $^3J_{FF} = 25$ Hz)

$\delta_{CF3} = -62$

cis

$\delta_{CF3} = -66$ (dtd, 3F)

$^3J_{FaCF3} = 17$ Hz
$^3J_{FbCF3} = 23$ Hz $\Big\}$ Could be switched
$^3J_{FcCF3} = 10$ Hz

$\delta_{Fa} = +23$ (ddq 2F)
$\delta_{Fb} = +85$ (dtq 1F)
$\delta_{Fb} = +66$ (dtq 1F)

$^2J_{FaFb} = 106$ Hz $\Big\}$ Could be switched
$^2J_{FaFc} = 87$ Hz
$^2J_{FbFc} = 184$ Hz

$F_3C-S-CF_3$ $\quad$ –65

trans $\quad$ $^3J_{FCF3} = 23$ Hz

+19

cis

$\delta_{CF3} = -64$
$^3J_{FaCF3} = 18$ Hz
$^3J_{FbCF3} = 17$ Hz

$\delta_{Fa} = +12$ (th, 2F)
$\delta_{Fb} = +50$ (th, 2F)
$^2J_{FaFb} = 94$ Hz

Scheme 7.26

$Cl-TeF_5$

+6.1 (d, 4F)
+45.1 (p, 1F)
$^2J_{FF} = 170$

SeF_5

$\delta_{Fe} = +46$ (d, 4F)
$\delta_{Fa} = +87$ (p, 1F)

$^2J_{FF} = 191$

TeF_5

$\delta_{Fe} = -54$ (d, 4F)
$\delta_{Fa} = -37$ (p, 1F)

$^2J_{FF} = 148$

TeF_4

trans

–58

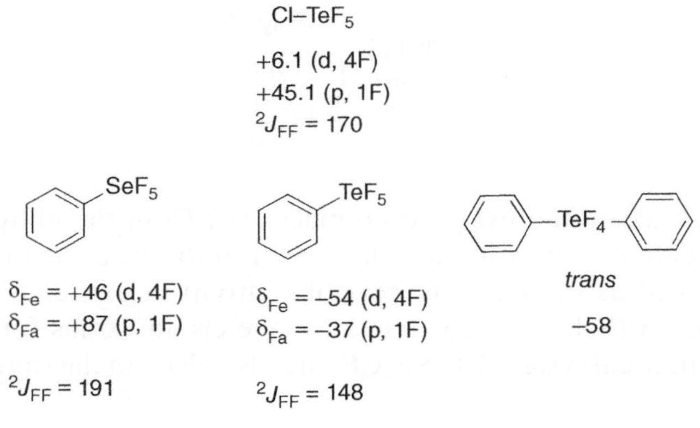

7.7.7. Organic Sulfinyl and Sulfonyl Fluorides

Typical examples of fluorine chemical shifts of sulfinyl and sulfonyl fluorides are given in Scheme 7.27.

Trifluorosulfinylbenzene (Scheme 7.28) has a trigonal bipyramid structure with one axial and two equatorial fluorines.

Scheme 7.27

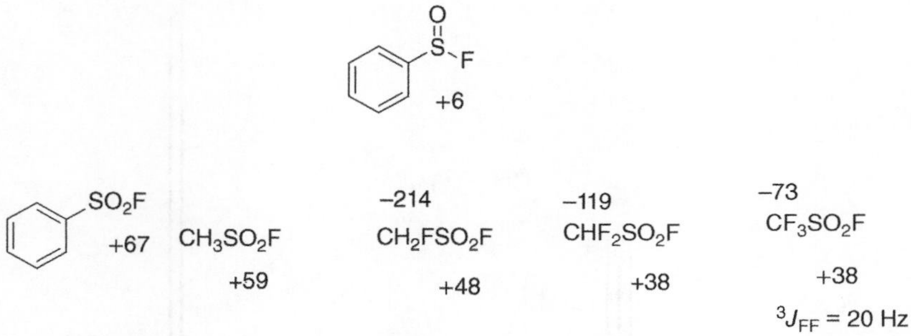

Scheme 7.28

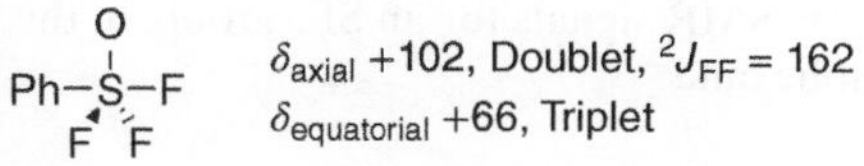

δ_{axial} +102, Doublet, $^2J_{FF}$ = 162
$\delta_{equatorial}$ +66, Triplet

7.8. THE PENTAFLUOROSULFANYL (SF₅) GROUP IN ORGANIC CHEMISTRY

In the past 10–15 years, the SF_5 group has emerged as a new, potentially interesting substituent with respect to its possible influence on biological activity.[6] That combined with the fact that there are now quite a large number of organic compounds bearing an SF_5 group justifies the particular attention that is given to it in this chapter.

Compared to a trifluoromethyl group, it has been shown that the SF_5 group has a similar but often enhanced effect on the lipophilicity of compounds and on essentially any physical or chemical property that derives from its strong polar influence. It also has the advantage of greater hydrolytic stability than does a trifluoromethyl group.

To a significant extent, the interest in compounds bearing an SF_5 group is a direct result of the increased availability of SF_5-containing building blocks, as well as by the appearance in the literature of new and convenient methods for incorporating SF_5 into aliphatic and aromatic compounds.

The SF5 group gives rise to highly characteristic signals in the fluorine NMR. First, these signals always appear at fields significantly *downfield* of internal standard $CFCl_3$. Thus, they have *positive* chemical shifts generally in the range of 55–90 ppm. Second, because the SF_5 group has fluorines existing in two different magnetic environments, one axial

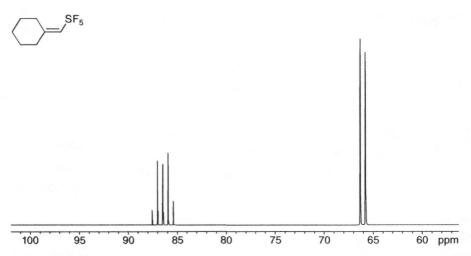

FIGURE 7.5. Typical ^{19}F NMR signals for an SF$_5$ group, in this case for (pentafluoro-sulfanylmethylene)cyclohexane

fluorine and four equatorial fluorines, it is always represented by an AB$_4$ pair of signals, a doublet integrating to four fluorines and a pentet integrating to one fluorine. For organo-SF$_5$ compounds, the pentet usually appears at a lower field than the doublet, but that is not always the case. The two-bond F/F coupling constant between the two types of fluorines of the SF$_5$ group is generally between 140 and 160 Hz. A typical AB$_4$ spectrum for the SF$_5$ group is given in Figure 7.5, with the spectrum of (pentafluorosulfanylmethylene)cyclohexane. In this spectrum, the four equatorial fluorines appear as a doublet at +66.0 ppm, with the single axial fluorine appearing as a pentet at +86.5 ppm. The two-bond F–F coupling constant between them is 147 Hz. If one looked more carefully, expanding the doublet, one would also be able to see the 9 Hz three-bond coupling to the vinylic hydrogen. This coupling is more apparent in the proton spectrum (Figure 7.6).

The influences of SF$_5$ groups upon ^{1}H and ^{13}C NMR spectra are also quite characteristic, and information about proton and carbon chemical shifts and the respective coupling constants will be included along with the ^{19}F data where they are available. The proton and carbon NMR spectra for (pentafluorosulfanylmethylene)cyclohexane are provided in Figures 7.6 and 7.7 as typical examples.

In the proton spectrum, the vinylic hydrogen appears as a pentet at δ 6.14 with a three-bond F–H coupling constant of 9 Hz, which means that *only* the equatorial fluorines couple to the hydrogen. There appear to be no examples of the axial fluorine of an SF$_5$ group exhibiting

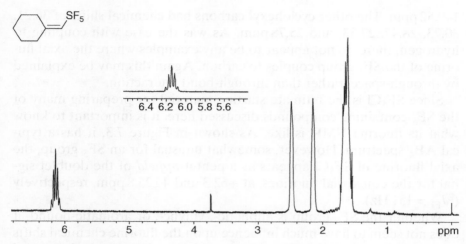

FIGURE 7.6. Proton spectrum of (pentafluorosulfanylmethylene)cyclohexane

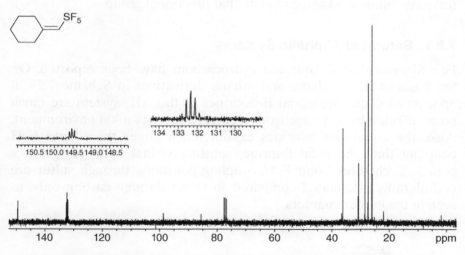

FIGURE 7.7. Carbon spectrum of (pentafluorosulfanylmethylene)cyclohexane

three-bond coupling to a C—H hydrogen. No explanation has thus far been proposed for this, other than that the coupling of the equatorial fluorines may be through-space. The allylic protons are broad singlets at 2.49 and 2.15 ppm, with the other six methylene hydrogens appearing as a broad multiplet centered at 1.65 ppm.

In the above ^{13}C NMR spectrum, the vinylic carbon bearing the SF$_5$ group appears as a pentet ($^2J_{FC} = 17.0$ Hz) at δ 132.34, with the other vinylic carbon also exhibiting coupling ($^3J_{FC} = 5.0$ Hz) and appearing at

149.52 ppm. The other cyclohexyl carbons had chemical shifts of 36.21, 30.73, 28.47, 27.33, and 25.78 ppm. As was the case with coupling to hydrogen, there do not appear to be any examples where the axial fluorine of the SF_5 group couples to carbon. Again this may be explained by through-space, rather than through-bond interaction.

Since SF_5Cl is the ultimate starting material for preparing many of the SF_5-containing compounds discussed here, it is important to know what its fluorine NMR is like. As shown in Figure 7.3, it has a typical AB_4 spectrum. However, somewhat unusual for an SF_5 group, the axial fluorine of SF_5Cl appears as a pentet *upfield* of the doublet signal for the equatorial fluorines, at +62.3 and +125.8 ppm, respectively ($^2J_{FF} = 151$ Hz).

Whether the SF_5 group is aliphatic-, vinylic-, or aromatic-bound, it does not seem to have much influence upon the fluorine chemical shifts observed. However, an SF_5 group that is proximate to a functional group can have its chemical shifts, particularly those of the equatorial fluorines, influenced somewhat by that functional group.

7.8.1. Saturated Aliphatic Systems

Few SF_5-substituted saturated hydrocarbons have been reported. On the basis of the methane and ethane derivatives in Scheme 7.29, it appears as if the equatorial B-fluorines of the AB_4 system are much more affected by a change from methyl to primary alkyl environment. Also, the equatorial fluorines exhibit much larger three-bond F–H coupling than the axial fluorine, splitting vicinal hydrogens into a pentet. Such three-bond F–H coupling constants through sulfur are considerably attenuated compared to those through carbon only, as seen in the earlier chapters.

Scheme 7.29

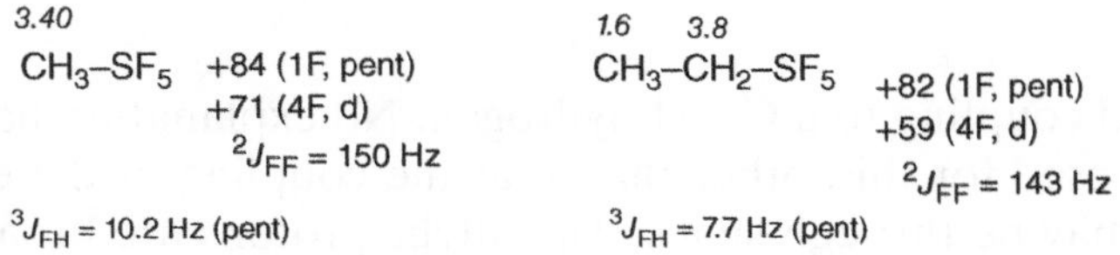

3.40
CH_3–SF_5 +84 (1F, pent)
          +71 (4F, d)
          $^2J_{FF} = 150$ Hz

$^3J_{FH} = 10.2$ Hz (pent)

1.6   3.8
CH_3–CH_2–SF_5 +82 (1F, pent)
             +59 (4F, d)
             $^2J_{FF} = 143$ Hz

$^3J_{FH} = 7.7$ Hz (pent)

Regarding the effect of an SF_5-group upon *proton chemical shifts*, its deshielding effect is significantly greater than that of a CF_3 group, but not as great as direct substitution by a single fluorine (Scheme 7.30).

Scheme 7.30

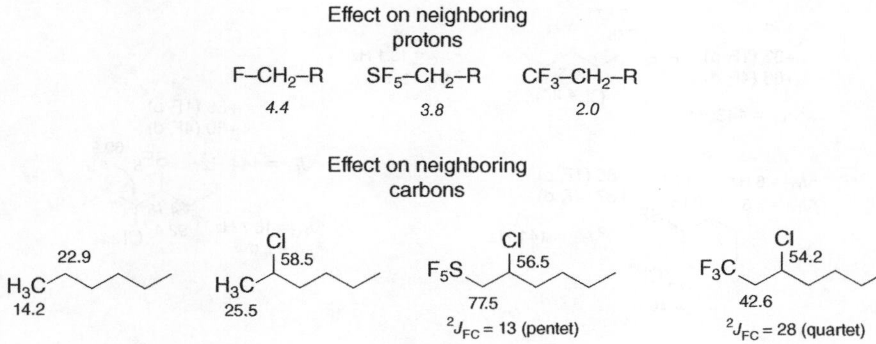
Effect on neighboring protons
F–CH2–R
SF5–CH2–R
CF3–CH2–R
4.4
3.8
2.0
Effect on neighboring carbons
22.9
H3C
14.2
Cl
58.5
H3C
25.5
F5S
Cl
56.5
77.5
2JFC = 13 (pentet)
F3C
Cl
54.2
42.6
2JFC = 28 (quartet)

Three-bond coupling to vicinal hydrogens generally leads to a pentet signal, meaning that coupling to the equatorial fluorines of the SF_5 substituent, as mentioned earlier, is much greater than to its axial fluorine.

With respect to *carbon* NMR, the SF_5 substituent also exerts a much greater deshielding influence on carbons to which it is attached than does a CF_3 substituent, as shown in the same Scheme 7.30. The SF_5 substituent deshields by about 50 ppm, whereas a CF_3 group only deshields by less than 20 ppm. As was the case for coupling to vicinal hydrogen, the equatorial fluorines of SF_5 also couple much more efficiently to carbon, which leads to the carbon signals appearing as pentets, as seen in Figure 7.6.

7.8.1.1. Impact of Halogen Substituents. Again, on the basis of the few published examples available, an α-halo substituent appears to shield both types of fluorines of the AB_4 system significantly and about equally (Scheme 7.31).

Scheme 7.31

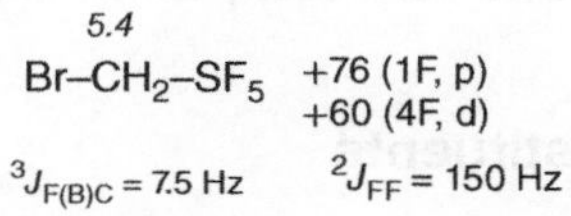
5.4
Br–CH2–SF5
+76 (1F, p)
+60 (4F, d)
3JF(B)C = 7.5 Hz
2JFF = 150 Hz

On the other hand, β-halogen substituents appear not to affect the axial fluorine much, but deshield the four equatorial fluorines (Scheme 7.32).

Scheme 7.32

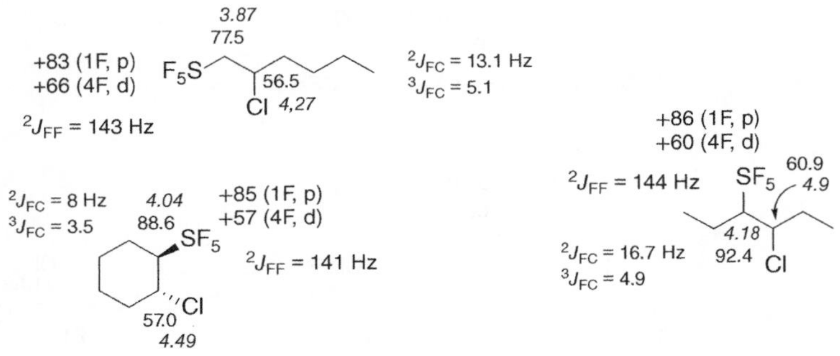

3.87
77.5

+83 (1F, p)
+66 (4F, d) F_5S 56.5

$^2J_{FF}$ = 143 Hz   Cl 4,27

$^2J_{FC}$ = 13.1 Hz
$^3J_{FC}$ = 5.1

$^2J_{FC}$ = 8 Hz   4.04   +85 (1F, p)
$^3J_{FC}$ = 3.5   88.6   +57 (4F, d)
SF$_5$ $^2J_{FF}$ = 141 Hz

57.0 Cl
4.49

+86 (1F, p)
+60 (4F, d)

$^2J_{FF}$ = 144 Hz   SF$_5$ 60.9
4.9

4.18

$^2J_{FC}$ = 16.7 Hz   92.4   Cl
$^3J_{FC}$ = 4.9

Many perfluoroalkyl SF$_5$ compounds are known and are exemplified by the data given in Scheme 7.33 for CF$_3$SF$_5$. Note that, in this case, a smaller but observable $^3J_{FF}$ coupling to the SF$_5$ *axial* fluorine is observed.

Scheme 7.33

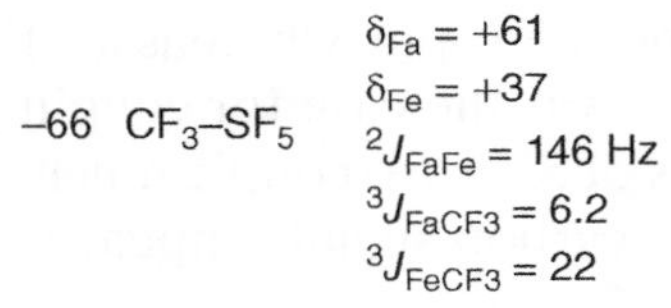

–66 CF$_3$–SF$_5$

δ_{Fa} = +61
δ_{Fe} = +37
$^2J_{FaFe}$ = 146 Hz
$^3J_{FaCF3}$ = 6.2
$^3J_{FeCF3}$ = 22

7.8.1.2. Impact of Carbonyl Groups. SF$_5$ bound directly to a carbonyl function is not a chemically stable situation (SF$_5$ being too good a leaving group). Thus, the few available examples of this class of compounds involve species that have the SF$_5$ group bound to a CH$_2$ bearing the carbonyl function of an acid, an ester, or a ketone (Scheme 7.34). The effect of such a carbonyl function is to shield the axial fluorine but have little effect on the equatorial fluorines. NMR data for the *perfluoro* acetic acid compound is also provided.

7.8.2. Vinylic SF$_5$ Substituents

The SF$_5$ fluorines of SF$_5$-ethane and SF$_5$-ethylene are almost indistinguishable, but the respective ^{13}C NMR spectra would certainly allow their distinction (Scheme 7.35). Fluorine, proton, and carbon NMR spectra of (pentafluorosulfanylmethylene)-cyclohexane have been provided as examples of such spectra (Figures 7.5–7.7, respectively).

Scheme 7.34

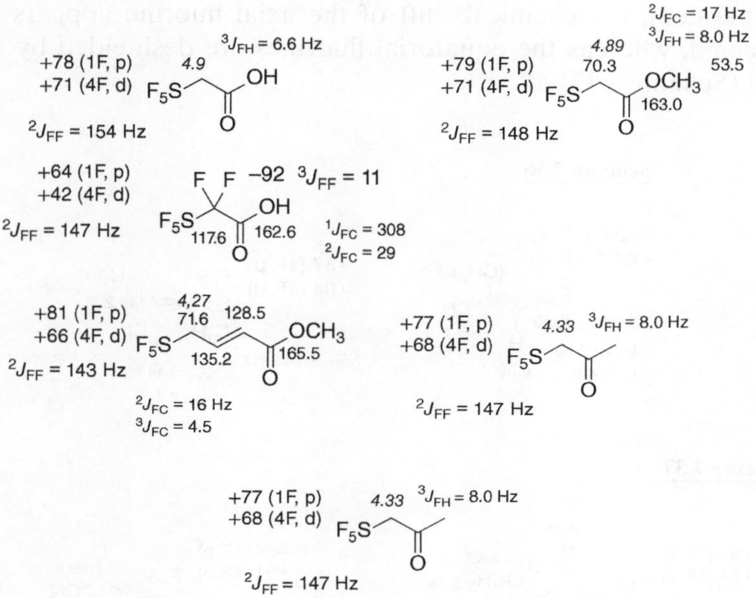

+78 (1F, p)
+71 (4F, d)

$^3J_{FH} = 6.6$ Hz

4.9

F$_5$S, OH

$^2J_{FF} = 154$ Hz

+79 (1F, p)
+71 (4F, d)

$^2J_{FC} = 17$ Hz
$^3J_{FH} = 8.0$ Hz

4.89
70.3

F$_5$S, OCH$_3$

53.5

163.0

$^2J_{FF} = 148$ Hz

+64 (1F, p)
+42 (4F, d)

$^2J_{FF} = 147$ Hz

F F   −92 $^3J_{FF} = 11$

F$_5$S, OH
117.6 162.6

$^1J_{FC} = 308$
$^2J_{FC} = 29$

+81 (1F, p)
+66 (4F, d)

$^2J_{FF} = 143$ Hz

4,27
71.6 128.5

F$_5$S, OCH$_3$
135.2 165.5

$^2J_{FC} = 16$ Hz
$^3J_{FC} = 4.5$

+77 (1F, p)
+68 (4F, d)

4.33 $^3J_{FH} = 8.0$ Hz

F$_5$S

$^2J_{FF} = 147$ Hz

+77 (1F, p)
+68 (4F, d)

4.33 $^3J_{FH} = 8.0$ Hz

F$_5$S

$^2J_{FF} = 147$ Hz

Scheme 7.35

6.1

H$_3$ SF$_5$

5.6 H$_2$ H$_1$
6.7

+80 (1F, p)
+59 (4F, d)

$^2J_{FF} = 147$ Hz

$^3J_{FH1} = 6.6$ Hz
$^4J_{FH2} = 2.8$

$^3J_{H1H2} = 10$ Hz
$^3J_{H1H3} = 17$ Hz
$^2J_{H2H3} \leq 1$ Hz

6.5

H SF$_5$

H$_3$C H
6.5

+83 (1F,p)
+62 (4F, d)

$^2J_{FF} = 149$ Hz

$^3J_{FH} = 5$ Hz
$^4J_{FH} = 2.2$

149.5 SF$_5$
132.3

H

6.15

$^3J_{FH} = 9$ Hz

+86.5 (1F, p)
+66.0 (4F, d)

$^2J_{FF} = 147$ Hz

$^2J_{FC} = 17.0$
$^3J_{FC} = 5.0$

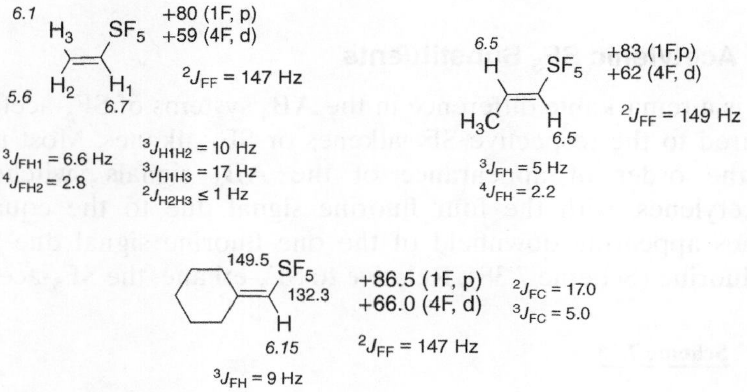

A β-chlorine on the alkene has little effect upon the chemical shift of the axial fluorine but acts to slightly deshield the equatorial fluorines of the SF$_5$ group (Scheme 7.36).

7.8.2.1. α,β-Unsaturated Carbonyl/Nitrile Systems.

7.8.2.1. α,β-Unsaturated Carbonyl/Nitrile Systems. When an SF_5 group is placed at the β-position of an α,β-unsaturated carbonyl or nitrile system, the chemical shift of the axial fluorine appears to be unaffected, whereas the equatorial fluorines are deshielded by about 5 ppm (Scheme 7.37).

Scheme 7.36

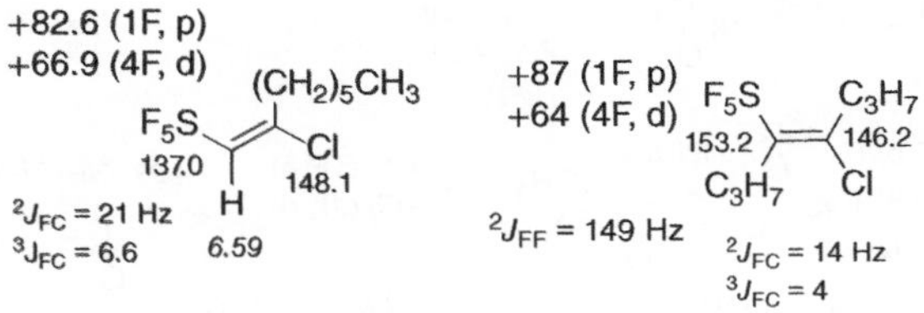

+82.6 (1F, p)
+66.9 (4F, d)

F_5S (CH$_2$)$_5$CH$_3$ Cl
137.0 148.1
H 6.59

$^2J_{FC}$ = 21 Hz
$^3J_{FC}$ = 6.6

+87 (1F, p)
+64 (4F, d)

F_5S C$_3$H$_7$
153.2 146.2
C$_3$H$_7$ Cl

$^2J_{FF}$ = 149 Hz

$^2J_{FC}$ = 14 Hz
$^3J_{FC}$ = 4

Scheme 7.37

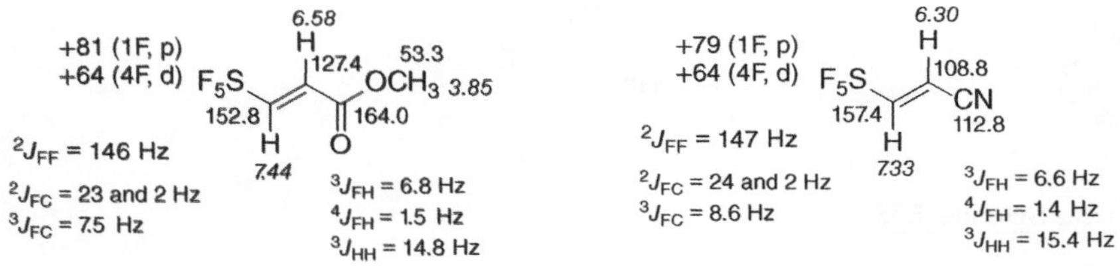

+81 (1F, p)
+64 (4F, d)

6.58
H
127.4 53.3
F_5S OCH$_3$ 3.85
152.8 164.0
H O
7.44

$^2J_{FF}$ = 146 Hz

$^2J_{FC}$ = 23 and 2 Hz
$^3J_{FC}$ = 7.5 Hz

$^3J_{FH}$ = 6.8 Hz
$^4J_{FH}$ = 1.5 Hz
$^3J_{HH}$ = 14.8 Hz

+79 (1F, p)
+64 (4F, d)

6.30
H
108.8
F_5S CN
157.4 112.8
H
7.33

$^2J_{FF}$ = 147 Hz

$^2J_{FC}$ = 24 and 2 Hz
$^3J_{FC}$ = 8.6 Hz

$^3J_{FH}$ = 6.6 Hz
$^4J_{FH}$ = 1.4 Hz
$^3J_{HH}$ = 15.4 Hz

7.8.3. Acetylenic SF_5 Substituents

There is a remarkable difference in the AB_4 systems of SF_5-acetylenes compared to the respective SF_5-alkenes or SF_5-alkanes. Most noticeably, the order of appearance of the AB_4 signals switches for SF_5-acetylenes, with the four fluorine signal due to the equatorial fluorines appearing downfield of the one fluorine signal due to the axial fluorine (Scheme 7.38). Relative to SF_5-ethane, the SF_5-acetylene

Scheme 7.38

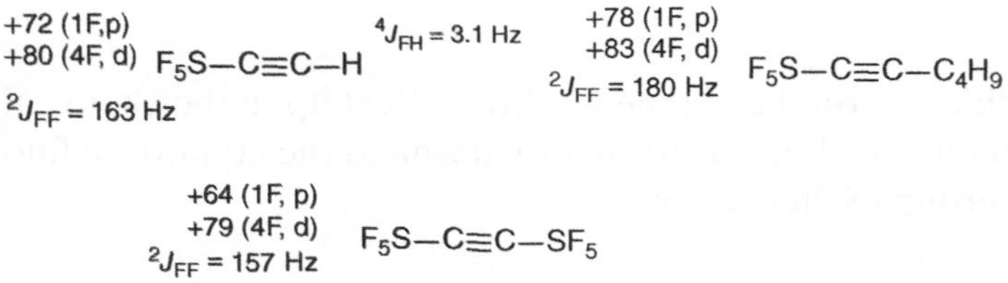

+72 (1F, p)
+80 (4F, d)

$F_5S—C≡C—H$ $^4J_{FH}$ = 3.1 Hz

$^2J_{FF}$ = 163 Hz

+78 (1F, p)
+83 (4F, d)

$F_5S—C≡C—C_4H_9$

$^2J_{FF}$ = 180 Hz

+64 (1F, p)
+79 (4F, d)

$F_5S—C≡C—SF_5$

$^2J_{FF}$ = 157 Hz

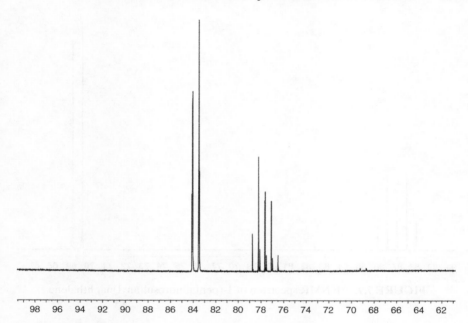

98 96 94 92 90 88 86 84 82 80 78 76 74 72 70 68 66 64 62

FIGURE 7.8. ^{19}F NMR spectrum of p-methylphenyl pentafluorosulfanyl acetylene

equatorial fluorines have shifted ~20 ppm downfield, whereas its axial fluorine is shifted ~10 ppm upfield compared to those of SF$_5$-ethane.

This switching of peaks can be seen in the fluorine spectrum of p-tolyl SF$_5$-acetylene (Figure 7.8).

The doublet for the four equatorial fluorines appears at +83.8 ppm, while the pentet for the single axial fluorine appears at 77.6 ppm. The two-bond coupling constant is 161 Hz.

7.8.4. Aromatic SF$_5$ Substituents

The ^{19}F NMR signals deriving from the AB$_4$ system of SF$_5$ benzenes appear in the same general region as those deriving from SF$_5$ alkenes (Scheme 7.39). The fluorine NMR spectrum of

Scheme 7.39

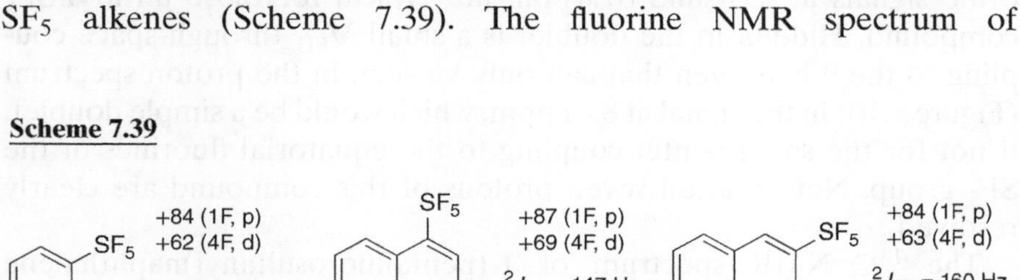

+84 (1F, p)
+62 (4F, d)

$^2J_{FF}$ = 149 Hz

+87 (1F, p)
+69 (4F, d)

$^2J_{FF}$ = 147 Hz

+84 (1F, p)
+63 (4F, d)

$^2J_{FF}$ = 160 Hz

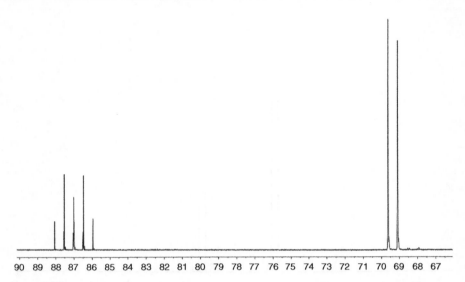

FIGURE 7.9. ^{19}F NMR spectrum of 1-(pentafluorosulfanyl)naphthalene

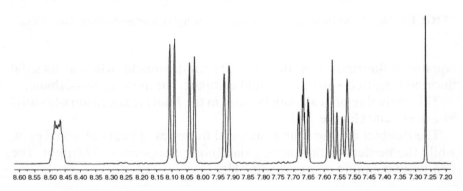

FIGURE 7.10. ^{1}H NMR spectrum of 1-(pentafluorosulfanyl)naphthalene

1-(pentafluorosulfanyl)naphthalene is shown in Figure 7.9. Its fluorine signals at 69.4 and 87.0 ppm are typical for those an aryl-SF$_5$ compound. Hidden in the doublet is a small $^5J_{FH}$ through-space coupling to the 9-hydrogen that can only be seen in the proton spectrum (Figure 7.10) in the signal at 8.47 ppm, which would be a simple doublet, if not for the small pentet coupling to the equatorial fluorines of the SF$_5$ group. Note that all seven protons of this compound are clearly resolved.

The ^{13}C NMR spectrum of 1-(pentafluorosulfanyl)naphthalene (Figure 7.11) is also typical, with all of its carbons being clearly resolved

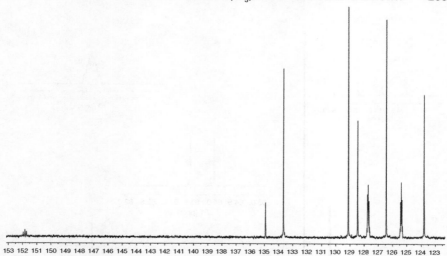

FIGURE 7.11. ^{13}C NMR spectrum of 1-(pentafluorosulfanyl)naphthalene

except for one of the quaternary carbons (at 127.74 ppm) that is obscured by the pentet at 127.76 ppm. The carbon bearing the SF$_5$ substituent appears at 151.8 ppm, with a two-bond coupling constant of only 17 Hz.

Comparing the effect of the SF$_5$ group on aromatic ^{13}C chemical shifts with that of a CF$_3$ group, one can see from the data in Scheme 7.40 that the two groups have subtle, but interestingly different effects. Both groups have a modest shielding influence upon the ortho carbons, while deshielding all other carbons, but the SF$_5$ group has a much stronger effect on the ipso carbon, a stronger effect on the para carbon, and a weaker effect on the meta carbon. As mentioned earlier, the two-bond coupling through the sulfur is much less efficient than through the carbon.

Scheme 7.40

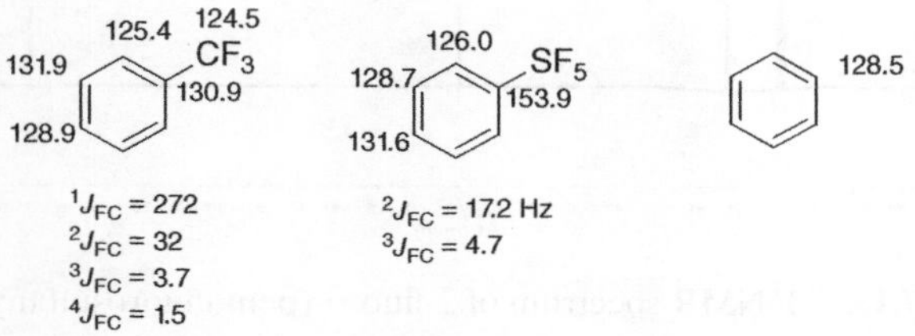

131.9　125.4　124.5　CF$_3$　130.9
128.9

$^1J_{FC}$ = 272
$^2J_{FC}$ = 32
$^3J_{FC}$ = 3.7
$^4J_{FC}$ = 1.5

128.7　126.0　SF$_5$　153.9
131.6

$^2J_{FC}$ = 17.2 Hz
$^3J_{FC}$ = 4.7

128.5

Fluorine spectra for 4-fluoro- and 2-fluoro-(pentafluorosulfanyl) benzene are given in Figures 7.12 and 7.13, respectively.

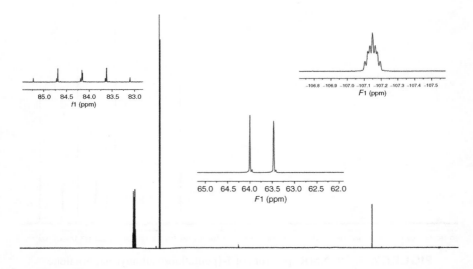

FIGURE 7.12. ^{19}F NMR spectrum of 4-fluoro-(pentafluorosulfanyl)benzene

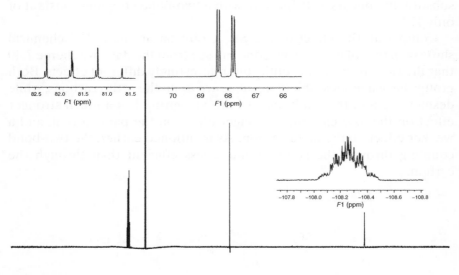

FIGURE 7.13. ^{19}F NMR spectrum of 2-fluoro-(pentafluorosulfanyl)benzene

TABLE 7.3. ^{19}F Chemical Shifts of the Equatorial and Axial
Fluorines of Aryl-SF$_5$ Compounds^a

Ar-SF$_5$	δ (ppm), Doublet	δ (ppm), Pentet	Solvent (CFCl$_3$)
Ph-SF5	62.3	84.1	CDCl$_3$
3-Br-PhSF$_5$	62.3	82.4	CDCl$_3$
4-Br-PhSF$_5$	62.5	83.0	CDCl$_3$
3-NO$_2$-PhSF$_5$	62.3	80.6	CDCl$_3$
4-NO$_2$-PhSF$_5$	62.2	80.6	CDCl$_3$
4-Me-PhSF$_5$	62.7	84.8	CDCl$_3$
2-NH$_2$-PhSF$_5$	64.4	87.8	CDCl$_3$
3-NH$_2$-PhSF$_5$	61.9	84.7	CDCl$_3$
4-NH$_2$-PhSF$_5$	64.1	87.4	CDCl$_3$
2-OH-PhSF$_5$	67.0	85.9	CDCl$_3$
3-OH-PhSF$_5$	62.1	83.8	CDCl$_3$
4-OH-PhSF$_5$	63.8	85.7	CDCl$_3$
2-F-PhSF$_5$	68.1	81.8	CDCl$_3$
3-F-PhSF$_5$	62.6	82.8	CDCl$_3$
4-F-PhSF$_5$	63.7	84.2	CDCl$_3$
3-B(OH)$_2$-PhSF$_5$	64.3	88.4	DMSO-d_6
4-B(OH)$_2$-PhSF$_5$	64.1	87.9	DMSO-d_6

^{a}The $^2J_{FF}$ coupling constants for all of the aryl-SF$_5$ compounds were
148–151 Hz.

There is nothing special in the fluorine spectrum of the 4-fluoro compound. The signal for the fluorine at the 4-position appears as the typical heptet that one obtains when the smaller coupling of the triplet of triplets is about half that of the larger coupling.

The spectrum for the 2-fluoro compound gives evidence of through-space coupling ($^4J_{FF} = 20$ Hz) between the 2-fluoro substituent and the equatorial fluorines of the SF$_5$ group.

The chemical shifts for the two fluorine signals of various substituted SF$_5$-benzenes are given in Table 7.3.

Notice that the chemical shifts for the four equatorial fluorines are not affected by other substituents on the ring when they are on the meta or para positions (±2.4 ppm). The outliers are the OH and F substituents in the 2-position, which exert a pronounced deshielding influence on the equatorial fluorines. The axial fluorines are much more affected by ring substitution (±7.8 ppm).

The carbon spectrum of 2-fluoro-(pentafluorosulfanyl)benzene (Figure 7.14) is also illustrative of the much greater long-range coupling influence of the fluorine substituent compared to the SF$_5$ group. The relative magnitudes of coupling to the single fluorine substituent allow unambiguous assignment of each of the aromatic carbons: δ 118.1

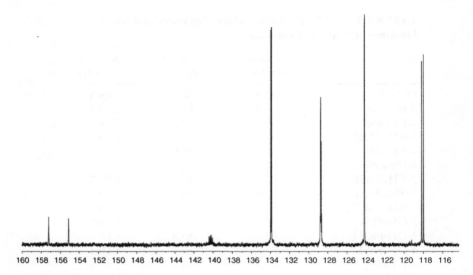

FIGURE 7.14. ^{13}C NMR spectrum of 2-fluoro-(pentafluorosulfanyl)benzene

(d, $^{2}J_{FC} = 24$ Hz, C3), 124.2 (d, $^{4}J_{FC} = 3.8$ Hz, C5), 128.7 (d, $^{3}J_{FC} = 4.9$ Hz, C6), 133.9 (d, $^{3}J_{FC} = 9.1$ Hz, C4), 140.2 (d, pent, $^{2}J_{FC} = 18.5$ and 11.2 Hz, C1), and 156.2 (d, $^{1}J_{FC} = 260$ Hz, C2).

7.8.5. Heterocyclic SF$_5$ Compounds

Fluorine, proton, and carbon spectra of 2-(pentafluorosulfanyl)pyridine are given in Figures 7.15–7.17.

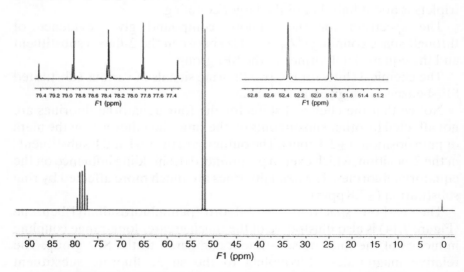

FIGURE 7.15. ^{19}F NMR spectrum of 2-(pentafluorosulfanyl)pyridine

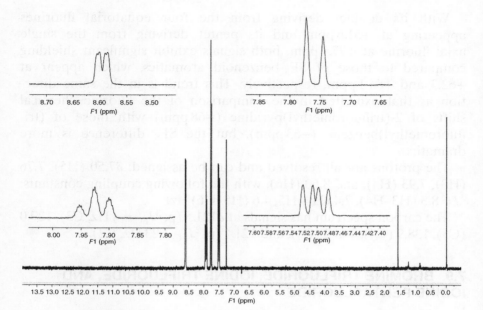

FIGURE 7.16. ^{1}H NMR spectrum of 2-(pentafluorosulfanyl)pyridine

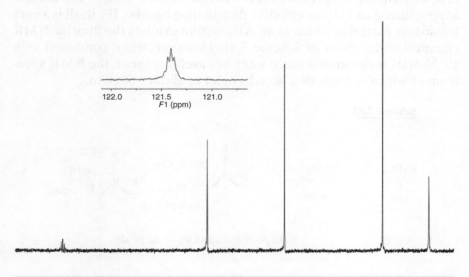

FIGURE 7.17. ^{13}C NMR spectrum of 2-(pentafluorosulfanyl)pyridine

With its doublet deriving from the four equatorial fluorines appearing at +51.6 ppm and its pentet deriving from the single axial fluorine at +77.9 ppm, both signals exhibit significant shielding compared to those of SF_5-benzenoid aromatics, which appear at +62.3 and +84.1 ppm, respectively. This trend is in the same direction as that exhibited in the comparison of the fluorine chemical shifts of 2-(trifluoromethyl)pyridine (−68 ppm) with those of (trifluoromethyl)benzene (−63 ppm), but the SF_5 difference is more dramatic.

The protons are all resolved and can be assigned: δ7.50 (H5), 7.76 (H3), 7.93 (H4), and 8.60 (H6), with the following coupling constants: $^3J = 8.3$ (H3–H4), 7.4 (H4–H5, 4.6 (H5–H6) Hz.

The carbon spectrum has signals at δ121.5 (p, $^3J_{FC} = 4$ Hz, C3), 127.0 (C5), 138.9 (C4), 148.2 (C6), and 121.4 (p, $^2J_{FC} = 23$ Hz, C2).

7.9. BROMINE TRIFLUORIDE, IODINE TRIFLUORIDE, AND IODINE PENTAFLUORIDE

BrF_3, IF_3, and IF_5 are highly reactive compounds that have received renewed attention in recent years because of their unique capabilities as specialized and highly effective fluorinating agents.[7] IF_5 itself is a very hazardous material, which as an AB_4 system exhibits the fluorine NMR chemical shifts given in Scheme 7.41. However, when combined with Et_3N-3HF, it becomes a much safer but useful reagent, the NMR spectrum of which consists of a broad singlet at about −50 ppm.[8]

Scheme 7.41

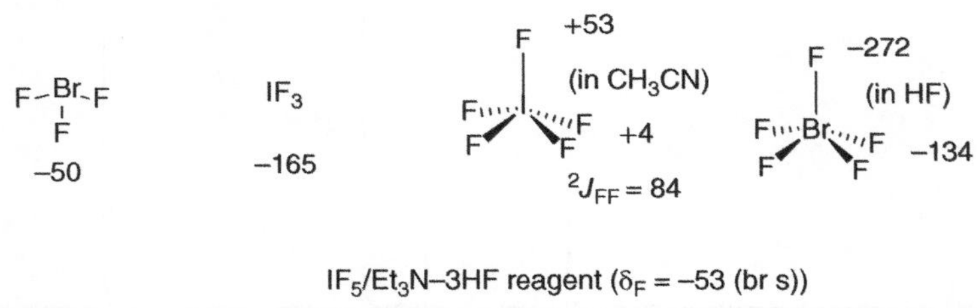

IF$_5$/Et$_3$N–3HF reagent (δ_F = −53 (br s))

7.10. ARYL AND ALKYL HALOGEN DIFLUORIDES AND TETRAFLUORIDES

The fluorines of various aryl chlorine, bromine, and iodine difluorides and tetrafluorides appear in their fluorine NMR spectra as singlets (unless there is a fluorine or trifluoromethyl group in the ortho position), which are sensitive to both the aryl substitution and the solvent (Scheme 7.42). The variations in chemical shifts, all given as reported, are distinctive.

Scheme 7.42

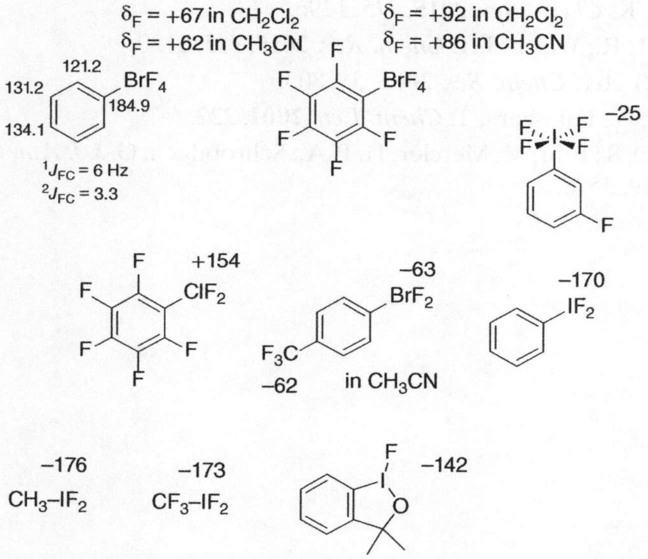

δ_F = +67 in CH$_2$Cl$_2$ δ_F = +92 in CH$_2$Cl$_2$

δ_F = +62 in CH$_3$CN δ_F = +86 in CH$_3$CN

121.2

131.2   BrF$_4$

     184.9

134.1

$^1J_{FC}$ = 6 Hz

$^2J_{FC}$ = 3.3

F BrF$_4$ −25

+154 CF$_2$ ... ClF$_2$

−63 BrF$_2$

F$_3$C

−62   in CH$_3$CN

−170 IF$_2$

−176   −173   −142

CH$_3$–IF$_2$ CF$_3$–IF$_2$

Data for two alkyl iodo difluorides and one example of a fluoroiodane electrophilic fluorination reagent are also provided.

7.11. XENON FLUORIDES

The chemical shifts of some xenon fluorides are given in Scheme 7.43.[9] Also shown is an organoxenon compound, Ph-XeF$_2$.

Scheme 7.43

XeF$_2$	XeF$_4$	XeF$_2$O
−179	−20.5	−49

PhXeF$_2$

−29.5

REFERENCES

1. Gombler, W.; Schaebs, J.; Willner, H. *Inorg. Chem.* **1990**, *29*, 2697.
2. Gillespie, R. J. *J. Chem. Ed.* **1963**, *40*, 295.
3. Ibbott, D. G.; Janzen, A. F. *Can. J. Chem.* **1972**, *50*, 2428.
4. Taha, A. N.; True, N. S.; LeMaster, C. B.; LeMaster, C. L.; Neugebauer-Crawford, S. M. *J. Phys. Chem. A* **2000**, *104*, 3341.
5. Seppelt, K. *Chem. Rev.* **2015**, *115*, 1296.
6. Savoie, P. R.; Welch, J. T. *Chem. Rev.* **2015**, *115*, 1130.
7. Rozen, S. *Acc. Chem. Res.* **2005**, *38*, 803.
8. Yoneda, N.; Fukuhara, T. *Chem. Lett.* **2001**, 222.
9. Brock, D. S.; Bilir, V.; Mercier, H. P. A.; Schrobilgen, G. J. *J. Am. Chem. Soc.* **2007**, *129*, 3598.

GENERAL INDEX

Guide to Fluorine NMR for Organic Chemists, Second Edition. William R. Dolbier, Jr.
© 2016 John Wiley & Sons, Inc. Published 2016 by John Wiley & Sons, Inc.

COMPOUND INDEX

Note: Compounds in bold indicate that there is a Figure associated with it.

Alkanes and haloalkanes

1-amino-4-fluorobicyclo[2.2.1]
heptane, F 14

1-amino-4-fluorobicyclo[2.2.2]
octane, F 14

1-amino-3-fluorobicyclo[1.1.1]
pentane, F 14

bromochlorofluoromethane, H 73;
C 73

bromochloroiodomethane, H 73;
C 73

bromochloromethane, H 73, C 73

1-bromo-1-chloro-2,2,2-
trifluoroethane, F 16; H 16

bromodichlorofluoromethane, F 73;
C 73

1-bromo-1,1-difluorohexane, F 149;
H 151; C 151

1-bromo-2,2-difluorohexane, F 151;
H 151; C 151

bromodifluoromethane, F 12, 149;
H 149

2-bromo-1,1-difluoro-1-
phenylethane, F 151

bromodiiodomethane, H 73; C 73

bromoethane, H 72; C 72

1-bromo-2-fluorobutane, H 74; C 74

bromofluorodiiodomethane, F 73;
C 73

1-bromo-1-fluoro-*trans*-2,3-
dimethylcyclopropane, F 71

1-bromo-2-fluoroethane, F 72; H 74

bromofluoroiodomethane, H 73;
C 73

1-bromo-1-fluorononane, F 71, 86;
H 74; C 74

2-bromo-1-fluoropropane, F 72

Guide to Fluorine NMR for Organic Chemists, Second Edition. William R. Dolbier, Jr.
© 2016 John Wiley & Sons, Inc. Published 2016 by John Wiley & Sons, Inc.

Alkynes

Benzenoid aromatics

Alcohols, ether and related compounds

Carboxylic acids, derivatives and nitriles

**Amines, imines, azides, nitro and
related compounds**

Hypofluorites (O-F) and N-F compounds

Phosphines, phosphonates and related compounds

Sulfides, sulfonates, selanes, tellanes and related compounds

Printed and bound by CPI Group (UK) Ltd, Croydon, CR0 4YY

16/04/2025

14658349-0002